Catalysis in Polymer Synthesis

ACS SYMPOSIUM SERIES **496**

Catalysis in Polymer Synthesis

Edwin J. Vandenberg, EDITOR
Arizona State University

Joseph C. Salamone, EDITOR
University of Massachusetts, Lowell

Developed from a symposium sponsored
by the Division of Polymeric Materials: Science and Engineering
at the 201st National Meeting
of the American Chemical Society,
Atlanta, Georgia,
April 14–19, 1991

American Chemical Society, Washington, DC 1992

Library of Congress Cataloging-in-Publication Data

Catalysis in polymer synthesis / Edwin J. Vandenberg, editor, Joseph C. Salamone, editor.

p. cm.—(ACS symposium series, ISSN 0097–6156; 496).

"Developed from a symposium sponsored by the Division of Polymeric Materials: Science and Engineering at the 201st National Meeting of the American Chemical Society, Atlanta, Georgia, April 14–19, 1991."

Includes bibliographical references and index.

ISBN 0–8412–2456–0

1. Polymerization—Congresses. 2. Catalysis—Congresses.

I. Vandenberg, Edwin J., 1918– . II. Salamone, Joseph C., 1939– . III. American Chemical Society. Division of Polymeric Materials: Science and Engineering. IV. American Chemical Society. Meeting (201st: 1991: Atlanta, Ga.). V. Series.

QD281.P6C27 1992
660′.28448—dc20 92–15608
CIP

The paper used in this publication meets the minimum requirements of American National Standard for Information Sciences—Permanence of Paper for Printed Library Materials, ANSI Z39.48–1984. ♾

PRINTED IN THE UNITED STATES OF AMERICA

ACS Symposium Series

M. Joan Comstock, *Series Editor*

1992 ACS Books Advisory Board

Foreword

THE ACS SYMPOSIUM SERIES was founded in 1974 to provide a medium for publishing symposia quickly in book form. The format of the Series parallels that of the continuing ADVANCES IN CHEMISTRY SERIES except that, in order to save time, the papers are not typeset, but are reproduced as they are submitted by the authors in camera-ready form. Papers are reviewed under the supervision of the editors with the assistance of the Advisory Board and are selected to maintain the integrity of the symposia. Both reviews and reports of research are acceptable, because symposia may embrace both types of presentation. However, verbatim reproductions of previously published papers are not accepted.

Contents

Preface .... **xi**

OVERVIEW

1. **Catalysis: A Key to Advances in Applied Polymer Science** **2**
   Edwin J. Vandenberg

TRANSITION METAL CATALYZED COORDINATION
POLYMERIZATION OF OLEFINS AND DIOLEFINS

2. **Most Advanced Magnesium Chloride Supported Ziegler–Natta Catalyst** **27**
   James C. W. Chien

3. **Polymerization of C_2–C_6 α-Olefins in the Presence of Modified Ziegler–Natta Catalysts Based on Titanium Halides and Organoaluminum Sulfates** **56**
   E. Chiellini, S. D'Antone, R. Solaro, F. Masi, F. Menconi, and L. Barazzoni

4. **Isotactic Olefin Polymerization with Optically Active Catalysts** **63**
   W. Kaminsky, S. Niedoba, N. Möller-Lindenhof, and O. Rabe

5. **1-Butene Polymerization with Ethylenebis-(1-indenyl)-zirconium Dichloride and Methylaluminoxane Catalyst System: Effect of Hydrogen Addition** **72**
   M. Kioka, A. Mizuno, T. Tsutsui, and N. Kashiwa

6. **Rate of Ethylene Polymerization with the Catalyst System (η^5-$RC_5H_4)_2ZrCl_2$–Methylaluminoxane: Effects of Cyclopentadienyl Ring Substituents** 78
Peter J. T. Tait, Brian L. Booth, and Moses O. Jejelowo

7. **Migratory Nickel(0)–Phosphorane Catalyst: α-Olefin Polymerization by 2,ω-Linkage** 88
G. Fink, V. Möhring, A. Heinrichs, and Ch. Denger

8. **Structure and Morphology of Highly Stereoregular Syndiotactic Polypropylene Produced by Homogeneous Catalysts** 104
Adam Galambos, Michael Wolkowicz, and Robert Zeigler

9. **Dicyclopentadiene Polymerization Using Well-Characterized Tungsten Phenoxide Complexes** 121
Andrew Bell

RING-OPENING COORDINATION POLYMERIZATION OF EPOXIDES

10. **Structural Effects and Reactivity of Polymer-Bound Functional Groups: Kinetics of Nucleophilic Displacement Reactions on Chlorinated Polyethers** 136
Douglas A. Wicks and David A. Tirrell

11. **Mechanism of Oxirane Coordination Polymerization of Soluble Polynuclear μ-Oxometal Alkoxide Aggregates** 149
Ph. Condé, L. Hocks, Ph. Teyssié, and R. Warin

12. **^{13}C NMR Analysis of Polyether Elastomers** 157
H. N. Cheng

13. **Calorimetric Study of Cure Behavior of an Amine–Epoxy Resin under Ambient Conditions** 170
Suk-fai Lau

14. **X-ray Fiber Diffraction Study of a Synthetic Analog of Cellulose: Poly[3,3-bis(hydroxymethyl)oxetane]** 182
J. M. Parris and R. H. Marchessault

ANIONIC POLYMERIZATION

15. **Metalloporphyrin Catalysts for Control of Polymerization** 194
Shohei Inoue

16. **Oligomerization of Oxiranes with Aluminum Complexes as Initiators** 205
V. Vincens, A Le Borgne, and N. Spassky

17. **Catalysts and Initiators for Controlling the Structure of Polymers with Inorganic Backbones** 215
Krzysztof Matyjaszewski

CATIONIC POLYMERIZATION

18. **Mechanisms and Catalysis in Cyclophosphazene Polymerization** 236
Harry R. Allcock

19. **Polyaddition of Epoxides and Diepoxides to the Acids of Phosphorus: Synthesis of Poly(alkylene phosphate)s** 248
S. Penczek, P. Kubisa, P. Klosinski, T. Biela, and A. Nyk

20. **Poly(methyl methacrylate)-*block*-polyisobutylene-*block*-poly(methyl methacrylate) Thermoplastic Elastomers: Synthesis, Characterization, and Some Mechanical Properties** 258
Joseph P. Kennedy and Jack L. Price

INDEXES

Author Index 280

Affiliation Index 280

Subject Index 281

Preface

UNUSUAL NEW ADDITION AND RING-OPENING polymerization methods for olefins, diolefins, and cyclic monomers (such as cyclic olefins and epoxides) have developed during the past 30 years from various metal catalysts. Many catalysts are based on transition metals and operate by completely new mechanisms, such as coordination or insertion polymerization. Many new polymers with unique properties, such as linear polyethylene and isotactic polypropylene, have been developed commercially into large-volume products for plastics, fibers, and films. Improved elastomers based on ethylene–propylene copolymers and on epichlorohydrin polymers and copolymers have been discovered and developed commercially from these new metal catalysts.

Research, which continues unabated in these areas, has resulted in supported, heterogeneous transition metal catalysts for isotactic polypropylene and polyethylenes. These catalysts have been extensively commercialized during the past 10 years because of simplified processes and lower costs. Very effective homogeneous catalysts have given unusual research products, such as syndiotactic polypropylene. Metathesis polymerization of cyclic olefins, discovered early in the transition metal catalyst studies, has been greatly extended during the past 10 years to show that it involves a ring-opening polymerization of an intermediate metallacyclobutane that can be modified to give living systems and that can be extended to more polar cyclic olefins. The commercialization of the metathesis polymerization of dicyclopentadiene is leading to useful plastic products by reaction injection molding.

This symposium was organized to recognize the achievements of Vandenberg in this field, especially in olefin and epoxide polymerization, at the occasion of his receiving the American Chemical Society Award in Applied Polymer Science. Because catalysis played a key role in his applied polymer science developments, we desired to present an up-to-date global view of the areas in which he contributed, emphasizing the newest and the most important research. We also tried to cover new areas in which catalysis is playing a key role, so we invited outstanding contributors from the United States, Europe, and Japan. Thus, this book provides up-to-date information on the recent research developments on metal catalysis for olefins, epoxides, oxetanes, cyclic phosphazenes, and cyclic silanes.

We greatly appreciate the excellent contributions and support of our authors. We also thank the Division of Polymeric Materials: Science and

Engineering of the American Chemical Society, Hercules Incorporated, Himont Research and Development Center, Zeon Chemicals U.S.A. Inc., and AKZO Chemicals Division for financial and other support that facilitated greatly the organization of the symposium upon which this book is based.

EDWIN J. VANDENBERG
Department of Chemistry
Arizona State University
Tempe, AZ 85287–1604

JOSEPH C. SALAMONE
Department of Chemistry
University of Massachusetts, Lowell
Lowell, MA 01854

January 13, 1992

Overview

A REVIEW OF E. J. VANDENBERG'S PERTINENT WORKS and related literature provides an overall background for the general theme of this book. The areas covered are (1) free radical redox emulsion polymerization catalysts in the late 1940s that were important to the development of improved SBRs, our largest volume synthetic rubber; (2) the development of new coordination polymerization catalysts in the 1950s based on Ziegler's revolutionary transition metal catalyst, which led to numerous new, improved, large-volume thermoplastics and elastomers; (3) the development of new coordination polymerization catalysts based on Al, Zn, and Mg for more polar monomers, especially cyclic monomers, such as the epichlorohydrin and propylene oxide elastomers; (4) the extension of transition metal catalysts to promising new areas, such as metathesis polymerization that is just now in early commercialization; (5) recent transition metal catalyst studies that have also led to new, more efficient heterogeneous catalysts offering future promise, such as for a greatly improved route to a new highly stereoregular syndiotactic polypropylene; and (6) the extension of the epoxide and oxetane developments to the preparation of new hydroxy polyethers that are analogs of natural polysaccharides.

Chapter 1

Catalysis: A Key to Advances in Applied Polymer Science

Edwin J. Vandenberg

Department of Chemistry, Arizona State University, Tempe, AZ 85287–1604

The author's contributions to Applied Polymer Science where catalysis played a key role are reviewed. The most important areas include: (1) olefin polymerization with Ziegler catalysts, especially as related to the independent discovery of: isotactic polypropylene, the first high yield catalysts for its preparation and the hydrogen method of controlling molecular weight; (2) the discovery of alkylaluminoxane catalysts for the ring opening polymerization of epichlorohydrin and propylene oxide to new, commercial polyether elastomers; and (3) oxetane polymerization with the alkylaluminoxane catalysts to some interesting new hydroxypolyethers which are analogs of poly(vinyl alcohol) and of cellulose, and to the first coordination copolymerization of tetrahydrofuran. The coordination polymerization of vinyl ethers at ambient temperatures with some unusual catalysts is described. Early studies are reviewed on the hydroperoxide redox emulsion polymerization of styrene-butadiene rubber, our largest volume synthetic rubber.

Catalysis is often, although not always, a key to advances in applied polymer science. The poineering developments in which I was always involved usually required new catalyst discoveries. Since one rarely knows the exact mechanism by which an achieved result is obtained in pioneering work, a broad definition of catalysts is needed. Therefore, in this paper, catalyst will be used in a very general sense as a minor component of a reaction mixture that enables a desired reaction to occur, independent of whether it is regenerated in active form as the purist requires. For example, the initiator of a free radical polymerization is a catalyst in this general sense even though it may be incorporated in the final product.

0097–6156/92/0496–0002$06.50/0

Early Polymerization Catalyst Studies

My first important catalyst discovery was in 1947 on redox free-radical polymerization, particularly for emulsion polymerization (1). After returning to the Hercules Research Center from an ordinance plant at the end of World War II, I had been assigned to longer range work on emulsion polymerization as it related to a Hercules emulsifier, disproportionated rosin soap (Dresinate 731), which had been accepted for some areas of the government rubber program (then, GR-S; now, SBR). One problem with this emulsifier was that reaction times were about 20% longer than with a fatty acid soap such as sodium stearate with potassium persulfate initiator. At this time, Dr. E. J. Lorand of Hercules was studying the air oxidation of activated hydrocarbons such as terpenes and cumene to hydroperoxides. My predecessor in the emulsion polymerization work, Dr. Arthur Drake, had tried cumene hydroperoxide (CHP,α,α-dimethylbenzyl hydroperoxide) as an initiator of SBR emulsion polymerization and found that it worked but was not better, perhaps somewhat inferior to persulfate, as I recall. We decided to do some work on this initiator and, at the time, became aware through PB reports coming out of Germany that the Germans had achieved very fast emulsion polymerizations with redox systems, i.e., the combination of an oxidizing agent such as a peroxide as hydrogen peroxide or benzoyl peroxide with a reducing agent such as iron pyrophosphate with a reducing sugar. We immediately tried this for emulsion polymerization on SBR and other vinyl monomers with disproportionated rosin soap emulsifier plus cumene hydroperoxide, iron pyrophosphate and a reducing sugar and obtained rates of polymerization which were 10-60 fold greater than with persulfate alone. Subsequently, we found that ordinary rosin soaps which were strong inhibitors with persulfate initiator could also give good rates with this new redox system. Thus, a unique aspect of these new redox systems was that they worked better in the presence of normal retarders or inhibitors of free radical polymerizations. We obtained five broad patents on this work (2,3). Kolthoff and Medalia (4) and Fryling et al (5) at Phillips Petroleum independently discovered these same systems. Phillips used them to make an improved SBR by low temperature (5°C) polymerization (6). Related redox systems are still used today for making SBR by 5°C emulsion polymerization. Hercules benefited by their emulsifier and hydroperoxide sales and in other ways as noted below.

In view of what appeared to be a unique behavior of these hydroperoxide redox systems, particularly with regard to inhibitors, we did some mechanism studies in order to understand why. We found that CHP reacted with ferrous iron to preferentially reduce the OH end of the molecule (Eq. 1), ultimately forming acetophenone and a methyl radical. In the actual redox system, the ferric iron is reduced back to ferrous iron by the reducing sugar so that little iron is required.

Presumably, the somewhat hindered and electronically favored RO· was less reactive to inhibitors giving it a better chance to add to monomer and initiate polymerization as compared to the more reactive radicals from persulfate, hydrogen peroxide or benzoyl peroxide. In the course of this study we ran a ferrous chloride

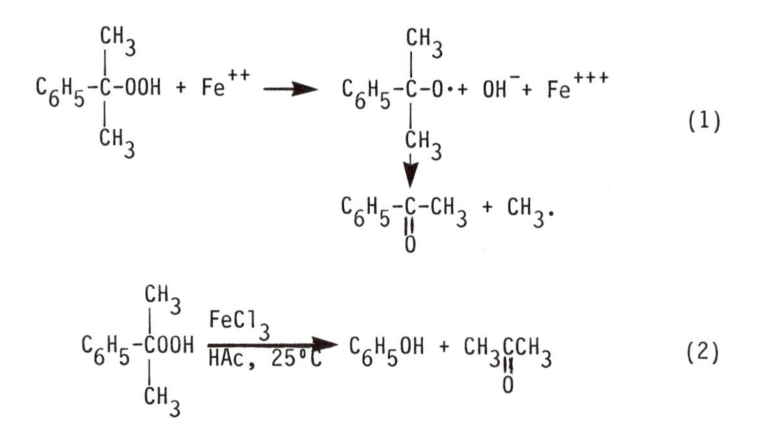

reduction in acetic acid and observed instead a very facile, acid-catalyzed ($FeCl_3$) cleavage of CHP to phenol and acetone (Eq. 2). This acid-catalyzed cleavage had been reported in the German literature by Hock and Lang (7) using aqueous H_2SO_4 during the war and by then (1947) was well known. We studied this reaction briefly and found that it went very readily with many acid catalysts in organic medium to give near quantitative yields of phenol and acetone. Although our work in this area was limited, our excellent results encouraged the Naval Stores operating department to aggressively pursue a new process for phenol and acetone. Subsequently, they joined with Distillers and developed a commercial process which was very successful (8) for these basic, very large volume raw materials. Indeed, almost all of the phenol in the world is made by this simple process. This achievement by Hercules and in a small but important way by my recognition of the possible value of this facile, high yield cleavage does, of course, relate to Applied Polymer Science since phenol and acetone are very important building blocks for important polymers as phenol-formaldehyde resins, epoxies via bis-phenol A, etc.

Another important development to come out of very limited work on the acid cleavage of CHP, which I recognized as a way of reacting hydroperoxides with other reactive molecules, was that of Boardman, a co-researcher at Hercules, who reacted CHP in the presence of an acid catalyst and α,α-dimethylbenzyl alcohol to give a new process for dicumyl peroxide (9), an important polymerization initiator and cross-linking agent for modifying polymers.

Thus, from our initial promising redox polymerization studies grew a number of important commercial developments related to polymers, some of which I had little detailed involvement in, except in the very early stage.

In the course of trying to find applications for our new hydroperoxide redox initiating systems, we discovered in 1951 systems which polymerized ethylene at low temperature (5°C) and low pressure (40 atm.), i.e., below the critical temperature (9.9°C). The best system found utilized an insoluble ferrous complex of ethylene diamine tetracetic acid with cumene hydroperoxide in 97.5%

tert-butyl alcohol (2.5% H_2O), but rates were low. The product, however, was more linear than the commercial high pressure product, since it contained much fewer CH_3 groups by IR, was higher melting (125°C), and was more crystalline by x-ray. In addition, the product was a brittle wax even though we did get the molecular weight up to 20,000 by raising the pressure to 300 atm. This is obviously the linear high density polyethylene which was unknown at that time but which was very easily made a few years later at much higher molecular weight from the Ziegler development and from the Phillips chromium oxide on silica catalyst (10). This polymer has become a very important, large volume commercial product. Our molecular weights were too low for us to recognize its value and, in any event, our process was too poor to be useful. In the course of this work we considered trying a combination of an organometallic such as a Grignard reagent with an organic chloride and a transition metal compound, such as $CoCl_2$, $NiCl_2$, $CrCl_3$, or $FeCl_3$, as a source of free radicals for initiating polymerization. This combination had been reported by Kharasch in his work on free-radical reactions (11), and it falls within the scope of the subsequently discovered Ziegler catalyst. But, the organometallic was not readily available, and we never tried it. So we missed what might have been a "logical" development of Ziegler catalysis. Indeed, there are examples in the literature where some element of this new coordination polymerization method was used, but in a very imperfect way, and the authors never recognized that it was a new and different method of polymerization (12).

Ziegler Catalyst Polymerizations

General. In mid-1954, Hercules acquired a license to Ziegler's new low pressure ethylene polymerization method which involved a new type of catalyst consisting of organometallics, such as alkylaluminums, combined with transition metal compounds, such as $TiCl_4$. In October of 1954, I was assigned to do scouting work in this new area of catalysis. Of course, we recognized the importance of Ziegler's discovery. Needless to say, these were very exciting times in the laboratory. We did not know very much about the mechanism of the polymerization method, although it was obvious that some new type of polymerization method was involved with growth of the polymer on the transition metal.

Propylene Polymerization. Just one week after I began to work in the field, I tried, on the same day, polymerizing propylene and also using hydrogen in an ethylene polymerization. In the case of propylene I obtained 92 milligrams of an unusual, insoluble, crystalline polymer which I found to be a fiber former. I concluded that it would be attractive for further study. In my experiments with hydrogen we ran three experiments plus a control and found that it definitely affected the molecular weight (Table I). The effect was really quite small and I suppose that a lot of people would have considered this inconsequential. However, we did pursue both of these leads.

In the case of propylene, ordinary Ziegler catalysts as described by Ziegler gave very low yields of crystalline polymer;

Table I. **First Experiments on the Effect of Hydrogen on the Polymerization of Ethylene with a Ziegler Catalyst**

Mole % H_2	$\frac{\eta_{sp}}{C}$(1)	M_w x 10^{-6}
0	21.0	1.33
0.6	17.4	1.11
1.1	16.4	1.05
2.3	16.4	1.05
2.3	15.8	1.01

(1) 0.1% conc. in decalin at 135°C.
SOURCE: Reproduced with permission from ref. 33, copyright 1983 Plenum

indeed, most of the polymer was an amorphous rubber or even a viscous liquid. Very quickly in 1955, we found catalysts and conditions which gave high yields of the crystalline polymer (Figures 1 and 2). Shortly after, Natta and coworkers reported that the polymer was stereoregular and coined the term "isotactic" (13). We did file a patent application on crystalline polypropylene as a new composition of matter, and were involved in a subsequent interference (10). However, our dates were not early enough and ultimately we were removed from the interference. In the case of hydrogen, we tried it on propylene and found it to be a very effective way of controlling molecular weight. We filed a patent application which also got involved in an interference. In this case, we prevailed and obtained a patent on our discovery (14). This method of controlling molecular weight is the major method used in the world today to control the molecular weight of isotactic polypropylene. The production capacity for polypropylene has grown rapidly so that it is now one of the major synthetic polymers with a worldwide consumption of about 12 billion lbs./yr. in 1985 (15).

Catalysts Giving High Stereoregularity - $TiCl_3 \cdot nAlCl_3$. In the case of propylene polymerization, our new catalysts utilized crystalline $TiCl_3$ solids which contained large amounts of $AlCl_3$, up to one mole per Ti. These $TiCl_3$ solids were obviously mixed crystals of $TiCl_3$ and $AlCl_3$ since the x-ray pattern gave no evidence for $AlCl_3$ per se. Both the $TiCl_3$-1.0 $AlCl_3$ and the $TiCl_3$-0.33 $AlCl_3$ mixed crystals gave new and different x-ray patterns which appeared related to the known purple α-$TiCl_3$ rather than the known brown β-$TiCl_3$. Color varied from reddish purple to violet. The $TiCl_3$-1.0 $AlCl_3$ product from $EtAlCl_2$-$TiCl_4$ reaction was called "purple substance" by our x-ray specialist to differentiate it from the known α-$TiCl_3$. This type proved important in the commercial manufacture of polypropylene. We initially made this new form by the room temperature reaction of $TiCl_4$ with an excess (2-4 mole/Ti) of $EtAlCl_2$. By using this pre-reaction product with triethyl aluminum in certain ratios where the triethyl aluminum reacted with the excess ethyl aluminum dichloride to form diethyl aluminum chloride (Figure 1), we obtained high yields of stereoregular polypropylene (16,17). This was our first high yield method of making isotactic polypropylene and may indeed be the first high rate, high yield method discovered by anyone. This method did

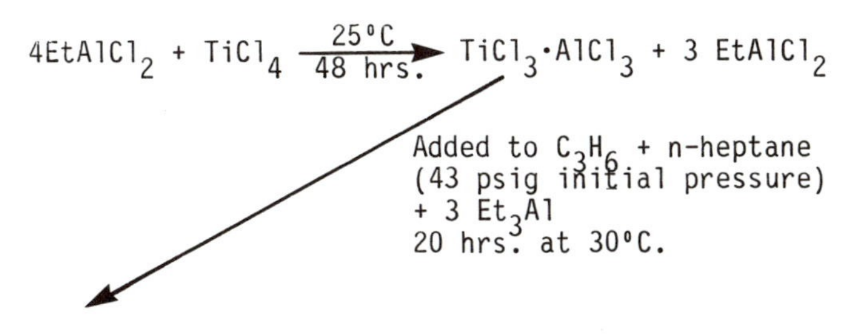

$TiCl_3 \cdot AlCl_3 + 6\ Et_2AlCl$

"Purple Substance"

δ Form

-95% yield, 86% conv. of heptane-insoluble polypropylene,

$$\frac{\eta_{sp}}{C} = 29$$

Figure 1. First High Yield Method for Isotactic Polypropylene (16,17) (Reproduced with permission from ref. 33, copyright 1983 Plenum)

involve the new so-called "purple substance" form of $TiCl_3$. Hercules obtained a composition of matter patent on this form (18). Years later, Natta and coworkers (19) reported their excellent crystallographic studies of the various modifications of $TiCl_3$ and called this form δ. It turns out that this δ form was the basis of all the early methods of making polypropylene commercially (20).

Our second and even higher yield (99%) and higher rate method of making isotactic polypropylene was discovered in 1955 as a logical development from our first high yield catalyst. If $TiCl_3 \cdot 1.0\ AlCl_3$ gave this excellent result, how would lower levels of $AlCl_3$ work? We prepared such a lower $AlCl_3$ catalyst by the stoichiometric reaction of trialkyl aluminum with $TiCl_4$, i.e., one alkyl aluminum bond per $TiCl_4$ to form $TiCl_3$-0.33 $AlCl_3$ as the sole product and then used this crystalline solid with Et_2AlCl as an activator (Figure 2) (21,22). This $TiCl_3$ solid was also a mixed crystal of $TiCl_3$ and $AlCl_3$ (no x-ray evidence for $AlCl_3$) and had a different x-ray pattern from the known "violet" α-$TiCl_3$ but is identical to that finally reported as the γ form by Natta (19), especially when R = iBu in Figure 2. This method for making isotactic polypropylene was very effective but the physical form of the product was less favorable and thus was not developed commercially by Hercules. Similar processes have been described in the patent literature (23).

However, this stoichiometric type of $TiCl_3$ has become of major importance in the commercial manufacture of polypropylene (12), based on the reduction of $TiCl_4$ by aluminum metal, as first reported by Ruff and Neumann (24). Improved methods of making this type of $TiCl_3$ were found and developed by Exxon (25), Stauffer (26), and others (12). In particular, Tornqvist et al (26) found that ballmilling was necessary for good polymerization rates. In addition, ballmilling converted the $TiCl_3$ to the δ form (19,27).

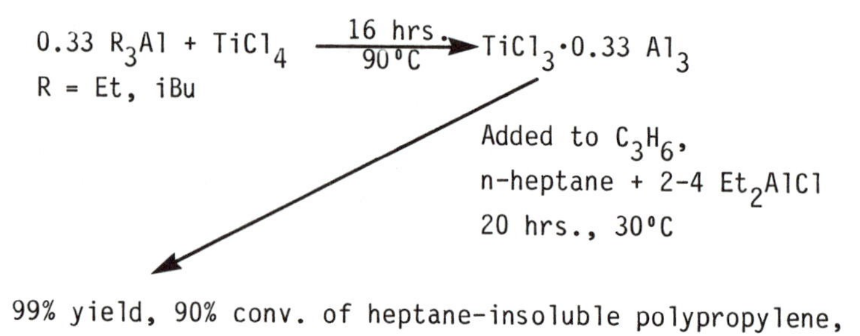

99% yield, 90% conv. of heptane-insoluble polypropylene,

$$\frac{\eta_{sp}}{C} = 29$$

Figure 2. Second High Yield High Method for Isotactic Polypropylene-Stochiometric Catalyst (21,22). (Reproduced with permission from ref. 33, copyright 1983 Plenum)

Also, the polypropylene was obtained in better physical form. Tornqvist et al also recognized the higher activity and other desirable aspects of the $TiCl_3 \cdot nAlCl_3$ mixed crystals, particularly $TiCl_3 \cdot 0.33\ ACl_3$, over $TiCl_3$, and ascribed this to a specific effect of $AlCl_3$. This was my belief too. This conclusion is somewhat suspect today since Solvay has reported $TiCl_3$ catalysts of four fold higher activity and with higher stereospecificity than with prior $TiCl_3 \cdot nAlCl_3$ catalysts using especially pure δ-type $TiCl_3$, prepared by special ether extraction and $TiCl_4$ treatment of a Et_2AlCl-$TiCl_4$ reaction product (28). We were not able to publish, except by patent, in this area of great commercial importance.

Commercial Development Aspects. We went on the make large quantities of isotactic polypropylene and found that it had an interesting combination of properties, although it was not immediately taken up by Hercules since much of Hercules effort at that time was directed toward developing a Ziegler-type process for high density polyethylene. I personally felt that polypropylene would ultimately be an important polymer based on the low cost of propylene and the properties of the polymer. Finally, after about a year or two, Hercules developed a commercial process in collaboration with Hoechst from our lead (29). As you know, Hercules, in late 1957, became the first U.S. producer and ultimately the world's largest producer of polypropylene. However, in this modern world of mergers and acquisitions, Hercules placed their polypropylene production and know-how in a 50-50 joint company with Montedison, called Himont, which, in turn, became a wholly-owned subsidiary of Montedison when Hercules sold its share.

This new Ziegler method of polymerization proved to be very useful commercially and is used on a very large scale for olefin (ethylene and propylene) and diolefin (butadiene and isoprene) polymerization. Some of the important discoveries discussed above for polypropylene were also of value to polyethylene. For example, the hydrogen method of controlling molecular weight is widely used for ethylene polymerization. It is also of interest that Example 53 in one of our basic olefin catalyst patents (16) reads directly on the very important linear low density polyethylene, showing a copolymer of ethylene with a small amount (5.8%) of octene-1. This example appears to predate the DuPont patent on linear low-density polyethylene (30). Extensive litigation has occurred between DuPont and Phillips on the validity of this DuPont patent (31,32). The issue has still not been completely resolved. In any event, we did not recognize the great potential value of the product of our early patent example.

There has been a tremendous amount of work done on catalysts for olefin and diolefin polymerization from both the practical and mechanistic point of view. Although many mechanism details are still under debate, we have previously published our general thoughts on this new method of polymerization, often referred to as coordination or insertion polymerization (33,34).

Polar Monomers. Early in scouting efforts on Ziegler catalysts, other monomers were studied, particularly polar momomers, to see whether we could obtain other unusual products. We had some success with vinyl chloride (35) and methyl methacrylate (36) and this work was explored by others at our laboratory (37,38). However, we had most success with vinyl ethers and found a variety of new catalysts which proved especially effective for making stereoregular, isotactic polymers from vinyl ethers (39). Some of these catalysts are shown in Figure 3. Initially, we devised a vanadium based catalyst for vinyl methyl ether, consisting of the stoichiometric reduction product from 0.33 Et_3Al plus VCl_4 (called SV catalyst) pretreated with iBu_3Al (PSV catalyst) used with additional R_3Al (40). Subsequently, we found better, new, non-transition metal catalysts (41-43) such as the combination of an aluminum alkoxide or alkyl with sulfuric acid (41). With these new catalysts, we were able to obtain much more highly stereoregular polymers than Schildknecht had obtained some ten years before with ordinary Lewis acid-type catalysts. We concluded that these new catalysts were examples of cationic insertion or coordination catalysts as contrasted with Ziegler catalysts which by that time were considered

Transition Metal Based

Pre-Treated Stoichiometric Vanadium (PSV) Catalyst

$$[2\ \underline{i}\text{-}Bu_3Al + VCl_3 \cdot 0.33\ AlCl_3] + R_3Al$$

Al Based

$$[R_3Al + H_2SO_4] + R_3Al$$

R = Alkyl or RO-

Figure 3. Vinyl Ether Coordination Catalysts (39). (Reproduced with permission from ref. 33, copyright Plenum)

as examples of anionic coordination catalysts. Figure 4 shows the type of mechanism that we proposed in a paper on that subject in

1963 (39). As you can see, the mechanism postulated is different than considered with olefins in that coordination involves the ether group of the last unit in the growing chain. We concluded that this was an example of coordination polymerization because the polymerizations were many orders of magnitude slower, even at or near ambient temperature, than typical cationic polymerizations, molecular weights were unusually high, and stereoregularity could be very high. Although this work was very interesting scientifically, it has not yet led to any important commercial developments. The catalysts discovered are unusual and still merit further study on monomers which polymerize readily by cationic methods.

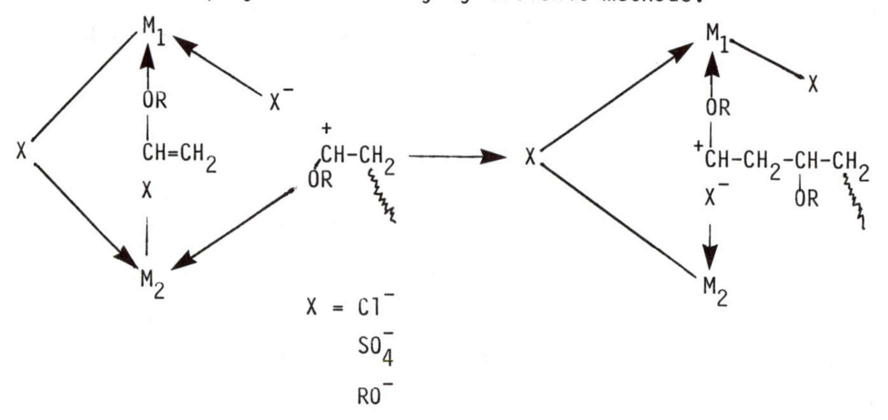

Figure 4. Mechanism of Vinyl Ether Coordination Polymerization (39). (Reproduced with permission from ref. 33, copyright 1983 Plenum)

Epoxide Polymerization

New Catalysts and Polymers. In 1957, we noted that some of our newly developed transition metal-based vinyl ether catalysts were somewhat similar to a new epoxide polymerization catalyst, an aluminum isopropoxide-$ZnCl_2$ combination, which Price and Osgan (44) had just found for polymerizing propylene oxide. Thus, we tried polymerizing an epoxide, specifically epichlorohydrin, with some of our new vinyl ether catalysts and some Ziegler catalysts and obtained a small amount, 9 mg., of crystalline polyepichlorohydrin, a new polymer, at least to us. Then through exploring this development, with a catalyst consisting of triisobutyl aluminium and vanadium trichloride, we found that if we left out the transition metal component we got even better results. In addition, the results were different in that the major product was now an amorphous rubber. This turned out to be another new polymer. In the course of exploring this lead, we found that we could not repeat our previous work with a new batch of monomer. Indeed in our first work, we had used a nearly empty bottle of epichlorohydrin that had been around the laboratory for some time. We immediately suspected that there was probably water in the epichlorohydrin. Thus, we reacted triisobutylaluminum with water and found it to be very effective catalytically. In this way, we discovered this very important class of new catalysts based on the reaction of alkyl aluminums with water, as shown in Figure 5 (45). In the course of studying the mechanism of polymerization with this new catalyst, we speculated that it was a coordination catalyst, i.e., the epoxide coordinated with the metal prior to its insertion in the chain. We thought that we could prove this hypothesis by adding a chelating agent such as acetylacetone to block the coordination site on the aluminum and thus prevent polymerization. Much to our surprise, when we tried this, we obtained an even better catalyst. How this improved catalyst works, especially with 1.0 acetylacetone per Al, is still not thoroughly understood, except to propose that aluminum becomes 5 or 6 coordinate in the propagation step. This catalyst, which I will refer to as the chelate catalyst, has indeed proven to be one of the most versatile catalysts in the field of epoxide polymerization as well as for the polymerization of some other monomers such as aldehydes (46), especially for oxetanes (47), and for episulfides (48).

The catalysts shown in Figure 5, i.e., the trialkylaluminum-H_2O and the trialkylaluminum-H_2O-chelate catalysts, were covered by a composition of matter patent (49). The chelate catalyst has been used commercially to produce the epichlorolydin-ethylene oxide copolymer rubber (50). The trialkyl aluminum-H_2O catalyst has been used to produce isotactic and atactic polyepichlorohydrin (51). The alkylaluminum-H_2O catalysts, or more correctly, called

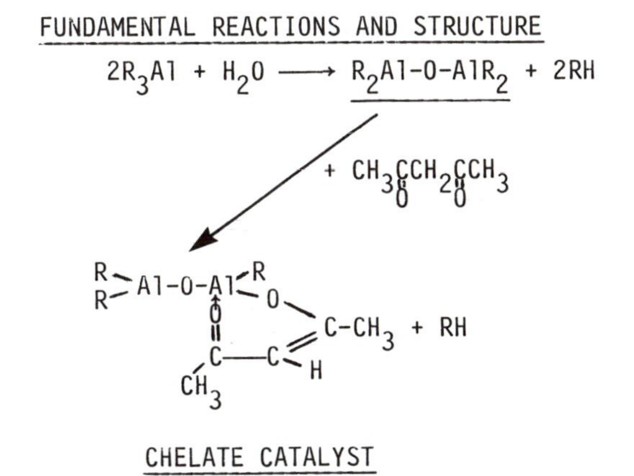

Figure 5. R_3Al-H_2O Catalysts (33). (Reproduced with permission from ref. 33, copyright 1983 Plenum)

alkylaluminoxanes, have been used in other research areas, i.e., (1) recently in some outstanding work by Kaminsky et al (52) (see also Kaminsky et al., this book) and others for some very unique transition metal based catalysts for propylene polymerization which required specific alkylaluminum-H_2O products as co-catalysts and (2) by a number of workers for the polymerization of cyclic esters such as β-butyrolactone (53,54). Specific alkylaluminum-H_2O catalysts have recently been offered for sale by Texas Alkyls, Ethyl Corporation and Schering Berlin Polymers, Inc.

While studying the alkylaluminum-H_2O catalysts, I also discovered a wide variety of other useful catalysts for epoxides. Many of these were based on the reaction of organometallics such as zinc or magnesium with di-or polyfunctional additives such as water, polyols, ammonia, amines, diacids, diketones, etc., (55). As a result of the discovery of these catalysts, particularly the alkylaluminum-water-chelate catalyst, we were able to make a large number of new high molecular weight polymers, particularly in the epoxide (55,56) and oxetane (47) areas.

Commercial Aspects. A few of the more interesting monosubstituted epoxides that were polymerized to both amorphous and crystalline high molecular weight polymers with coordination catalysts are illustrated in Table II. The epichlorohydrin amorphous homopolymers (57) and amorphous copolymers with ethylene oxide (58) are especially interesting oil resistant rubbers. The commercial polyether elastomers derived from epichlorohydrin or propylene oxide have been described in detail (50). The epichlorohydrin elastomers are being produced by Zeon Chemicals Company, an affiliate of Nippon

Table II. High Polymers from Monosubstituted Epoxides (33)

	Amorphous	Crystalline (m.p., °C)
$-CH_2-CH-O-$ | CH_2Cl		120
$-CH_2-CH-O-CH_2-CH_2O-$ | CH_2Cl	Solvent-resistant rubbers	-
$-CH_2-CH-O-CH_2-CH-O-$ | (on CH) CH_3; | (on CH) CH_2 | $O-CH_2-CH=CH_2$	S-Vulcanizable rubber	-
$-CH_2-CH-O-$ | $CH=CH_2$	Rubber	74

Zeon, in the U.S. under the trade name of Hydrin and by Nippon Zeon and Osaka Soda (now Daiso) in Japan. World-wide production is 12,000-13,000 tons/yr. The Zeon Chemicals Company acquired the commercial interests of the B. F. Goodrich Chemical Company (first licensee of Hercules Incorporated) who previously had acquired the full interests of Hercules Incorporated. The properties and advantages of these rubbers make an interesting story. Briefly, the copolymer is especially unusual - having a unique combination of properties. It has the oil resistance of nitrile rubbers combined with the good rubber properties and environmental resistance of neoprene.

The copolymers of propylene oxide with small amounts of an unsaturated epoxide, such as allyl glycidyl ether, can be sulfur vulcanized to vulcanizates with excellent low temperature and dynamic properties similar to natural rubber combined with good ozone resistance and good heat resistance. This polyether elastomer is available commercially through Zeon Chemical Company under the registered trademark Parel 58 and found to be especially useful as a specialty elastomer in applications such as automotive engine mounts (47). This product is an amorphous elastomer and is characterized by adjacent propylene oxide units being only about 70% head-to-tail as a result of the chelate catalyst used. Research studies (59) indicate that strength and oil resistance of this elastomer can be greatly improved by making this copolymer with a catalyst that will give a high yield of isotactic, head-to-tail poly (propylene oxide) from propylene oxide. This result is a further example of the criticality of the catalyst to the properties and thus the probable value of a product in applied polymer science. Further details on the past and future of polyether elastomers are given in my Charles Goodyear Medal lecture (60).

Mechanism Studies. Our extensive mechanism studies on the polymerization of epoxides and oxetanes, generally and specifically, with the R_3Al-H_2O and R_3Al-H_2O-acetylacetone catalysts have been reviewed in detail (33). The important conclusions are reiterated below:

(1) The R_3Al-H_2O catalyst can behave as a superior cationic catalyst with epoxides which polymerize readily cationically or as a coordination catalyst with other epoxides such as epichlorohydrin.

(2) The chelate catalyst always gives coordination polymerization. This catalyst involves polymerization at sterically-hindered aluminum sites which vary in steric hindrance and in their copolymerization ability. Thus, in the copolymerization of epichlorohydrin (ECH) and trimethylene oxide (TMO) with ethylene oxide under conditions needed to make a uniform copolymer with largely ECH or TMO, one obtains a small amount of water-soluble copolymer which is very high in ethylene oxide.

(3) All polymerizations of mono and disubstituted epoxides with all known anionic, cationic, or coordination catalyst occur with complete inversion of configuration of the ring opening carbon atom (56). This result appears to be a very important first principle of epoxide polymerization (55).

(4) The mechanism of the coordination polymerization of epoxides must involve two metal atoms in order to make it possible

to obtain a rearward attack on the epoxide. In the mechanism proposed (Equation 3), the epoxide is coordinated to one aluminum atom prior to its attack by the growing chain on the other aluminum atom. The coordination bonds in the catalyst structure are needed to move the growing polymer chain from one metal to an adjacent one without altering the valence of the metal.

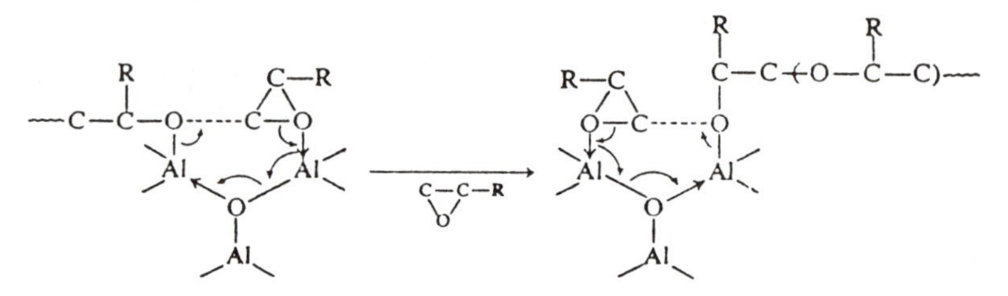

Hydroxy Polyethers

I previously reviewed our extensive work on making high molecular weight polyethers with reactive side chains such as esters, carbonyls, amines and hydroxyl (33). Since hydroxy polyethers are most intriguing in view of their obvious relation to polysaccharides, which are important both biochemically and commercially, I will review in depth our recent work in this area, since some of our products show promise of being of commercial importance.

Polyglycidol. Hydroxyl groups destroy our organometallic catalysts and thus eliminate direct polymerization of hydroxy monomers as a good route to high molecular weight polymers. Also, the nucleophilic replacement of chlorine in polyepichlorohydrin by hydroxyl is not acceptable since hydroxyl anion is a poor nucleophile and polyepichlorohydrin undergoes base cleavage very readily (61). Thus, in most of our work on the polymerization of hydroxy monomers, we have blocked the hydroxyl with an unreactive group which can be readily removed after polymerization. Most of our work has been done with trimethylsilyl ethers.

Our work on preparing water-soluble, high molecular weight atactic, and isotactic polyglycidol as either the racemic mixture or as the optically-active, pure enantiomer, has been reported in detail (33).

Hydroxyoxetane Polymers.

Poly(3-hydroxyoxetane), PHO - An Analog of Poly(vinyl alcohol). This work grew out of our efforts to make isotactic polyglycidol by polymerizing a pure enantiomer of glycidol or its trimethyl silyl ether with a base catalyst to form a rather low molecular weight, water-soluble polymer which was not polyglycidol. Instead, this product was partially branched PHO formed by a new rearrangement polymerization (62) as illustrated in general by Equation 4.

On checking the literature we found that Wojtowicz and Polak (63) had made 3-hydroxy oxetane (HO) and found that it spontaneously polymerized to a low molecular weight, water-soluble polymer, presumably PHO but which had not been adequately characterized. In

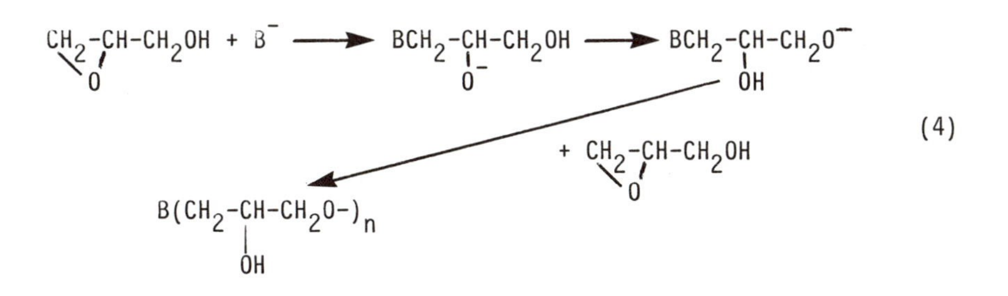

(4)

addition, the polymerization mechanism was not known. We undertook a study at Arizona State University of the Wojtowicz PHO, and the polymerization mechanism by which PHO was formed. Also, linear high molecular weight PHO was made and characterized by polymerizing the trimethylsilyl ether of HO using our organoaluminum-H_2O catalysts and the derived chelate, coordination catalyst, followed by removal of the blocking group (64). This work led to the following conclusions:

(1) The Wojtowicz spontaneous polymer was low molecular weight (Mn up to 2000), water-soluble, linear atactic PHO which was highly crystalline with a M.P. of 155°C.

(2) The polymerization mechanism for the Wojtowicz polymer involved the air oxidation of HO over several months at room temperature to carboxylic acids (as formic acid, hydroxycarboxylic acids, etc.) which polymerize HO to Wojtowicz PHO in an unusual cationic mechanism catalyzed by a proton acid. The rate of polymerization increased greatly with the acid strength of the proton acid, e.g., trifluoromethane sulfonic acid was explosive in pure HO. A detailed mechanism was proposed.

(3) High molecular weight, water-soluble, linear atactic, and highly crystalline PHO (mp = 155°C) was made by polymerizing the trimethylsilyl ether of HO with the i-Bu_3Al-0.7 H_2O cationic catalyst followed by hydrolysis. Its x-ray pattern was the same as the Wojtowicz PHO.

(4) Two ^{1}H-NMR methods for measuring the tacticity of PHO were developed based on finding two different types of methylene units at 400 MHz with the methine protons decoupled. Also, an ^{1}H-NMR method was developed for measuring branching in HO polymers.

(5) High molecular weight, linear PHO with enhanced isotacticity (89%) has been obtained in low yield as a water-insoluble fraction with mp of 223°C.

(6) The low molecular weight PHO prepared previously by the base-catalyzed, rearrangement polymerization of glycidol or its trimethylsilylether is highly branched.

High molecular weight, linear atacic PHO, is an interesting analog of poly(vinyl alcohol) since it can be considered to be a copolymer of vinyl alcohol and formaldehyde. The highly crystalline nature of this atactic polymer may be related to the crystalline nature of atactic poly(vinyl alcohol) (PVA). PHO has some obvious advantages over PVA. It is very easy to melt fabricate into a translucent, tough film with a Tg of -5 to 16°C depending on trace H_2O content. It is very soluble in H_2O and does not precipitate therefrom up to 150°C. A 4% aqueous solution is not affected by adding a saturated aqueous lime solution, and does not precipitate from H_2O by boric acid addition as does PVA. PHO should have a variety of interesting applications if an economic route can be developed. U.S. composition of matter coverage has been obtained on high molecular weight, linear(65) and branched (66) PHO and copolymers of HO.

Poly[3,3-Bis(hydroxymethyl)oxetane], PBHMO - An Analog of Cellulose. PBHMO is an interesting analog of cellulose since it has similar hydroxyl and ether contents, and based on its symmetrical structure would be expected to be very high melting and highly insoluble. There are important differences from cellulose, namely: the hydroxyls are all primary, the main chain oxygens are ordinary ethers and not acetals, and there are no asymmetric carbons. Thus, one might anticipate significant improvement in some properties compared to cellulose. This polymer was first reported in low molecular weight form by Farthing (67) from the polymerization of the cyclic acetone ketal of 3,3-bis(hydroxymethyl)oxetane with BF_3-etherate catalyst, followed by hydrolysis. However, this product was not adequately characterized. We have made PHBMO in high molecular weight (η_{inh} up to 5.2) by polymerizing the trimethylsilylether of 3,3-bis(hydroxymethyl)oxetane with the $i\text{-}u_3Al$-0.7 H_2O cationic catalyst at low temperature, followed by hydrolysis. PBHMO is highly crystalline, very high melting (314°C), with a Tg of 99°C, and highly insoluble. It is soluble in 75% H_2SO_4 at 30°C, being 65% converted to the acid sulfate ester; these conditions are useful for viscosity measurement, since the degradation rate is low and at least an order of magnitude less than for cellulose in this solvent. PBHMO can be prepared as oriented films and fibers using the lower melting diacetate (184°C) which can be melt or solution ($CHCl_3$) fabricated, oriented and then saponified to give oriented PBHMO. The PBHMO prepared by the Farthing (67) published method was too low in molecular weight to make oriented films or fibers by these same methods and thus would not be useful in such applications. Parris and Marchessault (this book) have analyzed the x-ray fiber diffraction pattern of PBHMO.

In view of the HO polymerization results with protonic acids as well as the economic advantage of not using a removable hydroxyl blocking group, we polymerized BHMO directly with trifluoromethane sulfonic acid at room temperature and obtained low molecular weight, perhaps somewhat branched, PBHMO (η_{inh} 0.1), m.p. 275°C. The limited branching indicated by the high m.p. is encouraging, suggesting that further studies may permit the direct preparation of high molecular weight PBHMO.

Poly(3-methyl-3-hydroxymethyloxetane), (PMHMO), prepared in high molecular weight (η_{inh} up to 3.8) by the same method used for PBMO, soluble and lower melting (165°C) than PBHMO, appears to be atactic and can be compression molded at 195°C to a tough, clear film which is readily oriented. Copolymers of BHMO with MHMO are crystalline over the entire composition range with a linear variation of mp with composition, a new example of isomorphism in the polymer area.

Poly(3-ethyl-3-hydroxymethyloxetane), PEHMO, was also prepared in high molecular weight (η_{inh} up to 2) by the same method used for PBHMO. Surprisingly, its melting point (175°C) was higher than PMHMO. It is probable that copolymers of EHMO with BAMO and MAMO will exhibit isomorphism.

Commercial Aspects. The poly(hydroxyoxetanes) described above are intriguing polymers which merit commercial development from both the process and use aspects.

PHO and copolymers of HO should fit into many water-soluble polymer applications now filled by poly(vinyl alcohol) or other water-soluble polymers. The apparent ease of melt fabricating PHO could make it advantageous for films where its oxygen barrier properties should be outstanding. The excellent water solubility which appears independent of temperature and not affected, or less affected, by base, boric acid and perhaps salts could be advantageous. Biocompatibility may be favorable since it is obviously a polymer of glycerin and needs to be carefully assessed. If PHO is biocompatible, numerous applications such as for capsules, blood-extender, etc., become feasible. Biodegradability also may be favorable since poly(vinyl alcohol) is reported to biodegrade readily (68).

PBHMO and copolymers with MHMO and EHMO represent a unique class of polymers exhibiting isomorphism which makes it easy to vary mp, fabricability and properties as needed. The obvious similarity to cellulose but with improved stability opens up a host of possible applications in many areas, including film, fiber and molded articles. Film permeability needs to be assessed for possible unique opportunities in this area. Hereto, biocompatibility needs to be assessed and could be a key to premium applications.

We have broad patent coverage on high molecular weight PBHMO, PMHMO and PEHMO as well as a broad spectrum of copolymers (69).

Coordination Copolymerization of Tetrahydrofuran and Oxepane with Oxetanes and Epoxides (70)

The previously described, very useful chelate catalyst, as typified by the Et_3Al-0.5 H_2O-0.5 acetylacetone product, usually prepared with Et_2O or tetrahydrofuran (THF) present, has all the known characteristics of a coordination catalyst for polymerizing epoxides and uniquely for oxetanes. Prior work indicates that this catalyst has a variety of polymerization sites which differ, at least in the steric restrictions at the site (see also H. N. Cheng, this book). During a study of oxetane polymerization with chelate catalyst prepared with THF present, we discovered that THF will copolymerize with various oxetanes. Thus, we found that the chelate catalyst gives fairly good copolymerization of THF (54% in monomer charge) with 3-(trimethylsilyloxy)-oxetane which, after hydrolysis, is a water-soluble, moderate molecular weight copolymer of THF (36%) with 3-hydroxyoxetane (HO). This apparent coordination, copolymerization of THF has been extended to trimethylene oxide (TMO), 3,3-(bis-trimethylsilyloxymethyl)oxetane, 3,3-bis(chloromethyl)oxetane

(BCMO), trans-2,3-epoxybutane (TBO) and propylene oxide, listed in order of decreasing copolymerizability with THF. Presumably this is the first known coordination copolymerization of THF which hitherto has only been polymerized with cationic catalysts (70). Oxepane also copolymerizes coordinately with TMO and BCMO, but less readily than THF, with the chelate catalyst.

TBO polymerizes slowly with the chelate catalyst to form stereoregular polymer which can be separated into an acetone-insoluble, highly stereoregular fraction and an acetone-soluble, somewhat less stereoregular fraction. The soluble fraction can be eliminated by using 1.0 acetyl acetone per Al in the catalyst or by adding a small amount of a very strong base (0.09 quinuclidine per Al). The copolymerization of TBO with THF (39%) gives insoluble stereoregular homopolymer and soluble copolymer containing about 23% THF, reflecting the varied steric hindrance of the sites.

Coordination copolymers of THF with TMO and TBO have been obtained with 57 and 65% THF respectively, indicating consecutive sequences of THF units which has been confirmed by ^{13}C-NMR. These data suggest that the coordination homopolymerization of THF may be possible.

These results indicate that it is very important to examine a new catalyst in considerable detail since such work may lead to new areas of polymer chemistry, possibly new polymers or new approaches to prior polymers which may lead to new developments in applied polymer science.

Conclusions

Catalysis studies obviously will continue to be one important route to new developments in applied polymer science either for new polymer discovery and development or for improving existing polymers in many ways such as for modified properties, for economics of synthesis and/or for ease of fabrication.

My most important direct contributions were on polyolefins, especially isotactic polypropylene and on polyether elastomers, especially the epichlorohydrin elastomers. On isotactic polypropylene, I discovered high yield catalysts which were essential to its commercial development, leading eventually to it being an important, large volume polymer. These catalysts which were used worldwide have been replaced in the last ten years largely (70-80%) with magnesium chloride-supported catalysts which give very high mileage and higher yields of isotactic polymer so that the commercial process could be simplified and costs reduced. Recently, Kaminsky et al (52) have reported some very unusual, highly active homogeneous catalysts for olefin polymerization and for propylene polymerization to both atactic and isotactic polypropylene. This field is an important example where further study by many workers has led to great improvements in catalysts which have led in many cases to either new or lower cost products.

I also discovered the hydrogen method of controlling molecular weight in transition metal catalyzed olefin polymerization which has proved very useful in commercial production and is still the preferred method.

Ziegler catalyst polymerization is an example of where detailed catalyst studies led to the early discovery of the new metathesis polymerization method (71). Extensive studies in this field by many workers (72-74) has led to many new catalysts, new processes, and new product discoveries as well as the commercialization of poly(dicyclopentadiene) (75). The 2-ω polymerization of α-olefins with some nickel (0) catalysts by Fink et al. (this book) is another outstanding example of the unusual results that can be achieved with transition metal catalysts.

My own efforts to extend Ziegler catalysts to other monomers led to aluminum alkoxide-H_2SO_4 catalysts for making isotactic vinyl etner polymers by a new coordination method. Similar studies led to new catalysts for ring opening polymerization, especially for epoxides as typified by the alkylaluminoxanes and related catatysts. These catalysts led to the discovery and commericial development of the epichlorohydrin oxide elastomers and propylene elastomers. These catalysts, although relatively inefficient are still important today. Although much catalyst work has been done on epoxides and some very important contributions made, especially b Inoue (76), better, especially high mileage catalysts, are needed. The methylaluminoxanes were found by Kaminsky et al (12) to be an essential component for homogeneous transition metal olefin catalysts.

Recent work on using our alkylaluminoxane catalysts on oxetanes has led to some interesting hydroxy polyethers, i.e., poly(3-hydroxyoxetane), an analog of poly(vinyl alcohol) and poly[3,3-bis(hydroxymethyl)oxetane], an analog of cellulose. Both of these interesting polymers, including some related polymers and copolymers which are covered by composition of matter patents merit commercial development (65,66,69).

Our chelate coordination catalyst, e.g., Et_3Al-$0.5H_2O$-0.5 acetylacetone permits the hitherto unknown coordination copolymerization of THF and oxepane with a number of oxetanes and epoxides. The results suggest that ultimately it may be possible to homopolymerize THF with coordination catalysts.

Acknowledgments

Hercules Incorporated and my associates at the Hercules Research Center over 43 years were very helpful in my major research and development activities described herein. Hercules management, in particular, were very understanding and provided me with a relatively free hand in exploring numerous research opportunities. Recent work on hydroxy polyethers and the coordination copolymerization of THF and oxepane with oxetanes and epoxides at Arizona State University was done with the invaluable collaboration of Dr. J. C. Mullis and was supported by the National Science Foundation (Grant No. DMR-8412792), the U.S. Army Research Office (Grant No. MIPR-117-87), the DuPont Company; Hercules Incorporated, and the donors of the Petroleum Research Fund, administered by the American Chemical Society.

Literature Cited

1. Vandenberg, E. J.; Hulse, G. E. Ind. Eng. Chem. **1948**, 40, 932.
2. Vandenberg, E. J. (to Hercules Incorporated), U. S. Patent 2,648,655-6, filed November 24, 1948, issued August 11, 1953.
3. Vandenberg, E. J. (to Hercules Incorporated), U. S. Patent 2,648,657-8, filed April 12, 1947, issued August, 11, 1953, and U.S. Patent 2,682,528 (1954).
4. Kolthoff, I. M.; Medalia, A. I. J. Polym. Sci. **1959**, 5, 391.
5. Fryling, C. F.; Landes, S. H.; St. John, W. M.; Uraneck, C. A. Ind. Eng. Chem. **1949**, 41, 986.
6. Schulze, W. A.; Reynolds, W. B.; Fryling, C. F.; Sperberg, L. R.; Troyan, J. E. India Rubber World **1945**, 117, 739.
7. Hock, H.; Lang, S. Ber. **1944**, 77B, 257.
8. Thurman, C. in Encyclopedia of Chemical Technology; Editor, M. Grayson; Kirk-Othmer; Third Edition, Wiley: **1982**, Vol. 17; p. 373.
9. Boardman, H. (to Hercules Incorporated) U.S. Patent 2,668,180 (1954).
10. Vandenberg, E. J.; Repka, B. C.; "Ziegler-Type Polymerizations," in C. E. Schildknecht, Editor; Polymerization Processes, Wiley, 1977, Chapter 11.
11. Kharasch, M. S.; Lambert, F. L.; Urry, W. H. J. Org. Chem. **1945**, 10, 298.
12. Boor, J. Jr., "Ziegler-Natta Catalysts and Polymerizations," Academic Press, Inc., New York, **1979**.
13. Natta, G.; Pino, P.; Corradini, P.; Danusso, F.; Mantica, E.; Mazzanti, G.; Moraglio, G. J. Amer. Chem. Soc. **1955**, 77, 1708.
14. Vandenberg, E. J. (to Hercules Incorporated), U. S. Patent 3,051,690, filed July 29, 1955, issued August 28, 1962.
15. Lieberman, R. B.; Barbe, P. C. in Encyclopedia of Polymer Science and Engineering, Editor, J. I. Kroschwitz, Second Edition, Wiley, **1988**, Vol. 13; p. 518.
16. Vandenberg, E. J. (to Hercules Incorporated), U. S. Patent 3,058,963, filed April 7, 1955, issued October 16, 1962.
17. Vandenberg, E. J. (to Hercules Incorporated), U. S. Patent 2,954,367, filed July 29, 1955, issued September 27, 1960.
18. Vandenberg, E. J. (to Hercules Incorporated), U. S. Patent 3,108,973, filed February 16, 1961, issued October 29, 1963.
19. Natta, G.; Corradini, P.; Allegra, G. J. Polym. Sci. **1961**, 51, 399.
20. Galli, P. "Polypropylene - A Quarter of a Century of Increasing Successful Development," presented at IUPAC Macromolecular Symposium, Florence, Italy (September, 1980). Preprints 1, 10. Also, Galli, P. and Natta, G. in Struct. Order Polym.; Lect. Int. Symp., Editors, Ciardelli F. and Giusti, P.; Pergamon, Oxford, Engl., **1981**, p. 63.
21. Vandenberg, E. J. (to Hercules Incorporated), U. S. Patent 3,261,821, filed December 31, 1959, issued July 19, 1966.
22. Vandenberg, E. J., Isotactic Polypropylene, E. L. Wittbecker, Editor, in Macromolecular Syntheses, **1974**, 5, 95.

23. Gamble, L. M.; Langer, A. W., Jr.; Neal, A. H. (Esso Research and Engineering Company), U. S. Patent 2,951,045 (August 30, 1960); Winkler, D. E.; Nozaki, K. (Shell Oil Company), U. S. Patent 2,971,925, (February 14, 1961); Fasce, E. V.; Fritz, R. J. (Esso Research and Engineering Company), U. S. Patent 2,999,086 (September 5, 1961); Kaufman, D.; McMullen, B. H. (to National Lead Company), U. S. Patent 3,109,822, filed May 28, 1957, issued November 5, 1963.
24. Ruff, O.; Neuman, F. Z. Anorg. Chem. **1923**, 128, 81.
25. Tornqvist, E.; Seelbach, C. W.; Langer, A. W., Jr. (to Esso Research and Engineering Co.), U. S. Patent 3,128,252, filed April 16, 1956, issued April 7, 1964.
26. Preliminary Bulletin No. 56-6, July 9, 1957, and Manual of Procedure for the Evaluation of Olefin Polymerization Catalyst, Ziegler-Natta Type, 2nd ed., Stauffer Chemical Company, 380 Madison Avenue, New York, 17, New York.
27. (a) Tornqvist, E. G. M.; Richardson, J. T.; Wilchinsky, Z. W.; Looney, R. W. J. Catal. **1967**, 8, 189.
(b) Tornqvist, E. G. M. Ann. N. Y. Acad. Sci. **1969**, 155, 447.
(c) Wilchinski, Z. W.; Looney, R. W.; Tornqvist, E. G. M. J. Catal. **1973**, 28, 351.
28. Hermans, J. P.; Henrioulle, P. (to Solvay & Cie), U. S. Patent 4,210,738 filed March 21, 1972, issued July 1, 1980.
29. Hercules Mixer, 41: 1-10 (January, 1958).
30. Anderson, A. W.; Stramatoff, G. S. (to E. I. DuPont de Nemours & Co.) U.S. Patent 4,076,698 (1978).
31. Chemical Week, March 4, 1987, p. 9.
32. Oil Gas J., June 27, **1988**, 86(26), p. 26.
33. Vandenberg, E. J., "Coordination Polymerization," in Coordination Polymerization, C. C. Price and E. J. Vandenberg, Editors, Plenum Press, New York, **1983**, p. 11.
34. Vandenberg, E. J. In Encyclopedia of Polymer Science and Engineering, Editor, J. I. Kroschwitz, Second Edition, Wiley, **1986**, Vol. 4, p. 174.
35. Vandenberg, E. J. (to Hercules Incorporated), U. S. Patent 3,422,082, filed April 19, 1956 and July 30, 1956, issued January 14, 1969.
36. Vandenberg, E. J. (to Hercules Incorporated), U. S. Patent 3,313,229, filed August 12, 1966, issued Aril 25, 1967.
37. Breslow, D. S.; Christman, D. L.; Espy, H. H.; Lukach, C. A. J. Appl. Polym. Sci.**1967**, 11, 73.
38. Breslow, D. S.; Kutner, A. J. Polym. Sci. **1971**, 9B, 129.
39. Vandenberg, E. J. J. Polym. Sci., Part C, **1963**, 207.
40. Vandenberg, E. J. (to Hercules Incorporated), U. S. Patent 3,284,426, fled March 18, 1957, issued November 8, 1966.
41. Christman, D. L.; Vandenberg, E. J. (to Hercules Incorporated), U. S. Patent 3,025,282, filed September 14, 1959, issued March 13, 1962.
42. Heck, R. F.; Vandenberg, E. J. (to Hercules Incorporated), U. S. Patent 3,025,283, filed September 14, 1959, issued March 13, 1962.
43. Vandenberg, E. J. (to Hercules Incorporated), U. S. Patent 3,159,613, filed December 27, 1960, issued December 1, 1964.
44. Osgan, M.; Price, C. C. J. Polym. Sci. **1959**, 34, 153.

45. Vandenberg, E. J. (to Hercules Incorporated), U. S. Patent 3,135,705, filed May 11, 1959, issued June 2, 1964.
46. Vandenberg, E. J. (to Hercules Incorporated), U. S. Patent 3,208,975, filed Dec. 22, 1961, issued Sept. 28, 1965.
47. Vandenberg, E. J.; Robinson, A. E., "Coordination Polymerization of Trimethylene Oxide", in Polyethers, Editor, E. J. Vandenberg, American Chemical Society, Washington, D.C., **1975**, 101.
48. Vandenberg, E. J. J. Polym. Sci., Part A1, **1972**, 10, 329.
49. Vandenberg, E. J. (to Hercules Incorporated), U. S. Patent 3,219,591, filed originally May 11, 1959, issued Nov. 23, 1965.
50. Vandenberg, E. J. in Encyclopedia of Chemical Technology, Editor, M. Grayson, Kirk-Othmer, Third Edition, Wiley, Vol. 8, **1979**, p. 568.
51. Vandenberg, E. J. in Macromol. Syntheses, Editor, W. J. Bailey, Wiley, **1972**, Vol. 4, p. 49.
52. Simm, H.; Kaminsky, W.; Vollmer, H.-J.; Wolfe, R. Angew. Chem. Int. Ed. Engl. **1980**, 19, 390.
53. Agostini, D.; Lando, J. B.; Shelton, J. R. J. Polym. Sci., Polym. Chem. Ed. **1971**, 6, 2771.
54. Gross, R. A.; Zhang, Y.; Konrad, G.; Lenz, R. W. Macromolecules **1988**, 21, 2657.
55. Vandenberg, E. J. J. Polym. Sci., Part A1, **1969**, 7, 525.
56. Vandenberg, E. J. Pure and Applied Chem., **1976**, 48, 295.
57. Vandenberg, E. J. (to Hercules Incorporated), U. S. Patent 3,158,580, filed March 11, 1960, issued November 24, 1964.
58. Vandenberg, E. J. (to Hercules Incorporated), U. S. Patent3,158,581, filed March 11, 1960, issued November 24, 1964.
59. Vandenberg, E. J. J. Polym. Sci., Part A: Polymer Chemistry **1986**, 24, 1423.
60. Vandenberg, E. J. Charles Goodyear Medal Lecture, Presented at American Chemical Society, Division of Rubber Chemistry meeting, Toronto, Canada, May 22, 1991. Rubber Chemistry and Technology, **1991**, 64(3), G56.
61. Vandenberg, E. J. J. Polym. Sci. Polym. Chem. Ed. **1972**, 10, 2903.
62. Vandenberg, E. J. J. Polym. Sci. Polym. Chem. Ed., **1985**, 23, 915.
63. Wojtowicz, J. A.; Polak, R. J. J. Org. Chem. **1973**, 38, 2061.
64. Vandenberg, E. J.; Mullis, J. C.; Juvet, R. S., Jr., Miller, T.; Nieman, R. A. J. Polym. Sci. Polym. Chem. Ed. **1989**, 27, 3113.
65. Mullis, J. C.; Vandenberg, E. J. (to Arizona Board of Regents), U.S. Patent 4,965,342 (1990).
66. Mullis, J. C.; Vandenberg, E. J. (to Arizona Board of Regents), U.S. Patent allowed, Serial No. 203,262.
67. Farthing, A. C. J. Chem. Soc. **1955**, 1955, 3648.
68. Marten, F. L in Encyclopedia of Polymer Science and Engineering, Editor, J. I. Kroschwitz, Wiley, **1988**, Vol. 17, p. 180.
69. Vandenberg, E. J. (to Arizona Board of Regents), U. S. Patent 4,833,183 (May 23, 1989).
70. Vandenberg, E. J.; Mullis, J. C. J. Polym. Sci. Polym. Chem. Ed., **1991**, 29, 1421.
71. Eleuterio, H. C. (to E. I. DuPont de Nemours and Co., Inc.) U. S. Patent No. 3,074,918 (1963).
72. Ivin, K. J. Olefin Metathesis, Academic Press, London (1983).

73. Schrock, R. R.; Feldman, J.; Cannizzo, L. F.; Grubbs, R. H. Macromolecules, **1987**, 20, 1169.
74. Ginsburg, E. J.; Gorman, C. B.; Marder, S. R.; Grubbs, R. H. J. Amer. Chem. Soc., **1989**, 111, 7621.
75. Breslow, D. S. Polymer Preprints, **1990**, 31(2), 410.
76. Sugimoto, H.; Aida, T.; Inoue, S. Macromolecules, **1990**, 23, 2869.

RECEIVED November 15, 1991

Transition Metal Catalyzed Coordination Polymerization of Olefins and Diolefins

Heterogeneous Catalysts

The commercialization and production of isotactic polypropylene, for the first 20 to 25 years, was done with heterogeneous catalysts based on the δ form of titanium trichloride. In the past 10–15 years, a new technology has been developed based on heterogeneous catalysts consisting of titanium chlorides supported on $MgCl_2$. These catalysts are much more effective, giving very high catalyst mileages, high rates, and high stereospecificity. Thus, greatly improved, lower cost processes were developed because the very small amount of catalyst and by-product amorphous polypropylene present in these new systems could be left in the product. Thus, costly purification methods could be avoided. Today, isotactic polypropylene is made largely (probably >80%) with these new heterogeneous, supported catalysts. The two chapters in this subsection deal with various facets of these new catalysts.

Homogeneous Catalysts

Homogeneous Ziegler catalysts were examined in detail in the late 1950s by Breslow et al. for ethylene polymerization using bis(cyclopentadienyl)-titanium dichloride (titanocene) plus organoaluminum chlorides and were found to be somewhat unusual, but they were never exploited. During the past 15 years, Kaminsky and collaborators discovered that the zirconium metallocene combined with methyl aluminoxane co-catalyst polymerized α-olefins, gave extremely high polymerization rates (even higher than the new heterogeneous supported catalysts), and could be modified to prepare isotactic polypropylene in a homogeneous system. These new homogeneous systems have been used by Ewen et al. and others to prepare highly stereoregular syndiotactic polypropylene in good yield and at conventional polymerization temperature, as contrasted with the early work of Natta et al. when the product was of poor stereoregularity and −78 °C was required. This subsection includes recent work with these homogeneous catalysts by Kaminsky et al. on α-olefins, by Kashiwa et al. on 1-butene, and by Tait et al. on ethylene. Studies on the structure and morphology of highly stereoregular syndiotactic polypropylene are reported by Galambos et al. A homogeneous catalyst based on a nickel(0)–phosphorane combination giving a unique α-olefin polymerization by 2,ω-linkage is described in detail by Fink and co-workers.

Metathesis Systems

Metathesis polymerization was discovered in the late 1950s during transition metal catalyst studies. This unusual polymerization method has been studied extensively ever since, as evidenced by two books published in the mid-1980s. After much research, the mechanism has been established as being a ring-opening polymerization of an intermediate metallacyclobutane. Recently, metathesis catalyst systems have been modified to give living systems and extended to more polar cyclic olefins. Outstanding advances continue to be made in this area. One of the most important contributors (R. R. Schrock) gave an excellent lecture in the symposium related to this book but elected not to provide a manuscript.

Commercialization of metathesis polymers has been limited. The most promising commercialization is on dicyclopentadiene to prepare useful products by reaction injection molding. The chapter by Bell presents a detailed catalyst study in this area.

Chapter 2

Most Advanced Magnesium Chloride Supported Ziegler–Natta Catalyst

James C. W. Chien

Department of Polymer Science and Engineering, University of Massachusetts, Amherst, MA 01003

Most advanced CH-type Ziegler-Natta catalysts were prepared from $MgCl_2 \cdot 3ROH$ adduct, phthalic anhydride, $TiCl_4$ and phthalate esters. The best catalyst has the Mg/Ti ratio of 22, the ester/Ti ratio of ca. 1, very high specific surface area and pore volume and rotationally disordered $MgCl_2$. Alkyl phthalate esters are superior to terephthalate or monofunctional esters. Only the Ti(III) ions are catalytically active; the amounts of isospecific and nonspecific sites are 25 and 15% of the Ti atoms, respectively, with propagation rate constant values of 200 and $11 (Ms)^{-1}$. Solid NMR and FTIR showed that the Lewis base is complexed to $MgCl_2$; EPR provided evidence for the coordination of ester to Ti(III). The optically active catalyst containing (S)2-methyl-butylphthalate preferentially polymerizes the (S) antipode of 4-methyl-1-hexene and the (R) antipodes of 3-methylolefins stereoelectively suggesting site stereoselection by the ester. Detailed comparisons were made with an earlier type of Ziegler-Natta catalyst derived from crystalline $MgCl_2$.

The first stereospecific Ziegler-Natta (Z-N) catalyst (*1*) for propylene polymerization, based on α-$TiCl_3$ obtained by hydrogen reduction of $TiCl_4$ (*2,3*), has only low activity of *ca.* 10 g PP/(g Ti·atm·h). δ-$TiCl_3 \cdot 0.33AlCl_3$, obtained by aluminum alkyl reduction of $TiCl_4$ and ball-milling, had an increased surface area and catalytic activity of *ca.* 200 g PP/(g Ti·atm·h). This catalyst was employed exclusively in the manufacturing of polypropylene for more than two decades and is still producing much of the polypropylene worldwide today. The introduction of $MgCl_2$-supported $TiCl_3$ catalysts (*4-7*) has changed the manufacturing technology of polypropylene. Because of greatly enhanced catalytic productivity, the process of catalyst removal became unnecessary. The morphology of the polypropylene

0097–6156/92/0496–0027$08.25/0

produced is large, nonfriable, spherical granules thus obviating the need for pelletization.

We had investigated in great depth an early version of such a catalyst referred to as the CW-catalyst (*8,9*). It was prepared by ball-milling one mol of crystalline $MgCl_2$ with 0.15 mol of ethylbenzoate, treating the mixture with 0.5 mol of *p*-cresol, reacting it with 0.25 mol of $AlEt_3$, and then heating the product with an excess of $TiCl_4$ (*8*). Examinations of each intermediate stage were made by EPR (*10*), FTIR (*11*), BET (*12*), porosimety, X-ray diffraction and solid ^{13}C-NMR(*13*). Ball-milling in the presence of EB was found to reduce $MgCl_2$ crystallite to $\ll$ 100Å in size and caused the formation of a mixture of cubic close packing (*ccp* ABCABC...), hexagonal close packing (*hcp*, ABAB...), and rotationally disordered (*rd*) structures. Treatment with PC and TEA acted to break up the crystallite aggregates. It is often thought that the active sites in the $MgCl_2$ supported catalysts are formed by epitaxial growth of $TiCl_3$ on $MgCl_2$ to account for the stereoselectivity displayed by these sites. The optimum CW-catalyst was prepared using two Lewis base: internal Lewis base (B_i= ethylbenzoate) and external Lewis base (B_e = methyl-*p*-toluate). Using isotopic radiolabeling to count the active sites (*13*) we found (*14-17*) that the function of ethylbenzoate was to help create active sites of all kinds, whereas methyl-*p*-toluate acts to inhibit preferentially the nonstereo-selective (C_a^*) sites.

More advanced heterogeneous ZN-catalysts, referred to as the CH-type catalysts, were prepared from soluble magnesium compounds. They are several-fold higher in activity and much more stereospecific than the CW-catalysts. This chapter describes our findings about this catalyst (*18,19*), and compares it with the CW-catalyst.

Catalyst Preparation

A CH(EH, BP) catalyst of extremely high activity and stereo-specificity was prepared by the following procedure(*18*). $MgCl_2$ (15 mmol) was dehydrated by heating in the presence of HCl and suspended in 10 mL of decane. Three equivalents of 2 ethylhexanol (EH) was added and the slurry was heated for 2 h at 130°C to form the $MgCl_2 \cdot 3ROH$ adduct solution. Then 2.25 mmol of phthalic anhydride was added, which reacted with ROH:

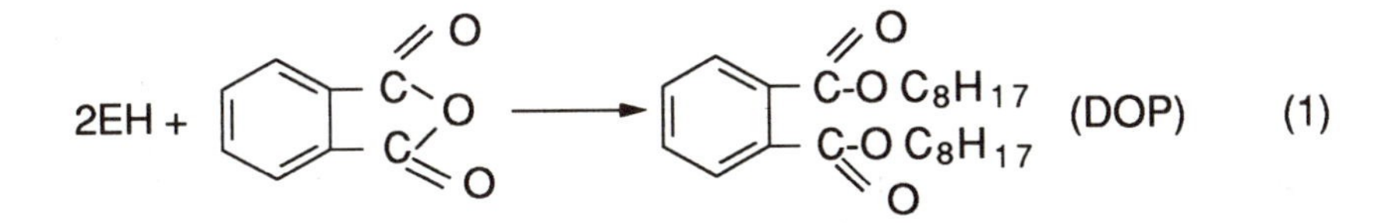

so that the "adduct" solution contains $MgCl_2 \cdot 3ROH$ with 0.15 mol eq. of DOP. This solution was cooled and added dropwise over 1 h to 70 mL $TiCl_4$ (0.63 mol, 42x$MgCl_2$) at -20°C. The reaction mixture was heated at a rate of 0.5°/min to 110°C and then 2.97 mmol of phthalate ester was added, and the mixture was stirred for 2 h at 110°C. After hot filtration, the solid was resuspended in 220 mL of $TiCl_4$ (1.8 mol), reacted again at 110°C for 2 h and hot filtered. Repeated washing with decane and heptane produced the CH-catalyst suspended in 50 mL of heptane. The amount of HCl liberated in the reaction,

$$MgCl_2 \cdot EH \cdot DOP + TiCl_4 \longrightarrow MgCl_2 \cdot DOP + Cl_3TiOC_8H_{17} + HCl \quad (2)$$

corresponded to > 89% of the EH present. Crystallization of $MgCl_2$ occured very slowly because of the slow addition of the $MgCl_2$ adduct at very low temperature and the presence of phthalate esters.

Table I Analysis of CH-catalysts

	CH-5	*CH-6*	*CH-7*	*CH(EH, BP)*
Mg (mmol/g)	6.56	7.56	6.64	7.96
Ti (mmol/g)	0.74	0.68	0.39	0.36
Cl (mmol/g)	16.3	16.6	15.8	18.0
Mg / Ti	8.9	11	17	22
EH[a]	0.27	0.31	0.009	0.006
BP[b]			0.44	0.29
DOP[c]	0.57	0.57	0.30	0.15
[[EBP]+DOP]/[Ti]	0.77	0.84	2.0	1.2

[a]2-Ethylhexanol, [b]di-*i*-butylphthalate, [c]di-octylphthalate.

In order to understand the process of the catalyst preparation, the composition of $MgCl_2$, and the role played by the ester, we had tried different preparative procedures and characterized the resulting catalysts (Table I). A CH-5 catalyst was obtained by reacting $MgCl_2$•3EH with $TiCl_4$, and washing. Additional treatment of CH-5 with $TiCl_4$ once more produced CH-6. The CH-7 catalyst was prepared similar to CH(EH, BP), except the second reaction with $TiCl_4$ was omitted. The assays for inorganic elements given in Table I showed [Cl] ≈ 2x[Mg] + 4x[Ti]. To analyze the organic components, the catalyst was hydrolyzed with 1N HCl, extracted with ether and analyzed by gas chromatographic method using *n*-hexanol as an internal standard.

CH-5 contains the highest amount of Ti and CH(EH, BP) the lowest; they differ by 0.38 mmol/g of Ti. Much of this difference may be attributed to the presence of $Cl_3Ti(OR)$ in the former catalyst analyzed as EH. The second treatment with $TiCl_4$ actually lowers slightly the [Ti] in CH-6. The ratio of phthalate ester to Ti is about 0.8. For the other two catalysts, di-*i*-butylphthalate (BP) was present during the reaction of $TiCl_4$ with $MgCl_2$•3EH. The hydrolyzates of these catalysts contained only negligible amount of EH. This suggests that BP can remove $Cl_3Ti(OR)$ or $TiCl_4$ physisorbed on $MgCl_2$. Conversely, there appears to be physisorbed phthalate esters in CH-7. The second $TiCl_4$ treatment removes the excess esters. In these catalysts the $MgCl_2$ is saturated with Ti upon the first reaction with $TiCl_4$. The second $TiCl_4$ treatment actually lowers the [Ti] slightly. The best catalyst, CH(EH, BP), contains the lowest amount of Ti and about one equivalent of phthalate esters per Ti.

The four different CH catalysts differ somewhat in specific surface areas (250-340 m^2/g), pore volume (0.23-0.37 cm^3/g) or pore radius (26-40Å) but without consistent trends of relationships with polymerization activity. The catalysts differ markedly in their X-ray diffraction pattens. Figure 1 shows the X-ray diffraction pattern of ball-milled $MgCl_2$ which contains reflections from both the *ccp* and *hcp* structures. In the case of CH-5 catalyst (Figure 2), the 15.3° reflection was considerably broadened. The crystallite size is estimated to be *ca.* 100Å according to Scherrer's formula (*19*). However, there is a shoulder indicating contribution to diffraction from rotational disorder (17.2°, 5.15Å). The most intense reflection is now found at 32.5° with shoulders on each side of it. This is consistent with the presence of more *hcp* than *ccp* structures and the presence of *rd* (33.8°, 2.65Å), i.e. 60° rotation of the Cl-Mg-Cl layers (*20-21*). The high intensity between 41° and 43° and 60.8° are consistent with a large population of *hcp* structure. The 2θ reflections at 20.3°, 22.1°, 37.6° and those

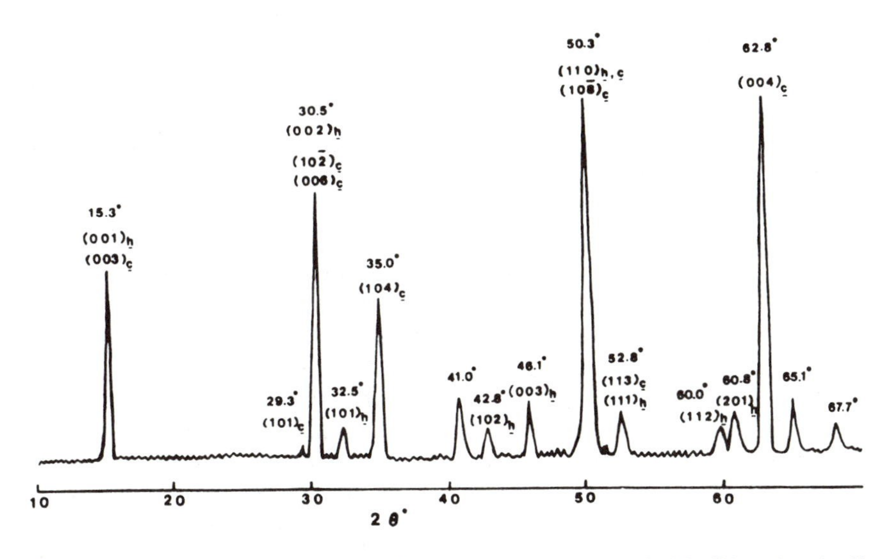

Figure 1. Powder x-ray diffraction pattern of $MgCl_2$ dry ball milled for 0.5 hr. *(18, 36)*

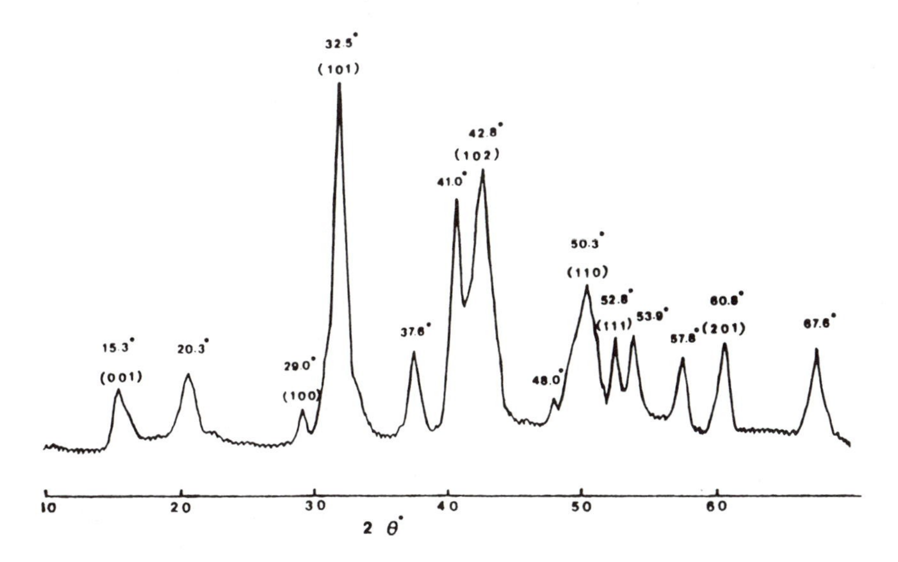

Figure 2. Powder x-ray diffraction pattern of CH-5 catalyst. *(18, 36)*

between 53° and 59° in figure 2 have not been noted in the previous reports, they may arise from the *rd*-$MgCl_2$. Additional treatment of CH-5 produces CH-6 whose X-ray diffraction pattern is shown in Figure 3. There is apparently annealing of the $MgCl_2$ crystallites as many reflections become narrower and resolved. The two catalysts do not differ appreciably in Ti, Cl_3TiOR, or DOP contents, nor in BET surface area and pore volume.

The X-ray diffraction patterns for CH-7 and CH(EH, BP) catalysts (Figure 4 and 5) are more diffuse. The $TiCl_4$ treatment of CH-7 did not have an annealing effect as in the case of CH-5. Instead, most of the reflections are broadened by the treatment, in particular the intensity distributions between 28° and 38° bears resemblance to that for extensively ball-milled $MgCl_2$ described by Galli et al.(*20*). It is interesting to compare the x-ray diffraction patterns of the CH(EH, BP) catalyst with that of the CW catalyst. The former, shown on the top of Figure 6, is that of Figure 5 reduced to the same scale for the angular coordinates of the CW catalysts. The 20.3°, 42.8° and 67.6° reflections appear clearly in the CH(EH, BP) catalyst whereas they are very weak or broad in the CW catalyst. On the other hand, the (100) reflection is resolved in the CW catalysts (Figure 6c) while it is only discernible as a shoulder in the CH-8 catalyst. Though the two catalysts have comparable crystallite dimensions, they differ in crystalline disorders (*21-23*).

The x-ray data helps to understand the catalyst formation processes. In equation 2 the adduct itself has 0.15 mol of DOP. The treatment with $TiCl_4$ dissociates some of the ester from $MgCl_2$,

$$MgCl_2 \cdot n\mathrm{DOP} + TiCl_4 \rightleftharpoons MgCl_2 + MgCl_2 \cdot m\mathrm{DOP} + TiCl_4 \cdot (n\text{-}m)\mathrm{DOP} \qquad (3)$$

and free $MgCl_2$ forms ordered crystallites (*s*),

$$n\text{-}MgCl_2 \longrightarrow (ccp \text{ and } hcp)\ MgCl_2\ (s) \qquad (4)$$

$MgCl_2$ complexed with DOP or BP precipitated slowly due to reduced enthalpy of crystallization and possible stabilization effect of the (*rd*)$MgCl_2$(*s*) by the ester. The protocol for the preparation of CH-8 was optimized to produce (*rd*)$(MgCl_2)_{22}$ $\cdot(TiCl_4 \cdot$ BP).

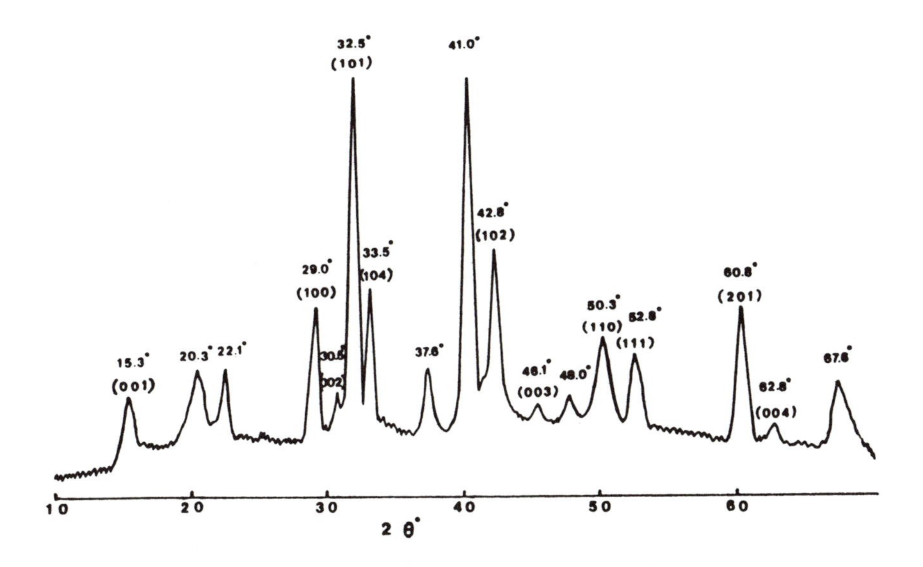

Figure 3. Powder x-ray diffraction pattern of CH-6 catalyst. *(18, 36)*

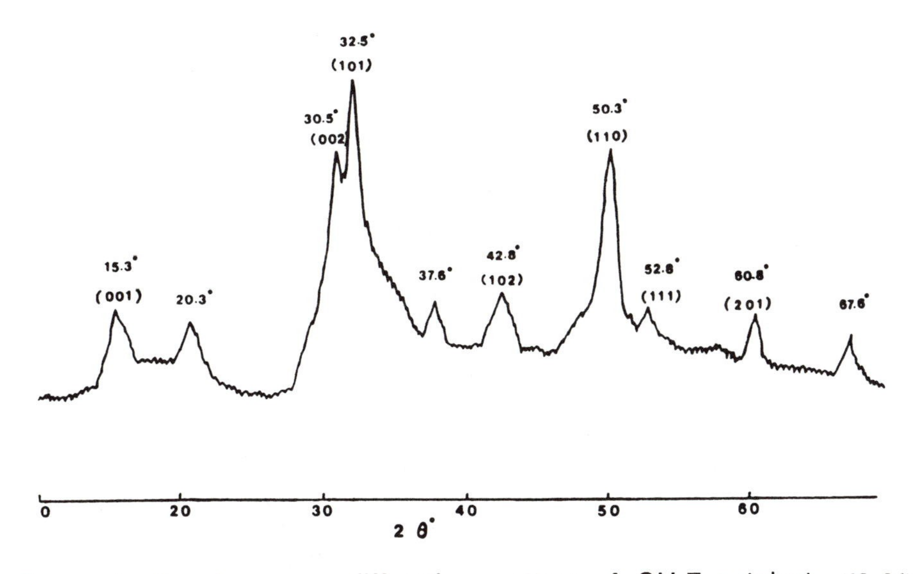

Figure 4. Powder x-ray diffraction pattern of CH-7 catalyst. *(18, 36)*

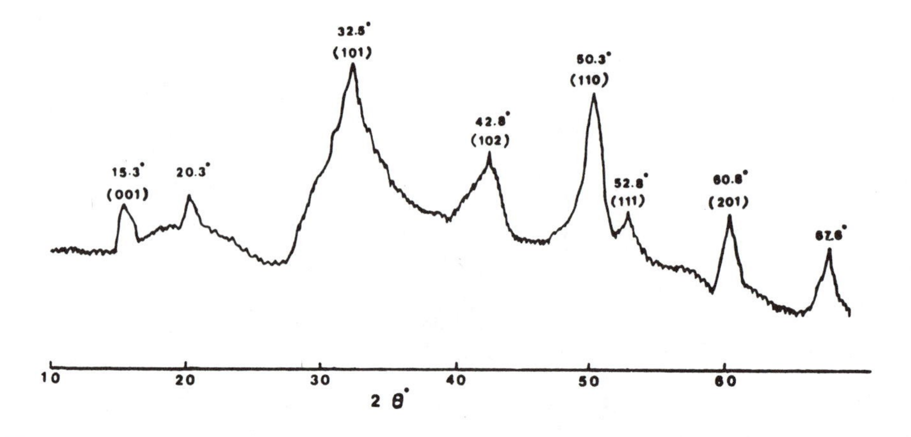

Figure 5. Powder x-ray diffraction pattern of CH-8 catalyst. *(18, 36)*

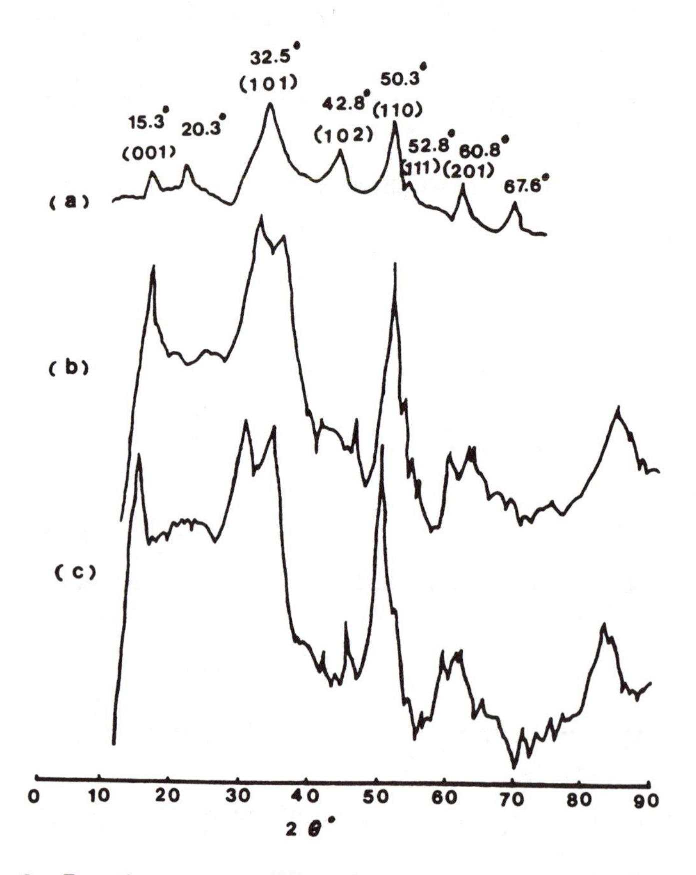

Figure 6. Powder x-ray diffraction pattern (a) CH-8 catalyst reduced from Figure 5; (b) CW catalyst support, $MgCl_2$/EB/PC /TEA; (c) CW catalyst, $MgCl_2$/EB/PC/TEA/$TiCl_4$. *(18, 36)*

The above CH catalysts differ appreciably in activity and stereospecificity. Arranged in the order of CH-5, CH-6, CH-7 to CH(EH, BP), the activity increases from 5.5 to 15.5 kg PP/(g Ti•atm•h) and the I.Y. (yield of refluxing *n*-heptane insoluble isotactic polypropylene, *i*-PP) increases from 96% to 99%.

Effect of Ester Structures

The action of phthalic ester occurs at the molecular level. The CW-catalyst utilizes aromatic monoester and the CH-catalyst needs a phthalic ester as B_i. For instance ball-milling of *c*-$MgCl_2$ with diesters did not produce any beneficial effect. A series of CW type catalysts were prepared by ball-milling of $MgCl_2$ in the presence of dimethyl phthalate, diethyl phthalate, di-*i*-butyl phthalate, or di-*n*-butyl phthalate followed by reactions with *p*-cresol, $AlEt_3$ and $TiCl_4$. There was no enhancement of either stereospecificity or productivity as compared to the CW-catalyst prepared with ethylbenzoate. The ball-milled product of $MgCl_2$ and di-2-ethylhexyl phthalate was too sticky to be useful for catalyst preparation.

On the other hand, the CH-type catalyst prepared with monoesters only is inferior in all respects compared to the phthalate containing CH-catalysts. An example is the catalyst obtained by reacting $MgCl_2$•3EH•0.3EB with $TiCl_4$ at 90°C. It was analyzed to contain 1.05 mmol/g of Ti, Mg/Ti = 5.7, 0.032 mmol/g of EH and 1.4 mmol/g of EB. It has an activity of only 3.3 kg PP/(g Ti•atm•h) and 86% I.Y. A second treatment of this catalyst with $TiCl_4$ had no beneficial effects. Other catalysts were prepared by reacting $MgCl_2$•3EH and $TiCl_4$ for 2h, then EB was introduced and stirred for 2 more h at 90°C. This catalyst has lower [Ti] of 0.47 mmol/g, Mg/Ti = 13.4, and 1.75 mmol/g of EB. The activity was only 8.5 kg PP/(g Ti•atm•h) and I.Y. = 90%. All these catalysts obtained with monoesters were characterized by low surface areas and pore volumes. Probably the monoester complex of $MgCl_2$ is less stable than the phthalate complexes and there is more free $MgCl_2$ in eq. 3 resulting in the formation of *ccp* and *hcp* $MgCl_2$ by eq. 4.

Other aromatic diesters were found to give inferior catalysts compared to BP (*24*). The catalytic activities for (-)-dimenthyl phthalate, di-*t*-butyltere-phthalate, and (-)-dimenthylterephthalate were 3.3, 6.2 and 2.4 kg PP/(g Ti•atm•h), respectively, and corresponding I.Y. values of 89, 90 and 85%.

Kinetics

The CH-type catalyst both more active and stereoselective than CW-type catalyst as shown in Table II. In order to understand these differences we determined the numbers of active sites, with and without stereochemical control, $[C^*_i]$ and $[C^*_a]$, respectively.

Table II Activities and Stereospecificities of the CH(EH, BP) and CW-catalysts in Propylene Polymerizations[a]

Catalyst	*Activator*	*Activity (kg PP/(g Ti atm hr))* *Total*	*Isotactic*	*I.Y. (%)*
CH(EH, BP)	$AlEt_3$/PES[b]	16.1	15.9	98.6
CH(EH, BP)	$AlEt_3$	20.3	14.6	72.0
CW	$AlEt_3$/MPT[c]	4.0	3.8	96.0
CW	$AlEt_3$	14.0	9.7	68.2

[a]Polymerization conditions: [Ti] = 0.24 M, A/T = 167, temp. = 50°C, $[C_3H_6]$ = 0.65 M, TEA/MPT = 3, TEA/PES-20; [b]PES = phenyltriethoxysilane; [c]MPT = methyl-*p*-toluate.

Active Site Determination. The active site concentrations were determined by quenching propylene polymerizations with CH_3OT, precipitating the polymers, extracting the stereo-irregular PP with boiling heptane, and radioassaying the fractions (*13-17*). The total metal-polymer-bond concentrations, $[MPB]_i$ and $[MPB]_a$ for the isotactic and atactic polymers, respectively, are calculated from the assays, the known specific activity of CH_3OT, and the kinetic isotope effect. Figure 7 and 8 gave the variation of [MPB] versus polymerization yields with and without PES as a modifier. Extrapolation to zero yield gave the initial active site concentration which are summarized in Table III. Both catalysts in the absence of external Lewis base have about the same $[C_t^*]$ of 70% to 75%, but half of the active sites in CH($-B_e$) are isospecific as compared to only one-fifth in the case of CW($-B_e$) (*16,17*). When the CW catalyst was prepared without the internal Lewis base (EB), i.e. CW($-B_i$, $-B_e$), there are fewer active sites of both kinds. Addition of B_i increases the number of both C_i^* and C_a^* to about the same extent. The roles of Lewis base in the CH-catalyst are

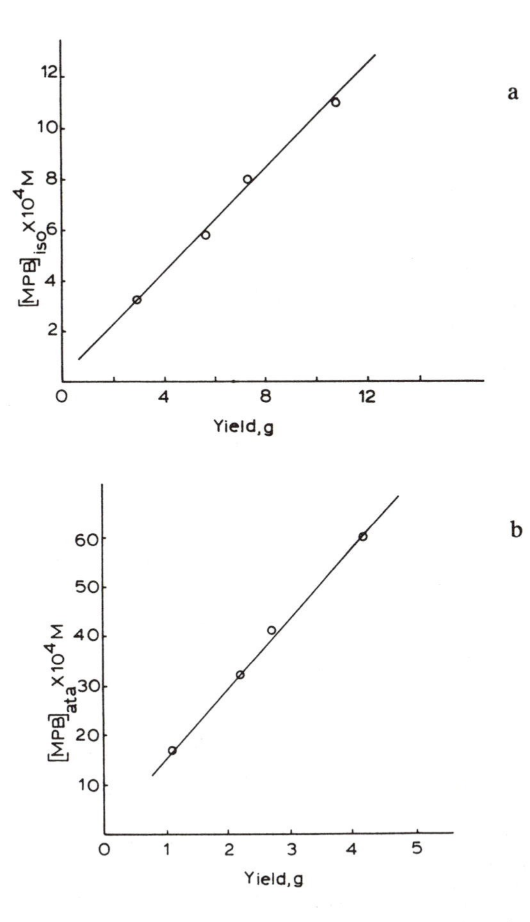

Figure 7. Variation of [MPB] versus polymerization yield at [Ti] = 0.24 mM, Al/Ti = 167, $[C_3H_6]$ = 0.65 M, temp. = 50°C, activated with $AlEt_3$ alone: (a) isotactic fraction; (b) atactic fraction. *(18, 36)*

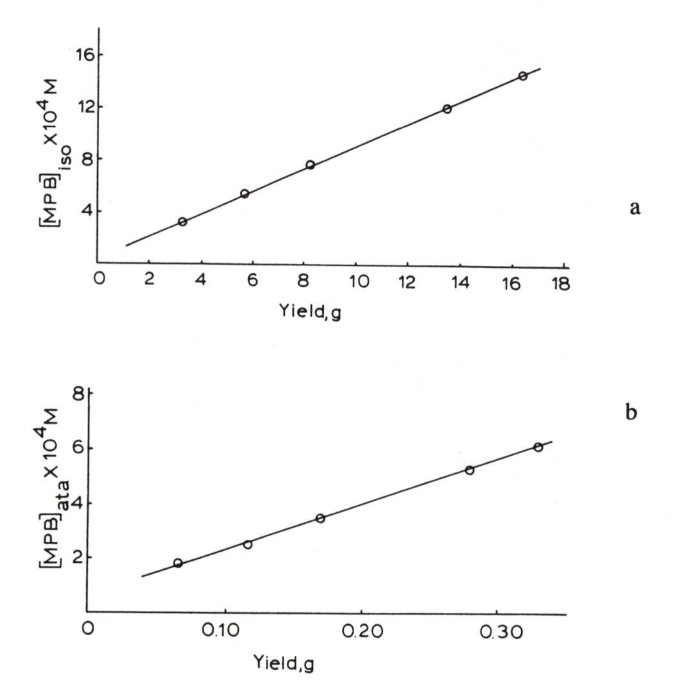

Figure 8. Variation of [MPB] versus polymerization yield, same conditions as in Figure 7 but activated with $AlEt_3$/PES (1:20): (a) isotactic fractions; (b) atactic fraction. *(18, 36)*

apparently more selective in favor of the creation of isospecific sites.

Table III Active Site Concentration

Catalyst	Activator	$[C^*]$ (mol% Ti)		
		isotactic	*atactic*	*total*
CH(EH, BP)	$AlEt_3$	33	42	75
CH(EH, BP)	$AlEt_3$/PES	25	15	40
CW	$AlEt_3$	15	55	70
CW	$AlEt_3$/MPT	6.7	6.3	13

The function of the B_e in the CW-catalyst is to inhibit about 90% of the C_a^* sites but unfortunately it also reduces C_i^* by 50% (*16,17*). By comparison PES reduces C_a^* and C_i^* by 64% and 24%, respectively, in the CH-catalyst. As a consequence, the CH-catalyst has 4-times more C_i^* sites but only 2 times more C_a^* sites than the CW-catalyst.

Chain Propagation. Polymerization by $MgCl_2$ supported catalysts reach their maximum rates ($R_{p,m}$) within two to three min. of activation. This was followed by rapid decay of R_p to a low stationary value ($R_{p,s}$). These values are given in Table IV. The rate constants of propagation were obtained from the maximum rates of polymerization and the active site concentrations,

$$k_{p,i} = R_{p,i} \; / \; ([C_i^*] \; [M]) \quad (5)$$

$$k_{p,a} = R_{p,a} \; / \; ([C_a^*] \; [M]) \quad (6)$$

where [M] is the propylene concentration.

The isospecific sites of CH- and CW-catalysts have the same $k_{p,i}$ values of 190 ± 20 $(M\ s)^{-1}$ with or without B_e and same $k_{p,a}$ values of 11 ± 1 $(M\ s)^{-1}$ in the presence of B_e. But the polymerization activity of the C_a^* sites for the CW- catalysts was doubled by the deletion of B_e; it is quadrupled for the CH-catalyst when PES was omitted. It appears that there is a distribution of nonspecific sites spanning a range of Lewis acidity. MPT can deactivate most of these sites, whereas PES is selective in poisoning those C_a^* sites which are more active in catalyzing nonspecific polymerizations.

Table IV Kinetic Parameters

		$CH\text{-}AlEt_3$	$CH\text{-}AlEt_3$/PES	$CW\text{-}AlEt_3$	$CW\text{-}AlEt_3/MPT$
$R_{p,m}$ x10^3(Ms^{-1})					
	total	12.0	8.3	7.2	2.2
	iso	8.6	8.1	4.9	2.1
	ata	3.4	0.25	2.3	0.12
k_p $(Ms)^{-1}$					
	total	102	126	66	109
	iso	170	206	209	200
	ata	51	11	27	11
$k^A{}_{tr}$ x10^3 (s^{-1})					
	iso	24.0	24.8	1.5	3.1
	ata	97.8	22.6	0.17	0.26
k_d (s^{-1})		$(1.81\pm0.42)\times10^{-3}$			
K'_d (s^{-1})		$(4.29\pm1.98)\times10^{-5}$			

Chain Transfer. The chain transfer rate with $AlEt_3$ can be measured directly (*25*) from the increase of [MPB] with yield (Y). The rate constant was calculated from the slope of the plot,

$$[MPB]_i = [C_i^*]_o + k^A{}_{tr,i}\, Y_i / k_{p,i}\, [M] \quad (7)$$

$$[MPB]_a = [C_a^*] + k^A{}_{tr,a}\, Y_a / k_{p,a}\, [M] \quad (8)$$

The values of the chain transfer rate constant (Table IV) are very different for the CH and CW catalysts. This can be explained by the fact that chain transfer is only with the free $AlEt_3$ but not with the $AlEt_3$ complexed with B_e. In the CH-catalyst with B_e, the $AlEt_3$/PES ratio is 20:1. It is $AlEt_3$/MPT = 3 for the CW-catalyst. In addition MPT is probably more Lewis basic than PES, consequently, there is a much higher concentration of free TEA in the CH(+B_e) system than in the CW(+B_e) system. This explaination can also account for the large decrease of $k^A{}_{tr}$ by B_e for the CW-catalyst and a small effect on the $k^A{}_{tr}$ of the CH-catalyst by PES.

Catalyst Deactivation. $MgCl_2$ supported catalysts are characterized by large decay of R_p, which was sometimes attributed to the encapsulation of catalyst by the semicrystalline polymer it produced. The rate of replenishment of consumed monomer by diffusion to reach the encapsulated active sites was said to become progressively slower as polymerization progresses. This postulate was shown to be unimportant (*19*). The R_p decay occurs during interrupted polymerizations in the absence of monomer (*26-28*). Furthermore, the R_p decay is the same for the CW-catalyst in both the propylene and decene polymerizations (*29-31*), even though polydecene is soluble in hydrocarbon solvent at polymerization temperature where as polypropylene is not.

The R_p decay for the CW-catalyst follows an apparent second order kinetics for propylene (*32*), ethylene (*14,15*) and decene (*29,31*) polymerizations. This relationship was interpreted by a pairwise deactivation of adjacent catalytic sites between adjacent $TiCl_3$ in the $MgCl_2$ surface (*31*). The R_p decays for all the CH-catalysts do not follow this apparent second order kinetics. Figure 9 contains the profiles for the CH-$AlEt_3$ and CH-$AlEt_3$ / PES catalyzed polymerizations. The plot of $(1/R_{p,t})-(1/R_{p,o})$ versus time is not linear (Figure 10). Instead, the deactivation kinetics is better analyzed as biphasic first order processes (Figures 11 and 12) indicating the absence of large clusters of Ti(III) ions. The first order rate constants of deactivation are summarized in Table IV. The average value of k_d for the rapid phase is $(1.81\pm0.42)\times10^{-3}\ s^{-1}$; the more stable sites have k_d value of $(4.3\pm2.0)\times10^{-5}\ s^{-1}$.

There are other observations which support the above kinetic analysis. The CW-catalyst is violet colored suggesting the presence of α, γ, or δ $TiCl_3$ phase thus the dominance of deactivation of two adjacent Ti(III) ions. The CH-catalysts, on the other hand, is beige colored even after activation indicating the absence of such $TiCl_3$ phases.

Involvement of Lewis Base at the Active Site.

The effects of Lewis bases on the creation of active sites and on the suppression of undesired nonselective sites were detailed above. But this does not constitute evidence for direct involvement of Lewis base as a part of the active site. Definitive evidences are provided by stereoelective polymerizations and EPR.

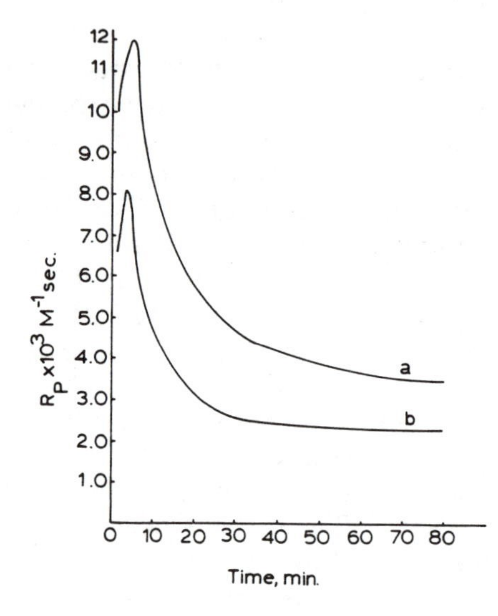

Figure 9. Propylene polymerization profiles for CH-catalyst at [Ti] = 0.24 mM, Al/Ti = 167, $[C_3H_6]$ = 0.65 M; temp. = 50°C: (a) activated with TEA; (b) activated with 20TEA/1PES. *(18, 36)*

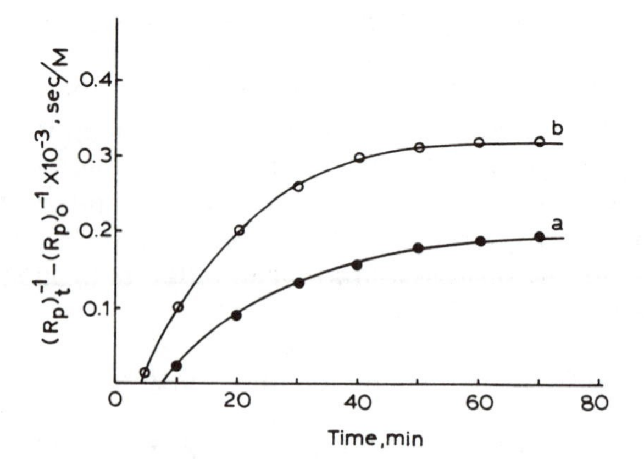

Figure 10. Plots of $(1/R_{p,t})-(1/R_{p,o})$ versus time data of fig. 9. *(18, 36)*

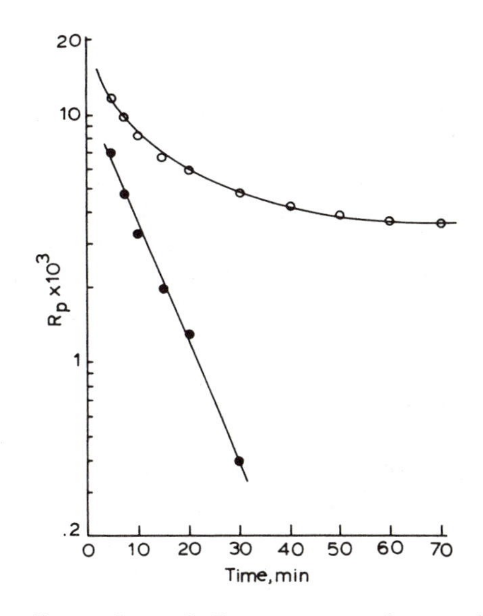

Figure 11. Semilog plot of R_p versus t for polymerization of fig. 9a. *(18, 36)*

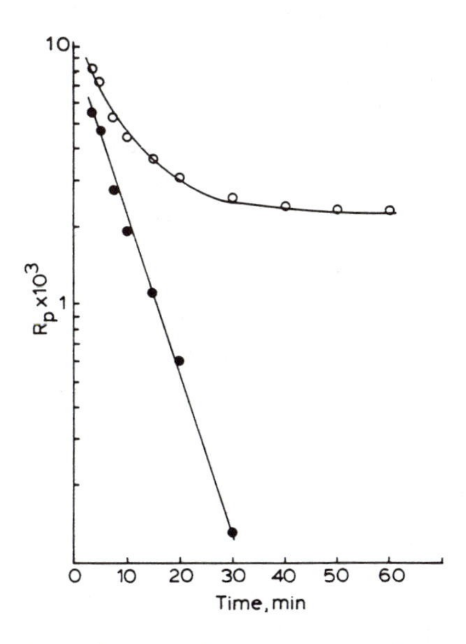

Figure 12. Semilog plot of R_p versus t for polymerization of fig. 9b. *(18, 36)*

Stereoelective Polymerization of Racemic Olefins. Chiral sites of an heterogeneous ZN-catalyst is a racemic mixture, polymerization occurs equally with both enantiofaces of a prochiral monomer. In the case of a *rac*-olefin, there is no enrichment or depletion of one antipode over the other in the polymer and unpolymerized monomer, respectively. However, stereoelective polymerization of one antipode takes place for an optically active catalyst. We have synthesized an optically active CH*((S)2MBP) catalyst using $MgCl_2$•3(S)2-methyl(butanol) and (S)2-methylbutyl-phthalate in place of $MgCl_2$•3EH and BP, respectively. For comparison purposes, an optically active CW* ((-)MBz) catalyst was prepared using (-) menthylbenzoate in place of ethylbenzoate. They were used to catalyze the polymerizations of racemic 4-methyl-1-hexene (4MH). Table V shows that the (S) antipode was preferentially polymerized with optical purity P_p = ((100-Y)/Y)P_m, where Y is the percent conversion. The recovered monomer was enriched in the (R) antipode with optical purity P_m = ($[\alpha]_m$ / $[\alpha_m{}^o]$)x100, where $[\alpha]_m{}^o$ is the molar optical rotation for the neat optically pure monomer = +3.03°. The stereoelectivity is given by the ratio of R_p for the two antipodes = R_p(R)/R_p(S) = (50+0.5P_p)/(50-0.5P_p). The CH* and CW* catalysts showed comparable stereoelectivity. In the polymerizations of *rac*-3MH (3-methyl-1-hexene) and *rac*-DMO (3,7-dimethyl-1-octene), the (R) antipodes were peferentially polymerized. The stereoelectivities are greater for the 3-methyl olefins than for 4MH. These results showed direct influence of the optically active Lewis base in stereoelective insertion of racemic monomers.

Site Stereoselection. The racemic CH((R,S)-2MBP) catalyst was found to be less active than the CH*((S)2MPP) catalyst in olefin polymerizations. For instance the former has activity for propylene polymerization ($[C_3H_6]$ = 0.73 M, $[C_3H_6]$ / [Ti] = 3000, time = 1 h, temp. = 50°C) of 24 kg PP / (g Ti • M • h) as compared to an activity of 49 kg PP / (g Ti • M • h) for the latter. The two catalysts have the same stereospecificity (IY = 98% for PP). The difference in activity is even greater for the polymerization of 4MH; the values are 9.3 and 70 g P4MH / (g Ti • M • h) for CH((R, S)2MBP) and CH*((S)2MBP) catalysts, respectively. Comparison of the propylene polymerizations kinetics with these two catalysts showed that the CH((R, S)2MBP) catalyzed polymerization was slow to reach $R_{p,m}$. The

Table V Stereoelective Polymerizations of Racemic α-Olefins

Catalyst	Monomer	$\frac{[M]^a}{[Ti]}$	Conv. (%)	Recovered monomer			Total polymer			$\frac{R_p(R)}{R_p(S)}$
				$[\alpha]^b$	P_m (%)	Prev.[c] chir.	$[\alpha]^d$	P_p (%)	Prev.[c] chir.	
CH*((S)2MBP)	4MH	3600	37	+0.02	0.66	R	+4.8	1.1	S	0.98
CW*((-)MBz)[b]	4MH	3600	27.4	+0.01	0.32	R	+3.7	0.85	S	0.98
CH*((S)2MBP)	3MH	1200	6.4	+0.066	0.33	S	_[e]	4.8	R	1.10
CH*((S)2MBP)	DMO	300	9.4	+0.157	0.76	S	-3.2	9.3	R	1.21
CH*((S)2MBP)	DMO	1200	8.4	+0.086	0.52	S	-4.3	5.6	R	1.12
CW*((-)MBz)[b]	DMO	3600	20.8	+0.13	1.10	S	-3.4	4.0	R	1.08

[a] [M] is monomer concentration; [b] at 25°C, sodium D line, neat; [c] prevailing chirality; [d] in cyclohexane; [e] P(3MH) is incompletely soluble in cyclohexane at 25°C.

rate of polymerization increased gradually during the first eleven min. to a maximum rate of $2x10^{-3}$ M s^{-1} followed by a modest 50% rate decrease. In contrast, the CH^*((S)2MBP) catalyzed polymerization reaches a maximum R_p of $6.4x10^{-4}$ M s^{-1} a couple of min. after the addition of the catalyst, after which R_p decayed to about 40% of the maximum rate. Active site determinations found the optically active catalyst to possess three-fold greater $[C_i^*]$ than the racemic catalyst (Table VI). The CH catalysts containing achiral Lewis bases are included for comparison, which has the identical $k_{p,i}$ values.

Table VI Kinetic Parameters of Propylene Polymerizations by Various CH Type Catalysts

catalyst	*CH(R,S)2MBP)*	*CH^*((S)2MBP)*	*CH(EH, BP)*
$[C_t]$ (mole % of Ti)	7.2	38	40
$[C_i]$ (mole % of Ti)	5.0	17	25
$[C_a]$ (mole % of Ti)	2.2	21	15
$k_{p,t}$ $(M\ s)^{-1}$	158	96	126
$k_{p,i}$ $(M\ s)^{-1}$	217	213	206
$k_{p,a}$ $(M\ s)^{-1}$	28	3.5	11
$k^A{}_{tr,i}$ X 10^3 s^{-1}	1.4	2.2	25
$k^A{}_{tr,a}$ X 10^3 s^{-1}	3.1	3.6	23

The differences between the optically active and racemic CH(2MBP) catalysts in olefin polymerizations may be explained by a difference in the relative stability of the diastereomeric complexes of C_l or C_d with either antipode of B_i^*,

$$C^*_l \cdot (S)B^* \underset{}{\overset{K_9\,(l,\ S)}{\rightleftharpoons}} C^*_l + (S)B^* \quad (9)$$

$$C^*_d \cdot (R)B^* \underset{}{\overset{K'_9\,(d,\ R)}{\rightleftharpoons}} C^*_d + (R)B^* \quad (9')$$

$$C^*_l \cdot (R)B^* \underset{}{\overset{K_{10}\,(l,\ R)}{\rightleftharpoons}} C^*_l + (R)B^* \quad (10)$$

$$C^*_d \cdot (S)B^* \underset{}{\overset{K'_{10}\,(d,\ S)}{\rightleftharpoons}} C^*_d + (S)B^* \quad (10')$$

The above equilibria may involve $AlEt_3$; all the species are attached to the $MgCl_2$ surface. The C^* species in the left-hand-side of the equilibria are in the resting states; the C^* species on the right-hand-side of the equilibria are catalytically active. One set of diastereomers is more stable than the other set; we can assume $K_9 >> K_{10}$ without loss of generality. In the CH^*((S)2MBP) system, the C^*_l is present largely in the dissociated state which is catalytically active, while the stable $C^*_d \cdot (S)B^*$ is not. In the racemic system both C^*_l and C^*_d form stable complexes with the corresponding preferred antipode $(R)B^*$ and $(S)B^*$. Dissociation of these stable complexes has to occur before they can initiate polymerization. The racemic catalyst has only one-eighth of the 4MH polymerization activity by CH^*((S)2MBP). The latter also has 5 time more active sites for propylene polymerization than the CH((R,S)2MPB) catalyst (Table VI). It seems that the *rac*-B_i complexes with both C^*_d and C^*_l, whereas the optically active B^*_i complexes preferentially complexes with either C^*_d or C^*_l. These experiments demonstrated site stereoselection by the chiral Lewis base.

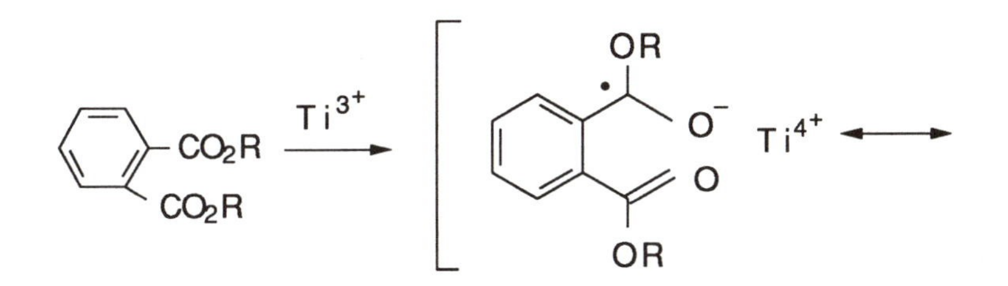

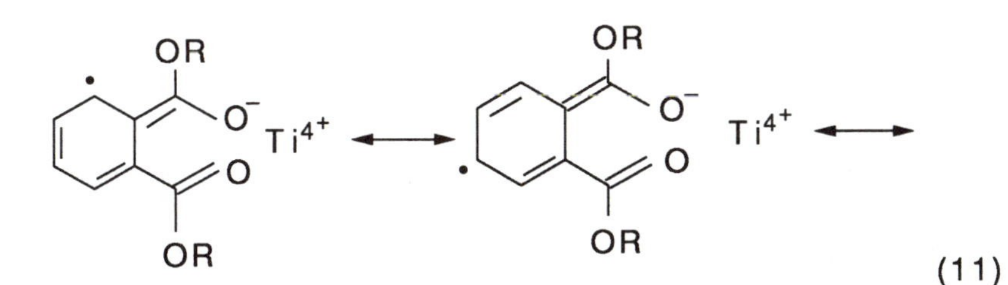

(11)

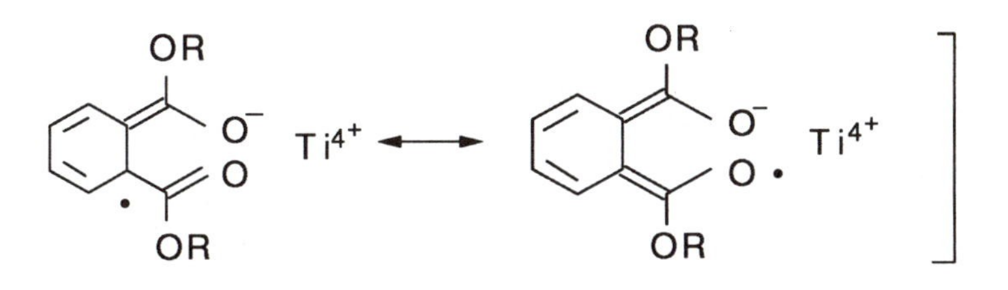

Electron Paramagnetic Resonance. The epr spectra of the CH(EH, BP) catalyst exhibits an intense and narrow $g \sim 2$ signal corresponding to 5-8% of the Ti(III) in the catalyst. This resonance was seen whether the catalyst was activated with $AlEt_3$ alone or with $AlEt_3$/PES both in the absence or in the presence of decene. This resonance was not observed in all those CW and CH type catalysts prepared with monoesters, i.e. EB, and activated either with $AlEt_3$ alone or with $AlEt_3$/MPT. It may be attributed to a radical anion of the phthalate ester as shown in eq. 11. This epr signal provides a direct evidence for the involvement of Lewis base in the active site.

Oxidation State and Epr of Titanium.

The amounts of Ti(II) and of Ti(III) ions have been determined by redox titration (*8,33,34*). The concentration of Ti(IV) ion was obtained by subtracting the former from the total Ti assay. The results are given in Table VII.

Table VII Oxidation State of Ti by Titration and Epr

	CH-AlEt₃	*CH-AlEt₃/*PES	*CW-AlEt₃*	*CW-AlEt₃/MPT*
Ti^{+2} (%)	19.3	33.4	28.7	37.9
Ti^{+3} (%)	73.7	38.5	65.3	28.2
Ti^{+4} (%)	7.0	28.1	6.0	33.9
Ti^{+3} (epr) (%)	34	24	19	8
Ti^{+3}(silent)(%)	39	15	53	17.4
$[C^*_i]$ (%)	33	25	15	6.5
$[C^*_a]$ (%)	42	15	55	6.5

A portion of the Ti(III) species were observable by epr, the remainder was not detected (epr silent). The spectra of CH(EH, BP) activated with $AlEt_3$ alone contains six resonances in the decreasing order of magnetic field: g_1 = 1.887, g_2 = 1.916, g_3 = 1.935, g_4 = 1.949, g_5 = 1.979 and g_e = 2.003. The last signal corresponds to the free electron signal (vide supra). In the epr spectra of this catalyst activated with $AlEt_3$/PES, the g_2 signal is indistinct, a g_6 = 1.966 signal takes the place of the g_4 and g_5 resonance on the higher and lower field sides of it, and the g_5 signal is more intense than that without PES. The amount of epr observable Ti(III) was obtained by double integration of the spectrum; the results are also contained in Table VII.

In the classical ZN-catalyst the epr observable Ti(III) concentration is only a small fraction of the total Ti(III), i.e. [Ti(III)(epr)] << [Ti(III)(total)]. Even in the case of the CW-catalyst [Ti(III)(silent)] = [Ti(III)(total)] - [Ti(III)(epr)] ≈ [Ti(III)(total)]. The present CH catalysts has [Ti(III)(epr)] comparable to [Ti(III)(silent)]. There is excellent agreement between $[C^*_i]$ and [Ti(III)(epr)] and between $[C^*_a]$ an [Ti(III) (silent)], regardless of whether $[C^*_i] >> [C^*_a]$ or vice verse. We concluded that the C^*_i sites are observable by epr and that the C^*_a sites are epr silent. The most likely reason for the latter to be epr silent is that they are bridged by chloride ions resulting in antiferromagnetism (*9,35*).

Differences between CH and CW catalysts.

Though the CH and CW systems are very similar in their polymerization of propylene except for the larger number of active sites in the former, they differ appreciably in the polymerization of bulky monomers. Table VIII gives the activities

Table VIII Polymerization of α-olefins by $MgCl_2$ Supported Ziegler-Natta Catalysts

Monomer	*Activity (g polymer/g Ti [M] h)*	
	CH((S)2MBP)*	*CW*((-)MBz)*
C_3H_6	22,000(1.0)[a]	15,000(1.0)[a]
4MP	3,800(0.17)	2,500(0.17)
4MH	55(0.0025)	230(0.015)
DMO	$0.60(2.7 \times 10^{-5})$	38(0.0025)
3MH	$0.60(2.7 \times 10^{-5})$	

[a]Activities normalized to propylene in parenthesis.

for the polymerizations of 4MP (4-methyl-1-pentene), 4MH, DMO and 3MH. They are normalized to the activity of propylene polymerization and given in parenthesis. The CH-catalyst is four and sixty times lower in activity for 4MH and DMO polymerizations, respectively, than the CW-catalyst. Since the activity is proportional to $k_p \cdot [C^*]$, the results indicate the CH-catalyst has relatively fewer active sites than the CW-catalyst for the polymerization of the branched monomer. However, the very low $[C^*]$ cannot be determined with any precision even by the radiolabeling method. If the active sites in the two catalysts differ only in amounts but are structurally very similar, then they should produce polymers having nearly the same molecular weights and distributions and other properties. Table IX compares these properties of polymers of DMO and of fractionated samples of P(4MH). The polymers obtained with the CH*((S)2MBP) catalyst have much lower M_n than the corresponding polymers derived from the CW*((-)MBz) catalyst. This is consistent with the greater $k^A{}_{tr}$ values for the former system (Table IV). However, the former have narrower distribution and greater enthapy of melting (ΔH_m) than the latter. This indicates that the dispersity of active structures which produces more stereoregular polymers is smaller in the CH-catalyst than the CW-catalyst.

Structures of Active Sites.

There are now sufficient knowledge about the isospecific propylene polymerization sites. It is a trivalent titanium species according to oxidation state determination, and is *not* bridged to an adjacent Ti(III) ion. The Lewis base is complexed to $MgCl_2$ according to FTIR (*10*) and solid state NMR (*12*), and to the Ti(III) ion according to epr (*36*). The ^{27}Al-shfs resolved in some of the epr spectra (*9*) indicates the complexation of some of the sites with an alkylaluminum compound. The possible structures for the isospecific sites in the CW-catalyst are shown in eqs. 12,

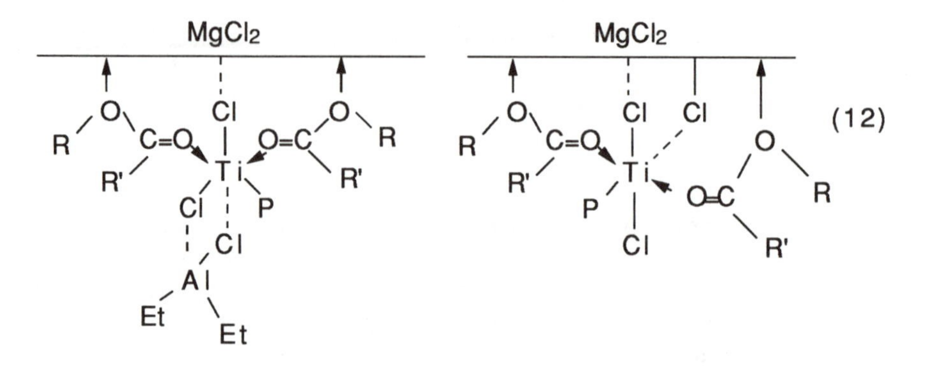

Table IX Properties of Polyolefins

Polymer	Fraction	Catalyst									
		CH*((S)2MBP)					CW*((-)MBz)				
		M_n x10^{-4}	M_w x10^{-4}	P.D.[a]	T_m (°C)	ΔH_m (cal/g)	M_n x10^{-4}	M_w x10^{-4}	P.D.	T_m (°C)	ΔH_m (cal/g)
P4MH	Acetone	0.18	0.75	4.1			-	-	-		
P4MH	Ethylacetate	0.525	1.72	3.3	97-121	1.02	2.55	9.20	3.6		
P4MH	Diethylether	0.99	2.77	3.0	186-191	2.37	7.13	29.9	4.2	186-194	0.54
P4MH	Cyclohexane	1.12	5.26	4.9	209-214	7.29	7.27	82.3	11.3	194-208	2.72
PDMO	Cyclohexane	0.72	5.76	8.0			6.66	152	22.9		

[a]P.D. is polymer dispersity.

In the case of the CH-catalyst the monoesters in the above structures are replaced by the phthalate esters. These active sites are characterized by low Lewis acidity, low affinity toward B_e, epr with rhombic symmetry, chirality, and long life time. The catalytic species which have much shorter life times, probably have neighboring species of similar structure but not bridged to one another. The mechanism of their deachvation had been elucidated (*33*).

The nonstereospecific sites are epr silent and more Lewis acidic than the stereospecific species. They may have structure such as (eq. 13),

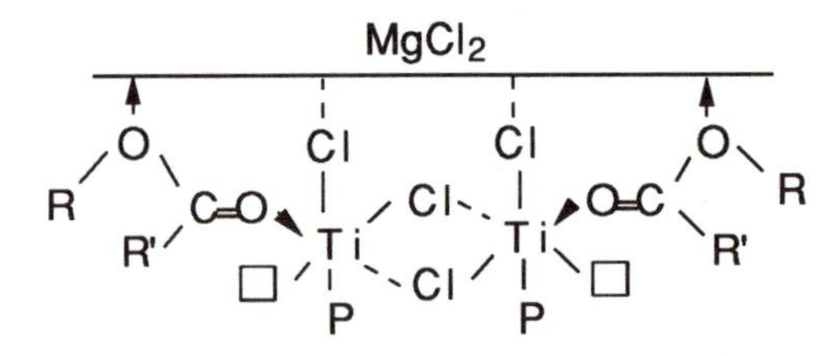

(13)

Many questions remains about the nature of the active sites which polymerizes 4- or 3-methyl olefins, such as the oxidation state, coordination vacancy, chirality, etc. The fact that there is stereoelection in the production of all fractions of P(4MH) indicates the involvement of optically active esters for all the active sites. But the esters are probably weakly complexed and dissociates easily to permit the monomers of appreciate steric bulk to be polymerized. The possible structure are,

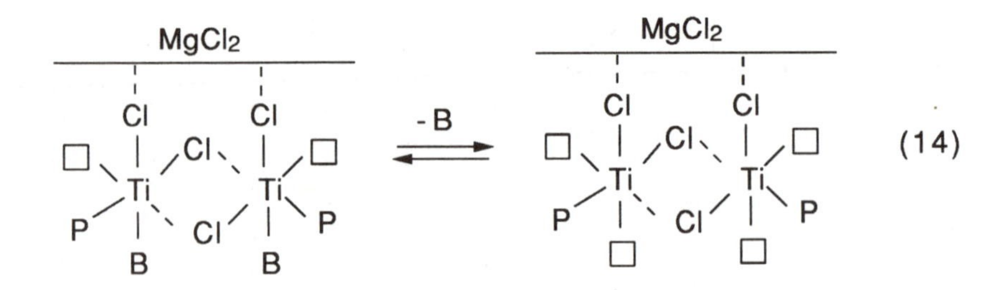

(14)

We know little or nothing whether a particular site is capable of polymerizing only one or more than one kind of monomer. For instance are the active sites for 4MH or DMO also active in aspecific polymerization of propylene.

Even the most advanced CH catalyst has much room for improvement. Only a quarter of the Ti in this catalyst with $AlEt_3$/PES as the cocatalyst are active in stereoselective polymerization of propylene. Furthermore the catalyst has $k_{p,i}$ value of 200 $(Ms)^{-1}$, which is about a factor of ten smaller than the $k_{p,i}$ of the ethylenebis(4,5,6,7-tetrahydro-1-indenyl) zirconium dichloride/methylaluminoxane catalyst (*37*). Therefore, order of magnitude improvement in heterogeneous ZN-catalyst is a reasonable goal to strive for.

Literature Cited

1. Abbreviations;
   B_e, external Lewis base; B_i, internal base; BP, di-*i*-butylphthalate, C^*_a, nonspecific active site; C^*_i, isospecific active site; C^*_t, total active site; *ccp*, cubic close packing; *c*-$MgCl_2$, crystalline $MgCl_2$; DMO, 2,3-dimethyl-1-actene; DOP, dioctyphthalate; EB, ethylbenzoate; EH, 2-ethylhexanol; *hcp*, hexagonal close packing;IY, isotactic polypropylene yield (% insoluble in refluxing *n*-heptane); $k_{p,i}$, isospecific propagation rate constant; $k_{p,a}$, nonspecific propagation rate constant; 3MH, 3-methyl-1-hexena; 4MH, 4-methyl-1-hexene; MPT, methyl-p-toluate; PES, phenyltriethoxy-silane; R_p, rate of polymerization; $R_{p,m}$, maximum rate of polymerization; *rd*, rotational disorder; ZN, Ziegler-Natta.
2. Natta, G. *J. Polym. Sci* **1959**, *34*, 21.
3. Chien, J. C. W. *J. Polym. Sci. A1* **1963**, *1*, 425, 1839.
4. U. S. Pat. 4,107,413; Ger. Pat. 2,029,992, Montedison.
5. Jap. Pat. 75-126, 590; Jap. Pat. 77-151, 691, Mitsui Petrochemicals.
6. Belg. Pat. 737,778, Hoechst; Belg. Pat. 743,325, Solvay & Cie.
7. Belg. Pat. 751,315, Stamicarbon; Ger. Pat. 2,003,075, Shell.
8. Chien, J. C. W.; Wu, J. C.; and Kuo, C. I. *J. Polym. Sci. Polym. Chem. Ed.* **1982**, *20*, 2019. (This is the first paper on the CW-catalyst).
9. Chien, J. C. W.; Dickinson, L. C.; and Vizzini, J. *J. Polym. Sci. Part A* **1990**, *28*, 2321.
10. Chien, J. C. W.; Wu, J. C. *J. Polym. Sci. Polym. Chem. Ed.* **1982,** *20*, 2461.

11. Chien, J. C. W.; Wu, J. C.; and Kuo, C. I. *J. Polym. Sci. Polym. Chem. Ed.* **1983**, *21,* 725.
12. Chien, J. C. W.; Wu, J. C.; Kuo, C. I. *J. Polym. Sci. Polym. Chem. Ed.* **1983**, *21,* 737.
13. Chien, J. C. W.; Kuo, C. I. *J. Polym. Sci. Polym. Chem. Ed.* **1985**, *23*, 731.
14. Chien, J. C. W.; Bres, P. *J. Polym. Sci. Polym. Chem. Ed.* **1986**, *24,* 2483.
15. Chien, J. C. W.; Bres, P. *J. Polym. Sci. Polym. Chem. Ed.* **1986**, *24*, 1967.
16. Chien, J. C. W.; Hu, Y. *J. Polym. Sci. Polym. Chem. Ed.* **1987**, *25,* 2847.
17. Chien, J. C. W.; Hu, J. *J. Polym. Sci. Polym. Chem. Ed.* **1987**, *25*, 2881.
18. Hu, Y.; Chien, J. C. W. *J. Polym. Sci. Part A* **1988**, *26,* 2003. This is the first paper on the CH-catalyst.
19. Chien, J. C. W. *J. Polym. Sci. Polym. Chem. Ed.* **1979**, *17,* 2555.
20. Galli, P.; Barbe, P.; Guidetti, G.; Zannetti, R.; Mortorana, A.; Marigo, A.; Bergozza, M.; and Fichera, A. *Eur. Polym. J.* **1983**, *19,* 19.
21. Allegra, G. *Nuovo Cim.* **1962**, *23*, 502.
22. Wilchinksy, Z. W.; Lovney, R. W.; and Tornqvist, E. G. M. *J. Catal.* **1973**, *28*, 351.
23. Allegra, G.; Bassi, I. W. *Gazz. Chim. Ital.* **1980**, *110*, 437.
24. Chien, J. C. W.; Hu, Y. and Vizzini, J. C. *J. Polym. Sci. Part A* **1990**, *28*, 273.
25. Chien, J. C. W. and Kuo, C. I. *J. Polym. Sci. Polym. Chem. Ed.* **1986**, *24,* 1779.
26. Keii. T.; Suzuki, E.; Tamura, M.; Murata, M. and Doi, Y. *Makromol. Chem.* **1982**, *18*, 2285.
27. Busico, V.; Corradini, P.; de Martine, L.; Proto, A.; Savino, V. and Alibizatti, E. *Makromol. Chem.* **1985**, *186*, 1279.
28. Galli, P.; Luciani, L. and Cecchin, B. *Angew. Makromol. Chem.* **1981**, *94,* 63.
29. Chien, J. C. W.; Ang, T. and Kuo, C. I. *J. Polym. Sci. Polym. Chem. Ed.* **1985**, *23,* 723.
30. Chien, J. C. W. and Ang, T. *J. Polym. Sci. Polym. Chem. Ed.* **1987**, *25,* 919.
31. Chien, J. C. W. and Ang. T. *J. Polym. Sci. Polym. Chem. Ed.* **1987**, *25*, 1011.
32. Chien, J. C. W. and Kuo, C. I. *J. Polym. Sci. Polym. Chem. Ed.* **1985**, *23*, 761.

33. Chien, J. C. W.; Weber, S. and Hu, Y. *J. Polym. Sci. Part A* **1989**, *27* 1499.
34. Weber, S.; Chien, J. C. W.; Hu, Y. in *Transition Metals and Organometallics as Catalysts for Olefin Polymerization*; Kamminsky, W.; Sinn, H. Ed.; Spinger-Verlag, Berlin, 1988 PP45-53.
35. Griffith, J. H. E.; Owen, J.; Park, J. G. and Partridge, M. F. *Proc. R. Soc. London Ser.* **1957**, *A, 250*, 84.
36. Chien, J. C. W.; Hu, Y. *J. Polym. Sci. Part A* **1989**, *27,* 897.
37. Chien, J. C. W.; Sugimoto, R. *J. Polym. Sci. Part A* **1991**, *29*, 459.

RECEIVED November 15, 1991

Chapter 3

Polymerization of C_2–C_6 α-Olefins in the Presence of Modified Ziegler–Natta Catalysts Based on Titanium Halides and Organoaluminum Sulfates

E. Chiellini[1], S. D'Antone[1], R. Solaro[1], F. Masi[2], F. Menconi[2], and L. Barazzoni[2]

[1]Department of Chemistry and Industrial Chemistry, CNR Center of Stereoordered Optically Active Macromolecules, University of Pisa, Via Risorgimento 35, 56100 Pisa, Italy
[2]Enichem S.p.A., via Maritano 26, S. Donato Milanese, Italy

The preparation of unconventional Ziegler-Natta catalysts is reported and their catalytic activity in the polymerization of ethylene, propylene and 4-methyl-1-pentene is compared with that of the corresponding conventional counterparts. The catalysts were obtained by reacting monomeric or oligomeric alkylaluminum sulfates with transition metal halides, either as such or supported on magnesium dichloride. In spite of their high oxygen content and saline features, the prepared unconventional catalysts were found comparably active as conventional systems in the polymerization of C2-C6 olefins. Essentially linear poly(ethylene) of fairly high molecular weight was obtained. Poly(propylene) and poly(4-methyl-1-pentene) with high isotacticity and molecular weight were also obtained.

In recent years a great relaunch of interest in fundamental and applied investigations of anionic coordinate polymerization of α-olefins is taking place *(1,2)*. The demand for high efficiency catalytic systems suitable to provide polyolefins with properties tailored to specific application and economically competitive with catalysts of the most recent generation is behind this new research and development impulse *(3,4)*.

In this context, modified Ziegler-Natta catalysts less chemically and thermally vulnerable appear exciting for the several implications connected with the polymerization of olefins containing functional groups and polymerization of hybrid comonomer mixtures. In previous papers we described the catalytic activity displayed in the polymerization of alkyl vinyl ethers by soluble stechiometrically defined organoaluminum sulfates or by their insoluble counterparts attained by addition of sulfuric acid to trialkylaluminum or trialkoxyaluminum derivatives *(5,6)*.

Indeed organoaluminum sulfates are structurally reminiscent of alumoxane derivatives that recently have been widely investigated and found to be active alone in the polymerization of lactones and epoxides *(7,8)* and of olefins when combined with transition metal halides *(9,10)*.

As a part of our continuing interest in the stereospecific polymerization of functional monomers *(11-14)* and in the chemical modification of reactive polymers *(15-18)*, in the present contribution we report on the investigation of the catalytic activity of unconventional Ziegler-Natta catalysts based on the use of monomeric and oligomeric organoaluminum sulfates in combination with transition metal halides in the

0097–6156/92/0496–0056$06.00/0

polymerization of ethylene, propylene and 4-methyl-1-pentene. Their activity is compared with that shown by analogous conventional Ziegler-Natta systems.

Experimental

Reagents. Bis(diisobutylaluminum) sulfate (DBAS) and bis(diisopropoxyaluminum) sulfate (DPOAS) were prepared as previously reported *(5,6)* by reacting dimethyl sulfate with diisobutylaluminum chloride and diisopropoxyaluminum bromide, respectively (Scheme I). The heterogeneous systems $(i.C_4H_9)_3Al/H_2SO_4$ (TBAS) and $(i.C_3H_7O)_3Al/H_2SO_4$ (TPOAS), were prepared by adding pure sulfuric acid to a two-fold excess of the corresponding aluminum derivative in hydrocarbon solution at -15°C under vigorous stirring, according to the procedure reported by Vandenberg *(19,20)*.

Scheme I. Preparation of bis(diisobutylaluminum) sulfate (DBAS) and bis(diisopropoxyaluminum) sulfate (DPOAS).

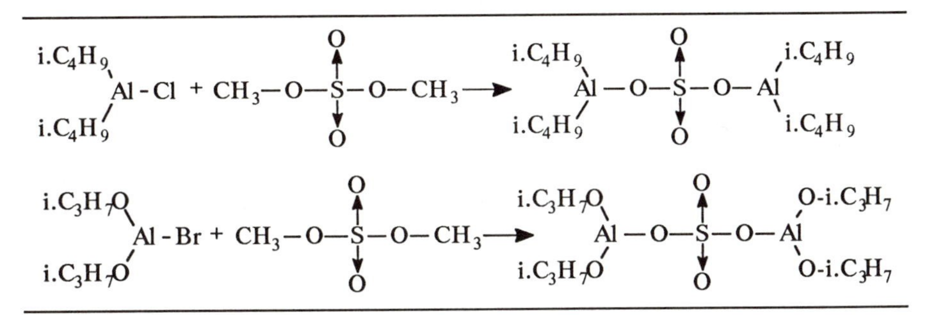

Commercial titanium trichloride "ARA", titanium tetrachloride, triisobutylaluminum (TIBA) and diethylaluminum chloride (DEAC) were purified by distillation under vacuum. Commercial grade $Ti(O\text{-}n.C_4H_9)_4$, $HfCl_4$, polymerization grade ethylene and propylene were used without further purification.

Research grade n-heptane and isooctane and commercial grade 4-methyl-1-pentene (4MP) were purified by refluxing over sodium/potassium alloy followed by distillation.

Magnesium-supported hafnium and titanium catalysts were prepared according to procedures similar to those previously described *(21)*.

Polymerization experiments. Polymerization experiments of ethylene, propylene, and 4-methyl-1-pentene were carried out according to common procedures. Three typical experiments (Runs E1 , P1, and MP1) are described in detail as typical examples, whereas data relevant to individual runs are summarized in tables I-III.

Polymerization of ethylene. In a 150 ml stainless steel pressure vessel was placed, under dry nitrogen atmosphere, a 30 ml glass vial containing a mixture of 10 ml of isooctane, 0.2 ml (0.81 mmol) of TIBA and 0.34 ml (0.23 mmol) of a 0.68 M solution of $TiCl_4$ in isooctane. The vessel was filled with ethylene up to 25 atm and then kept at 52°C for 8 h under mechanical shaking. The reaction product was washed with a 5% solution of HCl in methanol, extracted with boiling methanol and then dried under vacuum to give 4.0 g of polymeric product having m.p. 133°C and M_v 181,000. Data relevant to the series of polymerization experiments are collected in table I.

Table I. Polymerization of ethylene in the presence of conventional and modified Ziegler-Natta catalysts

Run	*Polymerization conditions*			*Conv.*[a]	*Polymer product*		
	$TiCl_n$ (n)	*Aluminum derivative*	*C_2H_4/Ti (mol/mol)*	*(%)*	*m.p. (°C)*	*[η]*[b] *(dl/g)*	$\overline{M}v \cdot 10^{-3}$
E1	4	TIBA	740	100	133	2.25	181
E2	3	TIBA	740	100	131	2.31	187
E3	4	DBAS	830	57	135	1.77	126
E4	3	DBAS	730	63	134	2.92	266
E5	4	TBAS	920	100	134	2.20	180
E6	3	TBAS	740	100	130	3.62	350
E7	4	DPOAS	710	0	-	-	-
E8	4	TPOAS	820	0	-	-	-

[a] Evaluated as (polymer weight/starting ethylene weight)•100. [b] In decalin at 135°C.

Polymerization of propylene. In a 150 ml stainless steel pressure vessel was placed, under dry nitrogen, a 30 ml glass vial containing a mixture of 10 ml of n-heptane, 2.08 ml (8.25 mmol) of TIBA and 0.27 ml (2.5 mmol) of a 9.26 M solution of $TiCl_4$ in n-heptane. The vessel was filled with propylene up to 8 atm and kept at 40°C for 8 h under mechanical stirring. The reaction product was refluxed with a 5% solution of HCl in methanol, extracted with boiling methanol and then dried under vacuum to yield 1.15 g of polymeric product. Data relevant to the series of polymerization experiments are collected in table II.

Table II. Polymerization of propylene in the presence of modified and conventional Ziegler-Natta catalysts

Run	*Polymerization conditions*			*Conv.*[a]	*Polymer product*			
	$TiCl_n$ (n)	*Aluminum derivative*	*Al/Ti (mol/mol)*	*(%)*	*T.I.*[b] *(%)*	*m.p.*[c] *(°C)*	*[η]*[c,d] *(dl/g)*	$\overline{M}v \cdot 10^{-3}$
P1[e]	4	TIBA	3.3	78	43	158	1.80	185
P2	3	TIBA	2.5	47	52	168	2.07	281
P3	4	DBAS	6.6	35	3	130[e]	1.25[e]	118[e]
P4	3	DBAS	3.0	95	63	168	2.30	252
P5	4	TBAS	3.3	39	20	152	1.79	180
P6	3	TBAS	2.5	47	62	163	3.69	356
P7	4	DPOAS	6.6	0	-	-	-	-
P8	4	TPOAS	3.3	0	-	-	-	-

[a] Evaluated as (polymer weight/starting propylene weight)•100. [b] Tacticity index, evaluated as percent fraction of polymeric product insoluble in n-heptane. [c] Fraction insoluble in n-heptane. [d] In decalin at 135°C. [e] Unfractionated polymer sample.

Polymerization of 4-methyl-1-pentene. A solution of 3.0 g (36 mmol) of 4-methyl-1-pentene in 30 ml of n-heptane was placed in a magnetically stirred 100 ml schlenke, then 1.2 ml (1.2 mmol) of 1 M solution of DBAS in n-heptane and 10.0 ml

(0.12 mmol Ti) of Ti/Mg catalyst suspension in n-heptane were added in that order. The mixture was stirred at 25°C for 5 hours, then the polymerization was interrupted with acidified methanol and the coagulated polymer was dried under vacuum to yield 10.2 g (29%) of polymeric product. Data relevant to the series of polymerization experiments are collected in table III.

Characterization of polymeric products. Viscosity measurements were carried out in decalin at 135°C by a Desreux-Bischoff dilution viscometer.

IR spectra were recorded on polymer films by a Perkin-Elmer 1600 FT-IR spectrophotometer. The D_{1378}/D_{1370} and D_{1378}/D_{1304} ratios were evaluated from the integrated absorbances of the 1378, 1370, and 1304 cm^{-1} bands.

NMR spectra were recorded by a Varian Gemini 200 spectrometer in 5 mm tubes on 10% (w/v) solutions in 1,2,4-trichlorobenzene/hexadeuterobenzene (9/1) at 110°C. ^{1}H NMR spectra were recorded at 200 MHz. Spectral conditions were as follows: size, 11,968 points; spectral width, 3 kHz; pulse, 30°; acquisition time, 2 s; number of scans, 1. ^{13}C NMR spectra were recorded at 50.3 MHz, under conditions of full proton decoupling. Spectral conditions were as follows: size, 23,936 points; spectral width, 15 kHz; pulse, 70°; relaxation delay, 2 s; acquisition time, 0.8 s; number of scans, 20,000. No weighing function was applied before Fourier transform.

Table III. Homopolymerization of 4-methyl-1-pentene by Ti/Mg catalyst in the presence of different aluminum derivatives

Run	*Polymerization conditions*				*Conv.*[a]	*Polymeric product*	
	Al derivative Type	*Al/Ti*	*Temp. (°C)*	*Duration (h)*	*(%)*	*[η]*[b] *(dl/g)*	*TI*[c] *(%)*
MP1-a	DBAS	10	25	22	23	16.1	80
MP1-b	DBAS	10	25	24	3	5.3	n.d.
MP2	DBAS	100	25	23	1	4.9	n.d.
MP3	DBAS	10	30	22	23	8.5	61
MP4	DBAS	10	40	22	71	8.3	67
MP5	DBAS	10	60	4	38	6.4	62
MP6	TIBA	10	25	1	85	2.0	51
MP7	DEAC	10	25	22	77	5.4	75
MP8	DEAC	10	60	4	85	2.1	52
MP9[d]	TIBA	15	25	1	9	20.3	94
MP10[d]	TIBA	15	40	1	16	18.7	93
MP11[d]	TIBA	15	50	1	20	16.8	90
MP12[d]	TIBA	15	60	1	25	19.1	88

[a] Evaluated as (polymer weight/4MP weight)·100. [b] In decalin at 135°C. [c] Tacticity index evaluated as percent fraction of polymeric product insoluble in diethyl ether. [d] Run carried out in the presence of a Hf/Mg catalyst.

Results and discussion

Homopolymerization of ethylene and propylene was investigated in the presence of modified Ziegler-Natta type catalysts obtained by reacting titanium trichloride and titanium tetrachloride with either soluble organoaluminum sulfates, such as bis(diisobutylaluminum) sulfate (DBAS) and bis(diisopropoxyaluminum) sulfate

(DPOAS), or with the heterogeneous slurries TBAS and TPOAS obtained by adding pure sulfuric acid to triisobutylaluminum and triisopropoxyaluminum, respectively (Tables I and II). For comparison polymerization experiments were also performed in the presence of two conventional catalysts prepared by reacting triisobutylaluminum (TIBA) with the same titanium halides.

Polymerization of ethylene was carried out under pressure (25 atm) in isooctane at 52°C for 8 h (Table I) by using molar ratio Al/Ti = 3.5. Fairly high conversions to polymeric products were observed in all the experiments performed in the presence of alkylaluminum derivatives, whereas the corresponding catalytic systems based on alkoxyaluminum derivatives did not display any ability to polymerize ethylene. This result confirms that only catalytic systems based on organoaluminum derivatives containing carbon-aluminum bonds are active as cocatalysts in the polymerization of ethylene and propylene, as reported in tables I and II.

Polymer samples obtained in runs E1-E6 are characterized by fairly high molecular weights (M_v 125,000-350,000) and exhibit melting point in the range 130-135°C, typical of linear polyethylene. The number of methyl groups per 1000 carbon atoms (Table IV), as evaluated from the relative optical density of the IR band at 1378 cm^{-1} (δ_s CH_3) referred to the band either at 1370 cm^{-1} (δ CH_2)or at 1304 cm^{-1} ($\delta_{t,w}$ CH_2) *(22-24)*, is generally lower than 10, as expected for conventional Ziegler-Natta catalysts and in keeping with the prevalent linear structure of the investigated samples.

Table IV. Number of methyl branching in polyethylene samples, as evaluated by IR spectroscopy

Sample	D_{1378}/D_{1370}	$CH_3/1000C$	D_{1378}/D_{1304}	$CH_3/1000C$
E1	0.275	< 10	0.818	4
E2	0.297	< 10	0.939	6
E3	0.327	< 10	1.016	7
E4	0.380	4	1.158	8
E5	0.493	14	1.845	17
E6	0.280	< 10	1.075	7

Polymerization of propylene was carried out at 5 atm in n-heptane at 40°C for 8 h by using Al/Ti molar ratios included between 2.5 and 6.6 (Table II) Also in this case, appreciable conversions to polymeric products were observed only when alkyl-aluminum derivatives were used. Fractionation with boiling n-heptane indicates that when using $TiCl_3$ a larger percentage of stereoregular polymer and higher molecular weights are obtained in the presence of aluminum sulfate derivatives as compared to TIBA. The opposite is true for samples prepared with $TiCl_4$, thus indicating that the reductive-alkylating activity of the organoaluminum sulfates is lower than that of TIBA.

The polymerization of 4MP was investigated in the presence of a $MgCl_2$ supported Titanium catalyst (Table III) *(21)*. By using DBAS as cocatalyst, independent of the Al/Ti molar ratio, very low conversions to polymeric products were observed at 25°C. By increasing the temperature up to 60°C, the polymer yield steadily increases, whereas the intrinsic viscosity reaches a maximum value of about 8.5 dl/g at 30-40°C and then decreases at 6.4 dl/g. The results obtained in runs carried out in the presence of TIBA and DEAC, clearly show that under comparable conditions the catalytic systems derived from these last organoaluminum derivatives are more active but give lower molecular weight poly(4MP) than the corresponding DBAS/Ti-Mg catalyst. These last systems can be therefore located, in terms of incidence of transfer reactions and hence of molecular weight, between the analogous catalytic systems based on Ti-Mg and Hf-Mg, respectively *(4)*.

Conclusions

The results reported in the present contribution have shown that novel unconventional Ziegler-Natta catalysts based on structurally defined or oligomeric alkylaluminum derivatives, as formally derived from sulfuric acid and trialkylaluminum, are able to polymerize ethylene and α-olefins such as propylene and 4-methyl-1-pentene.

The overall polymerization features and the properties of the polymers thus obtained do not substantially differ from analogous experiments carried out in the presence of the related conventional Ziegler-Natta systems.

Due to the presence of the sulfate moieties, unconventional catalysts appear less sensitive to moisture and oxygen than their conventional counterparts .

Moreover, the oligomeric organoaluminum sulfates, that have structural features somewhat reminiscent of alumoxanes, can be suitable in our opinion to produce a breakthrough to the exploitation of a new generation of Ziegler-Natta catalysts and in providing further constructive elements to the endless adventure of coordinate anionic polymerization.

References

1. "*Transition Metals and Organometallics as Catalysts for Olefin Polymerization*", Kaminsky, W.; Sinn, H., Eds., Springer Verlag, Berlin, 1988.
2. Ciardelli, F.; Carlini, C. in "*Comprehensive Polymer Science*", Allen, G.; Bevington, J.C., Eds., Pergamon Press, Oxford, 1992, Suppl. Vol. I
3. Masi, F.; Malquori, S.; Barazzoni, L.; Ferrero, C.; Moalli, A.; Menconi, F.; Invernizzi, R.; Zandonà, N.; Altomare, A.; Ciardelli, F. *Makromol. Chem. Suppl.* **1989**, *15*, 14.
4. Altomare, A.; Solaro, R.; Ciardelli, F.; Barazzoni, L; Menconi, F.; Malquori, S.; Masi, F. *J. Organomet. Chem.* **1991**, *417*, 29.
5. Chiellini, E.; Solaro, R.; D'Antone, S. *Makromol. Chem. Rapid Commun.* **1982**, *3*, 787.
6. Chiellini, E.; Solaro, R.; Masi, F. In "*Coordination Polymerization*", Price, C.C.; Vandenberg, E.J., Eds., Plenum Press: New York, 1983, pp. 141-165.
7. Bloemberg, S.; Holden, D.A.; Blum, T.L.; Hamer, G.K.; Marchessault, R.H. *Macromolecules* **1987**, *20*, 3086.
8. Gross, R.A.; Konrad, G.; Zhang, Y.; Lenz, R.W. *ACS Polym. Prep.* **1987**, *28*, 373.
9. Sinn, H.; Kaminski, W.; Vollmer, H.J.; Woldt, R. *Angew. Chem., Int. Ed.* **1980**, *19*, 390.
10. Kaminsky, W.; Schlobohm, M. *Makromol. Chem. Macromol. Symp.* **1986**, *4*, 103.
11. Chiellini, E. *Macromolecules* **1970**, *3*, 527.
12. Solaro, R.; D'Antone, S.; Orsini, M.; Andruzzi, F.; Chiellini, E. *J. Applied Polym. Sci.* **1983**, *28*, 3651
13. Chiellini, E.; Masi, F.; Senatori, L.; Solaro, R. *Chim. Ind. (Milan)* **1988**, *70*, 38.
14. Altomare, A.; Ciardelli, F.; Lima, R.; Solaro, R. *Polym. Adv. Techn.* **1991**, *2*, 3.
15. Chiellini, E.; Solaro, R.; D'Antone, S. *Makromol. Chem. Suppl.* **1981**, *15*, 82.
16. D'Antone, S.; Penco, M.; Solaro, R.; Chiellini, E. *Reactive Polymers* **1985**, *3*, 107.

17. Solaro, R.; D'Antone, S.; Chiellini, E. *Reactive Polymers* **1988**, *9*, 155
18. Chiellini, E.; D'Antone, S.; Solaro, R.; Callaioli, A.; Penco, M.; Galeazzi, L.; Petrocchi, E. *Chim. Ind. (Milan)* **1990**, *72*, 899.
19. Christman, D.L.; Vandenberg, E.J. (Hercules Inc.), *U.S. Pat. 3,025,281*, March **1962**; *Chem. Abstr.* **1962**, *57*, 2415e.
20. Heck, R.F.; Vandenberg, E.J. (Hercules Inc.), *U.S. Pat. 3,025,283*, March **1962**; *Chem. Abstr.* **1962**, *57*, 2415f.
21. Masi, F.; Malquori, S.; Barazzoni, L.; Invernizzi, R.; Altomare, A.; Ciardelli, F. *J. Mol. Catal.* **1989**, *56*, 143 .
22. Rugg, F.M.; Smith, J.J.; Wartman, L.H. *J. Polym. Sci.* **1953**, *1*, 1.
23. Slawinski, Jr., E.J.; Walter, H.; Miller, R.L. *J. Polym. Sci.* **1956**, *19*, 353.
24. Baker, C.; Maddams, W.F. *Makromol. Chem.* **1976**, *177*, 437.

RECEIVED February 18, 1992

Chapter 4

Isotactic Olefin Polymerization with Optically Active Catalysts

W. Kaminsky, S. Niedoba, N. Möller-Lindenhof, and O. Rabe

Institute for Technical and Macromolecular Chemistry, University of Hamburg, D–2000 Hamburg 13, Germany

With the homogeneous Ziegler-Natta catalyst based on chiral metallocenes/methylalumoxane highly isotactic polypropylene could be prepared. The catalysts can be separated into the optical active enantiomers. Even longer chained α-olefins as butene-1, pentene-1, or hexene-1, give isotactic polymers. Among the polymer features molecular weight and molecular weight distribution, solubility in toluene, melting point, density, and x-ray crystallinity as well as macrotacticity have been examined. Except for polyhexene all polymers investigated arrange in a stable helical conformation. Oligomers of butene-1 and pentene-1 show an optical rotation.

Among the great number of Ziegler-type catalysts, Vandenberg *(1)* examined at a very early stage the heterogeneous titanium trichloride system, especially for synthesizing isotactic polypropylene. Homogeneous systems have been preferentially studied in order to understand the elementary steps of the polymerization which is simpler in soluble systems than in heterogeneous ones. The situation has changed since in recent years a homogeneous catalyst based on metallocene and aluminoxane was discovered which is very active and also interesting for industrial uses *(2-3)*. In some cases special polymers could be synthesized only with soluble catalysts *(4)*.

Breslow *(5)* discovered that bis(cyclopentadienyl)-titanium(IV) compounds which are well soluble in aromatic hydrocarbons, could be used instead of titanium tetrachloride as the transition metal compound together with aluminum alkyls for ethylene polymerization. Subsequent research on this and other systems with various alkyl

0097–6156/92/0496–0063$06.00/0

groups has been conducted by many other research groups. With respect to the kinetics of polymerization and side reactions, this soluble system is probably the one that is best understood.

The use of metallocenes and aluminoxane as cocatalyst results in extremely high polymerization activities. The activity is higher than with heterogeneous catalysts. This system can easily be used on a laboratory scale. The chiral zirconocene ethylenebis(4,5,6,7-tetrahydroindenyl)zirconium dichloride ((S)-(+)-Et$(IndH_4)_2$-$ZrCl_2$ was the first initiator found to give highly isotactic polypropylene in a homogeneous system *(6,7)*. This catalyst allows separation into enantiomers by formation of diastereomers with S-(-)-1,1-bi-2-naphthol. The use of one enantiomer opens the opportunity to raise the uniformity among the polymer products as compared to those produced by means of conventional methods. With heterogeneous catalysts it is generally not possible to get only one type of isotactic polymer chain (Figure 1). Presuming that there is always a β-hydrogen transfer to form a double bond chain end, there are two different structures in the Fischer projection using heterogeneous catalysts and one using the optically active homogeneous system.

Isotactic Polyolefins

The catalyst used was (S)-(+)-Et$(IndH_4)_2$-$ZrCl_2$/MAO. Homopolymers from the prochiral 1-olefins propylene, 1-butene, 1-pentene and 1-hexene have been synthesized. This investigation is focused on the influence exerted by the polymerization temperature and the alkyl side chain length of the olefins on the reaction rate and the polymer features.

The reaction rate drops continuously from rather high initial values, according to the monomer type used, to rather low, almost constant values after several minutes (in the case of propylene) or even days (1-pentene). The mean activity values calculated over the whole polymerization period increase steadily with in-creasing temperatures from -50 °C up to +85 °C (in the case of propylene) and are highest for propylene (2,000 kg/mol m_{cat} x h x c monomer) as compared to those of the other monomers (Table I).

The most important result of this investigation was the finding that the soluble catalyst can polymerize even long-chain olefins like 1-pentene and 1-hexene in a highly isotactic manner.

Among the polymer features, the mean molecular mass and its distribution, the solubility in toluene at room temperature, the melting point, density and x-ray crys-

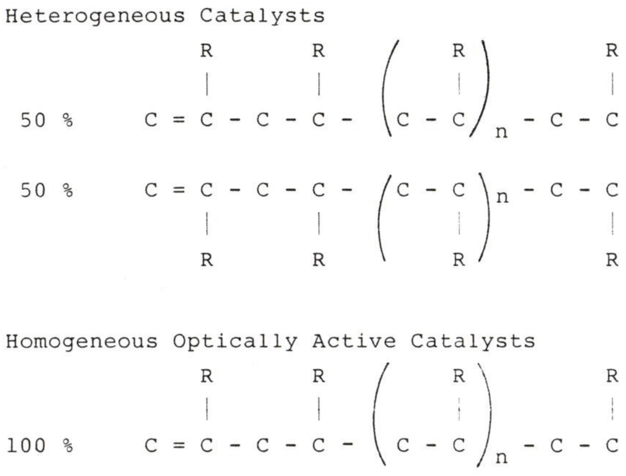

Heterogeneous Catalysts

```
               R     R   /   R \        R
               |     |  (    |  )       |
50 %       C = C - C - C - \C - C/n  - C - C

50 %       C = C - C - C - /C - C\n  - C - C
               |     |  (    |  )       |
               R     R   \   R /        R
```

Homogeneous Optically Active Catalysts

```
               R     R   /   R \        R
               |     |  (    |  )       |
100 %      C = C - C - C - \C - C/n  - C - C
```

Figure 1. Stereochemistry of isotactic polypropylene chains.

Table I. Polymerization of Olefins with (S)-(+)-Et$(IndH_4)_2ZrCl_2$/MAO in Toluene

Polymerization Temperature (°C)	Cat-Conc. (10^{-6}mol/l)	Monomer (mol/l)	Activity (kg/mol Zr x h x C_m
Polypropylene			
- 53	14,5	0,9	0,6
- 15	6,8	0,9	67,0
- 9	27,9	1,2	79,0
+ 16	19,9	1,9	295,0
+ 23	19,9	1,7	1154,0
+ 53	3,1	1,6	1649,0
+ 83	15,5	0,9	2038,0
Poly(1-butene)			
- 15	15,0	1,1	0,2
+ 10	81,0	8,1	0,4
+ 20	10,0	0,5	2,6
+ 48	22,3	0,9	9,0
+ 60	9,0	0,4	455,0
Poly(1-pentene)			
- 18	10,5	2,3	0,02
0	6,7	2,3	0,09
+ 12	10,5	2,3	0,4
+ 23	35,8	2,7	1,6
+ 42	4,5	2,3	3,6
Poly(1-hexene)			
+ 8	9,9	2,6	0,1
+ 11	10,5	0,9	6,5

C_m = Concentration of the monomer (mol/l)

tallinity as well as the macrotacticity (determined by IR-spectroscopy) have been examined.

These features are dependent on intermolecular forces. In all cases studied, a strong influence of the polymerization temperature on these properties is observed and leads to a decrease with rising temperature except for solubility and molecular mass distribution.

With polypropylene (Table II) the mean molecular masses decrease in the temperature range from -15 °C to +83 °C from 200,000 g/mol down to almost 1,300 g/mol (factor of 140). The melting point of 158 °C of a polypropylene prepared at -53 °C is very similar to that of a commercial product (162 °C). The density is even a little higher, at 0.9150 (- 15°C) compared to 0,9132. At higher polymerization temperatures the melting point goes down from 152 °C to a glass transition temperature of -35 °C., x-ray crystallinity from 67,7 % down to 14,4 %, and IR-spectroscopic macrotacticity from 72 % down to 29 %. In accordance with these values solubility in toluene at room temperature increases and the molecular mass distribution becomes even more narrow from Mw/Mn = 4,7 to 1,5.

The ^{13}C-NMR measured isotacticity (pentads) at the lowest polymerization temperatures are much higher than that of the commercial product (Table II).

The melting point is still lower, however, because there are some irregularities with the homogeneous catalyst (head to head and 1,3-insertion) and the mistakes are random (statistically) distributed over the whole chain.

Together with this temperature influence on intermolecular characteristics there is another influence working, i.e. the growing alkyl side chain length of the different monomers used. In both polybutene and polypentene, the isotacticity at low preparation temperatures is very high, decreases with increasing polymerization temperature (Table III). The influence of temperature is higher at the melting point. The molecular weight of poly(1-butene) goes down from 45 000 (prepared at -15 °C) to 5 000 (prepared at 60 °C). Table IV displays selected features of different polyolefins which have been synthesized with the chiral catalyst system at approx. 10 °C.

With respect to mean molecular masses - as observed over the whole temperature range studied (Tables II and III) - the difference is greatest between polypropylene and polypentene. With respect to crystallinity, however, the difference is most evident between polypentene and polyhexene.

By means of SAXS-investigations and IR-spectroscopy data this change of behaviour can be ascribed to the growing alkyl side chain length leading to a preferential formation of polymorphic modifications and a growing preponderance of the side chain to determine the molecular

Table II. Isotacticity (I), Melting Point (mp), Molecular Weight (Mη) and Density of Polypropylenes Catalyzed with (S)-(+)-Et$(IndH_4)_2ZrCl_2$/MAO

Polymer- ization Temp. (°C)	(I) (mmmm)	mp (°C)	Mη (g/mol)	ρ (g/m^3)	Mw/Mn
- 53	99,7	158			
- 15	99,4	152	200 000	0,9150	2,5
0	98,9	147	130 000	0,9075	4,3
15	97,7	148	60 000	0,9070	3,1
37	95,5	159	27 000	0,9056	2,7
53	91,5	95	10 500	0,9041	2,3
83	54,0	- 35	1 260	0,8700	1,5
(a)	95,8	162	350 000	0,9132	6,0

(a) = Commercial Product

Table III. Isotacticity (I), Melting Point (mp), Glass Transition Point (*), and Molecular Weight (Mη) of Polybutene, Polypentene, and Polyhexene Catalyzed with (S)-(+)-Et$(IndH_4)_2ZrCl_2$/MAO

Polymerization Temperature (°C)	(I) (mmmm)	mp (°C)	Mη (g/mol)
Poly(i-butene)			
- 15	98,0	119	45 000
+ 10	97,7	97	45 000
+ 20	95,5	94	34 000
+ 48	89,6	78	15 000
+ 60	72,2	-	5 000
Poly(1-pentene)			
- 18	99,0	69	18 000
0	99,0	66	12 900
+ 23	98,0	62	7 000
+ 42	85,6	-61*	1 900
Poly(1-hexene)			
+ 8	95,8	-48*	37 000

state of order. Except for polyhexene all polymers investigated build up stable helical conformations. In the case of polypropylene this preferred conformation can be observed in the IR-spectra for polymerization temperatures up to 66 °C. The lower limit for the detection of the helical segment length is 11 - 12 units.

From the point of view along a single polymer chain, the intra-molecular features exhibited were studied. Here the uniformity of bond formation is most important and was adequately investigated by means of ^{13}C-NMR-spectroscopy. Again a continuous decline from highly regularily attached units (microisotacticity > 99 %) at low polymerization temperatures down to predominantly atactic configurations at high temperatures (microisotacticity < 50 %) is observed. The influence of the alkyl side chain, however, is much smaller than in the case of intermolecular characteristics. Even completely amorphous polyhexene (polymerized at + 11 °C) turns out to be still very isotactic (> 95 %). The influence of the alkyl chain causes the β-hydrogen transfer reaction to be more preferred so that the chains become shorter.

From these two conclusions the following may be deduced: the tendency of the catalyst to build up polymers from 1-olefins in an isotactic manner is not seriously hindered by increasing the alkyl side chain length. On the other hand it must be admitted that very high regularity in the bond formation process does not *per se* lead to improved macroscopic features.

In conclusion, the polymerization of different olefins with optically active homogeneous catalyst gives polymers which are highly stereospecific but with no special physical properties. The polymers are not optically active. The existence of optical activity could happen only if there were formed a one-handed helix structure stable under conditions at which optical activity can be measured. This normally requires very bulky side groups, as in poly(tripehylmethyl methacrylate) *(8)* or polychloral *(9)*. Such a condition is not found in these polyolefins.

Oligomers

Solutions of high molecular weight isotactic polyolefins produced with the optically active catalyst are expected to show no or only very low optical activity, because gelices of these polymers rapidly change their screw sense in solution *(10)*. In contrast to this, solutions of oligomers or hydrooligomers should give a measurable optical rotation *(11,12)*. The oligomers can be synthesized at similar temperatures and lower monomer concentration. Table V shows the specific rotation for oligomers of 1-butene.

Table VI. Molar Rotation of Pentene Oligomers. Catalyst: (S)-(+)-$Et(IndH_4)_2ZrMe_2$

Polyolefin	Mol.Weight (g/mol)	Melting Point °C	Isotacticity % mmmm Pentads	X-Ray, % Crystallinity
Polypropylene	60 000	148	98,8	65,0
Poly(1-butene)	44 000	97	97,7	66,9
Poly(1-pentene)	10 000	66	98,3	37,9
Poly(1-hexene)	17 000	- 48	95,8	amorphous

Table V. Specific Optical Rotation at Different Wave Lengths of 1-Butene Oligomers Prepared with (S)-(+)-$Et(IndH_4)_2ZrCl_2$/MAO at 50 or 70 °C

Wave Length:	Φ 589 nm	Φ 365 nm
Dimer	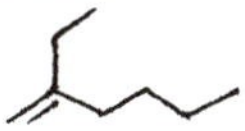 0	0
Trimers (50 °C)	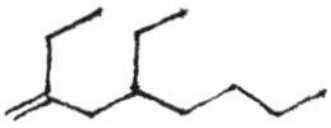 - 1,0	- 3,5
Trimers (70 °C)	- 0,3	- 0,8
Tetramers (50 °C)	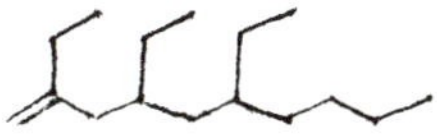 - 3,2	- 10,0
Tetramers (70 °C)	- 1,2	- 3,4

As predicted, the dimer shows no optical rotation. The optical rotation of the higher oligomers decreases with increasing oligomerization temperature. The rotation of the tetramers is higher than that of the trimers. The optical purity (ee) of the oligomers increases from 10 % by high oligomerization temperatures (60 °C) up to 90 % by low temperatures (0 °C).

Also the next homologue, 1-pentene, could be oligomerized with the asymmetric ethylene-bridged (S)-Et-$(IndH_4)_2ZrMe_2$/MAO-system. The product mixtures, as well as fractions of defined degree of oligomerization, show measurable optical activity. Specific optical rotation values are of the same order of magnitude as in the case of 1-butene oligomers. Table VI lists molar optical rotation values for pentene oligomers synthesized with the optically active (S)-$Et(IndH_4)_2ZrMe_2$ catalyst.

Table IV. Comparison of Properties of Polyolefins Prepared with (S)-(+)-Et$(IndH_4)_2ZrCl_2$/MAO at 10 °C

Component \ Wave length	589 nm	546 nm	365 nm
Dimers	-	-	-
Trimers	- 1,33	- 1,65	- 5,07
Tetramers	- 2,59	- 3,41	- 8,11
Product Mixture (M_n = 290 g/mol)	- 1,86	- 2,15	- 5,58

As expected, the optical rotation first increases with increasing number of chiral centers in the molecule and then declines again after going through a maximum. This behavior is due to the lower degree of asymmetry of the chiral carbon atoms in longer chained molecules.

It must be emphasized that the oligomers described above are, again, genuine olefins and still bear a terminal double bond. This explains the absence of optical activity for the dimers. Through this functionalization the oligomers become available for use in further organic syntheses where optically active hydrocarbon groups are to be introduced.

The product molecular weights were confirmed by GC-/MS and is in accord with the expected highly isospecific 1,2-insertion and termination by β-hydride transfer (Figure 2).

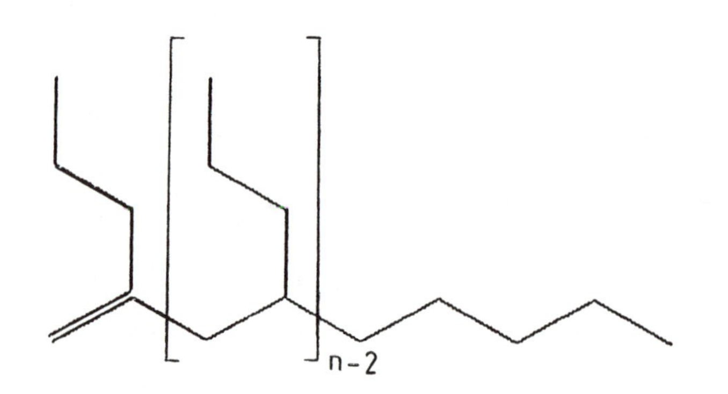

Figure 2. Structure of 1-pentene oligomers; n: degree of oligomerization.

Literature Cited

1. Vandenberg, E.J. (1963) US Pat. 3 108 973 (to Hercules Incorporated, C.A. (1964) 60: 8155 C
2. Sinn, H.; Kaminsky, W.; *Advances in Organometallic Chemistry*, Vol. *18*, Acad. Press, New York, **1980**, p. 99
3. Sinn, H.; Kaminsky, W.; Vollmer, H.-J.; Woldt, R.; *Angew.Chem.Int.Ed.Engl.* **1980**, *19*, 390
4. Ewen, J.A.; Jones, R.L.; Razari, A.; *J.Am.Chem.Soc.* **1988**, *110*, 6255
5. Breslow, D.S. (1958) U.S. Pat. 2 827 446 (to Hercules Incorporated)
6. Kaminsky, W.; Külper, K.; Brintzinger, H.H.;, Wild, F.R.W.P.; *Angew.Chem.Int.Ed.Engl.*, **1985**, *24*, 507
7. Kaminsky, W.; *Angew.Makromol.Chem.*, **1986**, *145/146*, 149
8. Okamoto, Y; Yashima, E; *Prog.Polym.Sci.*, **1990**, *15* (2), 263-298
9. Vogl, O.; Corley, L.S.; Harris, W.J.; Taycox, G.D.; Zhang, J.; *Makromol.Chem.Suppl.*, **1985**, *13*, 1
10. Pino, P.; *Adv.Polym.Sci.*, **1966**, *4*, 393
11. Pino, P.; Cioni, P.; Wei, J.; *J.Am.Chem.Soc.*, **1987**, *109*, 6189
12. Kaminsky, W.; Ahlers, A.; Möller-Lindenhof, N.; *Angew.Chem.Int.Ed.Eng.*, **1989**, *28*, 1216.

RECEIVED November 15, 1991

Chapter 5

1-Butene Polymerization with Ethylenebis-(1-indenyl)zirconium Dichloride and Methylaluminoxane Catalyst System

Effect of Hydrogen Addition

M. Kioka, A. Mizuno, T. Tsutsui, and N. Kashiwa

Mitsui Petrochemical Industries, 3–2–5 Kasumigaseki Chiyoda-ku, Tokyo, Japan

Butene-1 polymerizations were performed with and without hydrogen addition using ethylenebis(1-indenyl)zirconium dichloride and methylaluminoxane catalyst system. The catalyst activity was remarkably enhanced by 60-70 times by addition of hydrogen. The obtained polymers were characterized by ^{13}C NMR, DSC and intrinsic viscosity measurements. The reason for the activity enhancement by hydrogen addition is discussed on the basis of the analytical data.

Isotactic propylene polymerization with ethylenebis(1-indenyl) zirconium dichloride ($Et(Ind)_2ZrCl_2$) or ethylenebis(4,5,6,7-tetrahydro-1-indenyl)zirconium dichloride ($Et(TH\text{-}Ind)_2ZrCl_2$) in conjunction with methylaluminoxane (MAO) has been reported in the literature (1-3). The structural features (4-7 and Mizuno,A. et.al. Polymer,in press) of the polymers are high isotacticity and the presence of a small amount of irregular units based on 2,1- or 1,3-propylene insertion.

The effect of hydrogen on propylene polymerization catalyzed by $Et(Ind)_2ZrCl_2$ with MAO has been also reported (8). Namely, with the addition of hydrogen, the activity was enhanced three times. As one of the most plausible reasons for the activity enhancement, the chain transfer reaction with hydrogen at 2,1-inserted active chain ends was proposed

In this paper, butene-1 polymerizations with $Et(Ind)_2ZrCl_2$ and MAO catalyst system were performed with hydrogen addition (H_2 system) and without (non-H_2 system) in order to get further information concerning the hydrogen effect. On the basis of detailed information on the polymer microstructures, the effect of hydrogen on the activity was also discussed.

0097–6156/92/0496–0072$06.00/0

Experimental

Preparation of catalyst. $Et(Ind)_2ZrCl_2$ (6) and MAO (9) were prepared according to previous papers.

Polymerization of butene-1. In a 500ml glass reactor equipped with a stirrer, 250ml of toluene were placed and pure butene-1 or a mixture of butene-1 and hydrogen was bubbled through the system. Subsequently, 0.15g of MAO (=ca.2.5mmol of Al) and 2.5×10^{-3} mmol of $Et(Ind)_2ZrCl_2$ were added at 30℃, in this order. Polymerization was carried out under atmospheric pressure at 30℃ for 1 hr and stopped by the addition of a small amount of methanol. The whole product was poured into a large amount of methanol. The resulting powdery polymer was collected by filtration and dried in vacuum at 80℃ for 12hr.

Intrinsic viscosity measurement. Intrinsic viscosity was measured at 135℃ using decalin as solvent.

DSC measurement. DSC curves were recorded at a heating rate of 10℃ min^{-1} in order to measure melting temperature on a Perkin-Elmer 7 differential scanning calorimeter. The instrument was calibrated by the measurements of the melting points of indium and lead. The weight of the sample was ~5mg.

^{13}C NMR measurement. The polymer solution was prepared by dissolving ~ 100 mg of the polymer sample at 120 ℃ in ca. 0.6ml of hexachlorobutadiene, including ca. 0.05ml of deuterobenzene which was used for field stabilization, in a 5 mm o.d. glass tube. $^{13}C\{1H\}$ NMR spectra were recorded on a JEOL GX-500 spectrometer operating at 125.8MHz in a Fourier-transform mode. Instrumental conditions were as follows: pulse angle 45° ; pulse repetition time 5.0s; spectral width 20000 Hz; number of scans 20000 ~ 30000; data point 64k. The ^{1}H decoupled DEPT method was also used to discriminate the carbon species. The $^{13}C\{1H\}$ NMR spectra of the polymers dissolved in perdeuterocyclohexane at 60℃ were also measured in order to observe olefinic carbon resonances, which may be hidden under the solvent resonances in hexachlorobutadiene solution.

Results and discussion

In Table 1 are shown the results of butene-1 polymerizations with $Et(Ind)_2ZrCl_2$ and MAO catalyst system. In the H_2 system, the catalyst activity per unit butene-1 pressure was almost constant, not depending on the concentration of H_2 under these experimental conditions, but remarkably enhanced by 60-70 times in comparison with in the non-H2 system. This activity enhancement was much higher than that of propylene polymerization in our previous paper (8).

The molecular weight in terms of intrinsic viscosity of the polymerdecreased by the addition of 10% of hydrogen to butene-1, but further addition resulted in practically no change of the molecular weight.

Table 1. Effect of hydrogen on butene-1(C_4) polymerization

Run No.	C_4-Flow (l/h)	H_2-Flow (l/h)	Yield (g)	Activity (gPB-1/ mmol·Zr·h)	Intrinsic viscosity (dl/g)	Melting temperature (℃)
1	50	0	1.2	240	0.28	93.2
2	50	5	76.7	15340	0.19	98.4
3	50	10	82.0	16400	0.20	97.1
4	50	20	81.4	16280	0.18	97.6

The microstructures of the polymers obtained both in the non-H_2 and H_2 systems were investigated with the method of high-field ^{13}C NMR spectroscopy. The observed spectra of Run 1 and 2 are shown in Figure 1. The ^{13}C spectral features of the samples of Run 3 and 4 were almost same as that of Run 2. The peaks were assigned on the basis of DEPT measurement for determination of carbon species, the chemical shift additive rule of Lindeman-Adams (10) and the previously reported literature data (11). The assignment of the carbon peaks based on the terminal groups and the irregular units is added to Figure 1.

The structural features of the polymer obtained in the non-H_2 system was characterized with terminal groups having n-butyl group as the α end group and 2- and 3-pentenyl groups as the ω one, and also with 0.3 and 0.2 units of irregular structures formed by 2,1- and 1,4-insertions of butene-1 per one polymer chain. Moreover, it is noticeable that vinylidene group as ω end group was not detected unlike polypropylene with the same catalyst system. Thus, the above α and ω end groups of the polymer are considered to be formed by the chain transfer reaction in Scheme-1.

Scheme-1

```
     C    C                               C             C
     C    C  C4                           C             C
Zr-C-C-C-C-P → Zr-C-C-C-C + C-C=C-C-C-C-P + C-C-C=C-C-C-P
                        C
             C4         C
             → Zr-C-C-C-C-C-C

         C                     C
         C   C4                C
   Zr-C-C-P → Zr-C-C-C-C + C=C-P     (negligible)
             C4;butene-1,  P;polymer chain
```

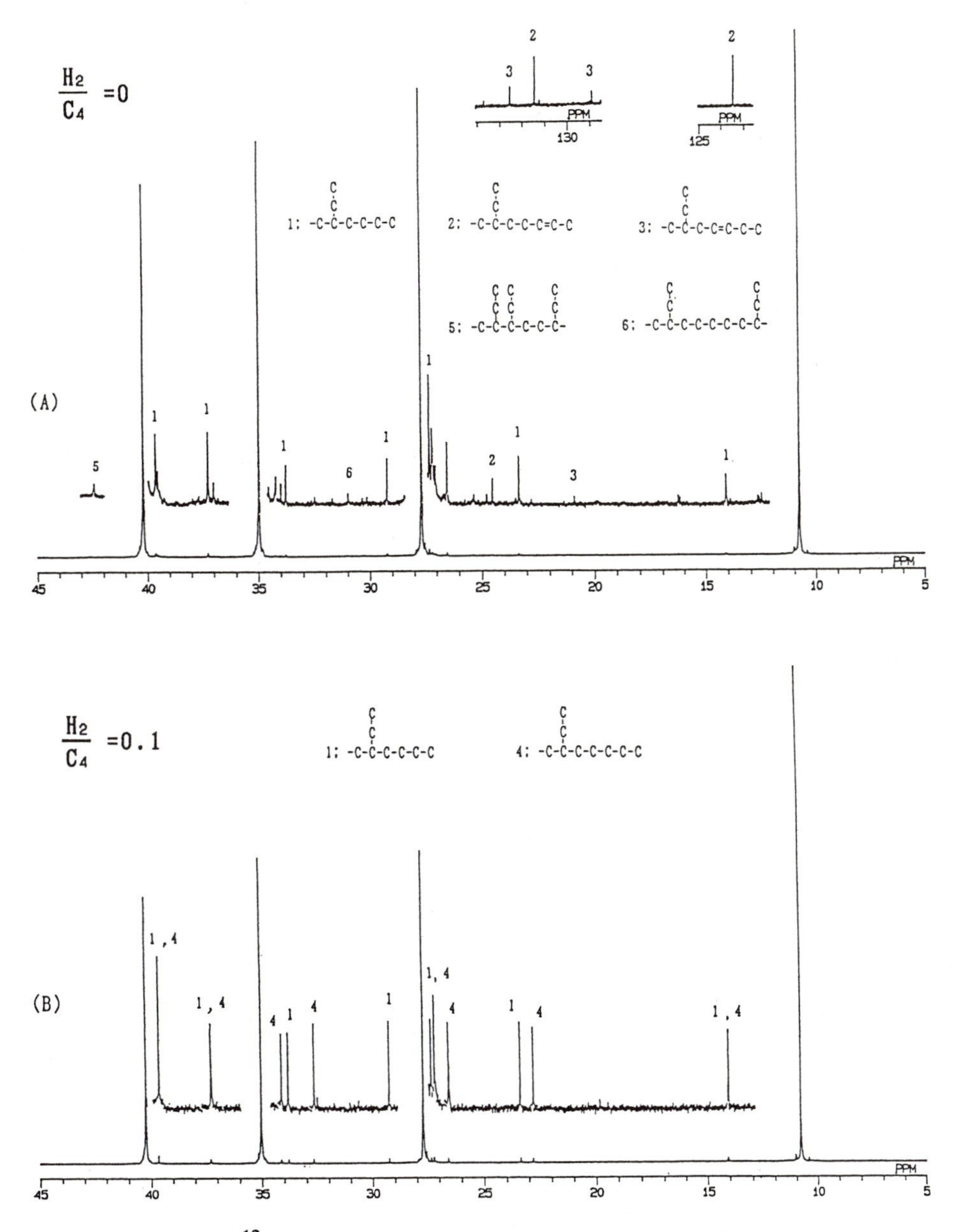

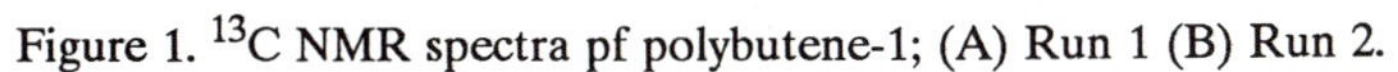
Figure 1. ^{13}C NMR spectra pf polybutene-1; (A) Run 1 (B) Run 2.

On the other hand, the spectrum of the polymer formed in the H_2 system shows a much simpler pattern in comparison with that in the non-H_2 system. From the spectral analysis, the polymer was shown to be composed of only head-to-tail units as the inner polymer chain structure and have n-butyl and n-pentyl groups as each terminal group. This microstructure reveals directly that the chain transfer reaction with hydrogen takes place preferentially at 2,1- or 1,4-inserted active chain end as shown in Scheme-2.

Scheme-2

```
     C       C                              C
     |       |                              |
     C       C   H2                         C
     |       |                              |
 Zr-C-C-C-C-P  →  Zr-H + C-C-C-C-C-C-P

        C                    C
        |                    |
        C       H2           C
        |                    |
  Zr-C-C-P     →  Zr-H + C-C-P            (negligible)

              P;polymer chain
```

In our previous paper (8) concerning the hydrogen effect on propylene polymerization, it was proposed that the chain transfer reaction with hydrogen is a plausible reason for the activity enhancement by assuming that propylene insertion is retarded at 2,1-inserted active chain end, compared with 1,2-inserted one. This proposal is likewise applied to the interpretation of the activity enhancement by hydrogen addition in butene-1 polymerization from the microstructural analysis. However, as mentioned above, the effect of hydrogen on activity in butene-1 polymerization is much higher than that in propylene polymerization. This remarkable difference is explainable from the stronger retarding effect at 2,1-inserted active chain ends in butene-1 polymerization due to larger steric hindrance.

This assumption is strongly supported by no detection of vinylidene group as a terminal structure of the polybutene-1 obtained in the non-H_2 system unlike the corresponding polypropylene.

The melting temperature was enhanced for 4-5℃ by addition of hydrogen. This enhancement of melting temperature may reflect the disappearance of the irregular structures since the considerable decrease of molecular weight, which tends to lower the melting temperature, was accompanied by the addition of hydrogen.

Literature Cited

1. Kaminsky,W.;Kulper.K;Niedoba.S. Makromol. Chem. **1986**,3,377
2. Kaminsky,W.;Kulper.K;Brintzinger,H.H.;Wild,F.R.W.P. Angew. Chem.,Int. Edn.Engl. **1985**,24,507
3. Kaminsky,W. Angew. Makromol. Chem. **1986**,145/146,149

4. Grassi,A.;Zambelli,A.;Resconi,L.;Albizzati,E.;Mozzocchi,R. Macromolecules **1988**,21,617
5. Tsutsui,T.;Mizuno,A.;Kashiwa,N. Makromol. Chem. **1989**,190,1177
6. Tsutsui,T.;Ishimaru,N.;Mizuno,A;Toyota,A,;Kashiwa,N. Polymer **1989**,30, 1350
7. Cheng,H.N.;Ewen,J.A. Makromol. Chem. **1989**,190,1931
8. Tsutsui,T;Kashiwa,N.;Mizuno,A. Makromol. Chem.,Rapid Commun. **1990**,11 565
9. Tsutsui,T;Kashiwa,N. Polymer, Commun. **1988**,29,180
10. Lindeman,L.A.;Adams,J.Q. Anal. Chem. **1971**,43,1245
11. Doi,Y.;Asakura,T. Macromolecules **1981**,14,69

RECEIVED December 4, 1991

Chapter 6

Rate of Ethylene Polymerization with the Catalyst System $(\eta^5\text{-}RC_5H_4)_2ZrCl_2$–Methylaluminoxane

Effects of Cyclopentadienyl Ring Substituents

Peter J. T. Tait, Brian L. Booth, and Moses O. Jejelowo

Department of Chemistry, University of Manchester Institute of Science and Technology, P.O. Box 88, Manchester, M60 1QD, United Kingdom

Ethylene was polymerized, at 60°C, using the homogeneous catalyst system $(\eta^5\text{-}RC_5H_4)_2ZrCl_2$-methylaluminoxane (where R = H, Me, n-Pr, i-Pr and t-Bu). A modest increase in the maximum rate of ethylene polymerization according to the order t-Bu < H < i-Pr < n-Pr < Me was observed. However active centre studies carried out using a ^{14}CO radio-tagging method demonstrated that active centre concentrations, C^*, remained more or less constant for all the catalysts systems, leading to the conclusion that there is some variation in the propagation rate constant, k_p, for the different catalyst systems investigated. Possible explanations are given in terms of the opposing influences of the electronic and the steric effects of the ring substituents.

Soluble zirconium complexes with aluminoxane co-catalysts have proved to be highly active and versatile homogeneous catalysts for the polymerization of ethylene and propylene(*1-3*). Recent work by Kaminsky(*4*) and Ewen(*5*) have shown that substituents in the η^5-cyclopentadienyl rings of $(\eta^5\text{-}C_5H_5)_2ZrCl_2$ affect both the rate of polymerization of ethylene and the molecular weight of the polymer. Kaminsky(*4*) has proposed that in the case of $(\eta^5\text{-}C_5Me_5)_2ZrCl_2$ and methylaluminoxane, the broad molecular weight distribution can be explained by supposing two different active centres C_1 and C_2 which differ not only in their reaction rates, but also in that different prereaction times are required for their formation. Ewen(*5*) has found that the number average molecular weights, M_n, decrease linearly with polymerization rates in the systems $(\eta^5\text{-}RC_5H_4)_2ZrCl_2$ (where R = H, Me and Et) and $(\eta^5\text{-}C_5Me_5)_2ZrCl_2$, and has stated that the observed

0097–6156/92/0496–0078$06.00/0

comparative rates of polymerization in these systems can be explained by a balance between the steric and the electronic effects of the substituents on the cyclopentadienyl rings. Previous work in this Department(6) has also shown that using $(\eta^5\text{-}RC_5H_4)_2TiCl_2$ {where R = H, Me, $(CH_3O)_3Si(CH_2)_3$} differences in the relative rates of hydrogenation of olefins were observed with variation in the nature of R.

In an attempt to investigate the effects of substituents on the cyclopentadienyl rings in more detail, a systematic study of the polymerization of ethylene using the complexes $(\eta^5\text{-}RC_5H_4)_2ZrCl_2$ (where R = H, Me, i-Pr, n-Pr and t-Bu) and methylaluminoxane has been carried out, and the results of this investigation are now reported.

Results and Discussion. Polymerizations of ethylene were carried out at 60 °C and 1 atm pressure using the catalyst system $(\eta^5\text{-}RC_5H_4)_2ZrCl_2$-methylaluminoxane, and typical rate-time profiles are shown in Figures 1, 2 and 3, where R is Me, n-Pr and t-Bu, respectively.

Except for the system where R = t-Bu, the maximum rate was achieved within 20 s of adding the last component of the polymerization mixture, and this rate was then maintained for the duration of the polymerization. Thus, the prereaction times were extremely short, and no significant differences in these times were observed when using the different $(\eta^5\text{-}RC_5H_4)_2ZrCl_2$ complexes. When R = t-Bu, however, the maximum rate was achieved only after 8 min and the rate then decreased to a steady low level.

From a comparison, however, of Figures 1, 2 and 3 it is apparent that there is a very significant difference in the nature of the rate-time plots when R is t-Bu. A schematic representation of the variation of the polymerization rate with variation in the R group is also shown in Figure 4. Variation of the cyclopentadienyl substitutent led only to a slight variation in the maximum rate of ethylene polymerization. Whilst the variation was not large it was nevertheless real and increase in catalyst activity followed the order t-Bu < H < i-Pr < n-Pr < Me. Comparative values of the polymerization rates are listed in Table I.

Table I
Polymerization System $(\eta^5\text{-}RC_5H_4)_2ZrCl_2$/Methylaluminoxane/Ethylene: Effect of Cyclopentadienyl Ring Substituent on Polymerization Rate

R	Rate /Kg PE $(mmol\ Zr)^{-1}\ h^{-1}$
H	9.1
Me	10.1
n-Pr	9.7
i-Pr	9.6
t-Bu	7.6*

[Zr] = 0.024 mmol dm^{-3}; Al/Zr = 1000; Toluene = 0.30 dm^{-3}; T = 60 °C
$[C_2H_4]$ = 0.0353 mol dm^{-3}
*Rate is the maximum level attainable.

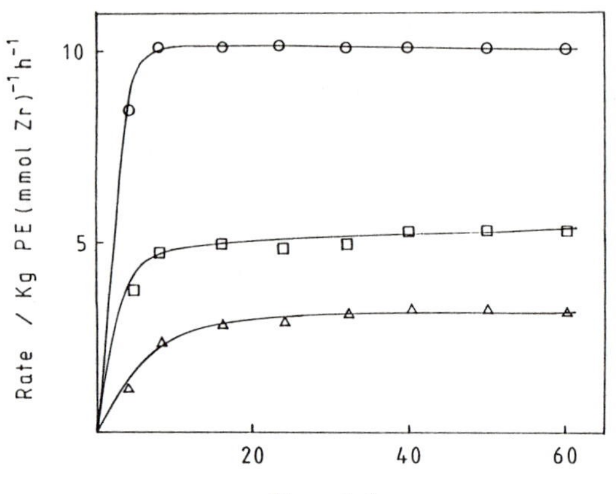

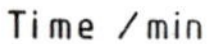

Figure 1. Plot of rate of polymerization of ethylene versus time using the catalyst system $(\eta^5\text{-MeC}_5H_4)_2ZrCl_2$/methylaluminoxane at 60 °C and 1 atm pressure.
$[Zr] = 0.024$ mmol dm^{-3}; toluene = 0.30 dm^3
Al/Zr : △ 500; □ 700; ○ 1000.

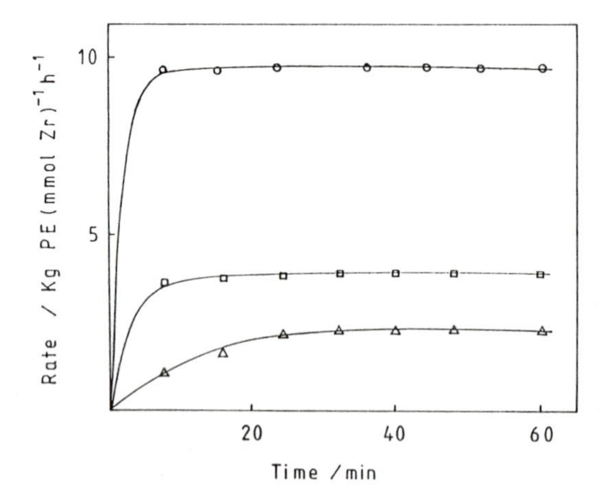

Figure 2. Plot of rate of polymerization of ethylene versus time using the catalyst system $(\eta^5\text{-n-PrC}_5H_4)_2ZrCl_2$/methylaluminoxane at 60 °C and 1 atm pressure.
$[Zr] = 0.024$ mmol dm^{-3}; toluene = 0.30 dm^3
Al/Zr : △ 500; □ 700; ○ 1000.

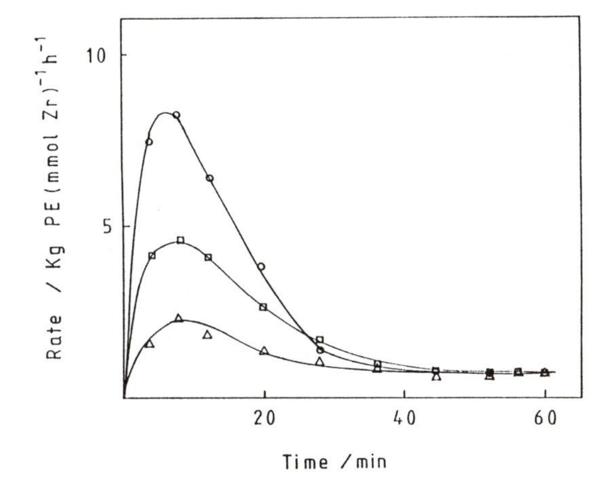

Figure 3. Plot of rate of polymerization of ethylene versus time using the catalyst system $(t\text{-}BuC_5H_4)_2ZrCl_2$/methylaluminoxane at 60 °C and 1 atm pressure.
[Zr] = 0.024 mmol dm^{-3}; toluene = 0.30 dm^3

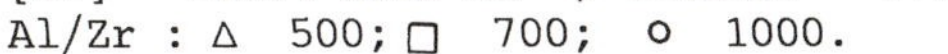
Al/Zr : Δ 500; □ 700; o 1000.

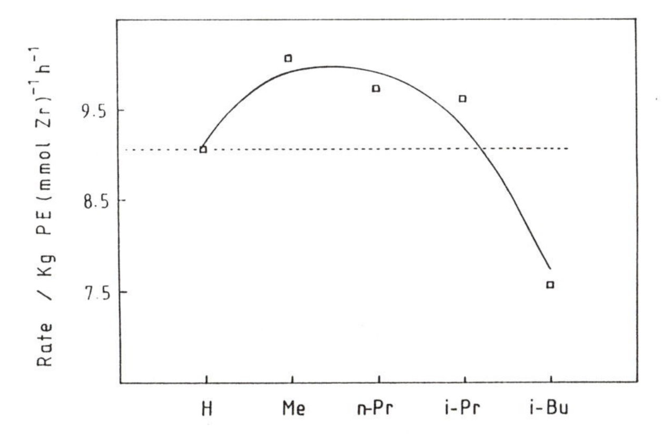

Figure 4. Schematic representation of effect of substituent R on the rate of polymerization of ethylene using the catalyst system $(\eta^5\text{-R-}C_5H_4)_2ZrCl_2$/methylaluminoxane at 60 °C and 1 atm pressure.

The use of ^{14}CO-radio-tagging as a method for the determination of active centre concentration has the great advantage that it allows determination of C* values as a function of polymerization time. Typical plots of C* as a function of polymerization time are shown in Figures 5 and 6. In all systems studied the value of C* remained constant throughout the duration of the polymerization. The use of the equation (8)

$$R_p = k_p C^* [M]$$

where [M] is the monomer concentration, kept constant throughout the duration of these polymerization experiments, allows the propagation rate coefficient, k_p, to be calculated at various times within a given polymerization. Typical values of k_p as a function of polymerization time are shown in Figures 7 and 8. The behaviour of the polymerization system where R=t-Bu is again very different from any of the other polymerization systems which were investigated in this study.

It has been suggested by Kaminsky(4) that, perhaps, in these reactions there are two types of active centres. The first type, C_1^*, is produced rapidly but inserts monomer units quite slowly, while the second, C_2^*, requires a longer prereaction formation time, but once formed inserts monomer units at a faster rate than those of the C_1^* type. If the nature of the substituent on the cyclopentadienyl ring can affect the ratio of the active centres, $C_1^*:C_2^*$, then an active centre study should be able to detect this phenomenon. We have shown previously(7) for the polymerization of ethylene using the homogeneous catalyst system $(\eta^5\text{-}C_5H_5)_2ZrCl_2$-methylaluminoxane that using a ^{14}CO radio-tagging method for active centre determination, all zirconium atoms are involved in active centres. In order to obtain accurate values of C^* it is, however, essential to use a short contact time. Using this same method, active centre studies were carried out on the representative systems where R = H, Me and t-Bu.

It can be seen from the results given in Table II that, within the limits of experimental error, the values of C^* are unaffected by the nature of R. However, these results do not provide any information on the types of centres which may be present.

As can be seen from an examination of Table II, the observed small changes in the relative rates of polymerization with the type of R substituent are due to a similar variation of the propagation rate constant. It can also be seen from Table II that the values of k_p decrease in the order Me > H > t-Bu, i.e., the same order as was found for the relative rates of polymerization. This lends support to the proposal first made by Ewen(5) that the main influences on k_p values in these systems, are the electronic and the steric effects of the R substituents. Thus, the inductive and hyperconjugative effects of the alkyl substituents will increase the electron density of the η^5-cyclopentadienyl ligand leading to an increased electron density at the zirconium atom which in turn leads to an increased rate of polymerization. This increase in polymerization rate may be due to a lowering of the stability of the η^2-alkene-zirconium bond resulting in a more weakly coordinated monomer, facilitating insertion into the growing polymer chain. Alternatively, or, in

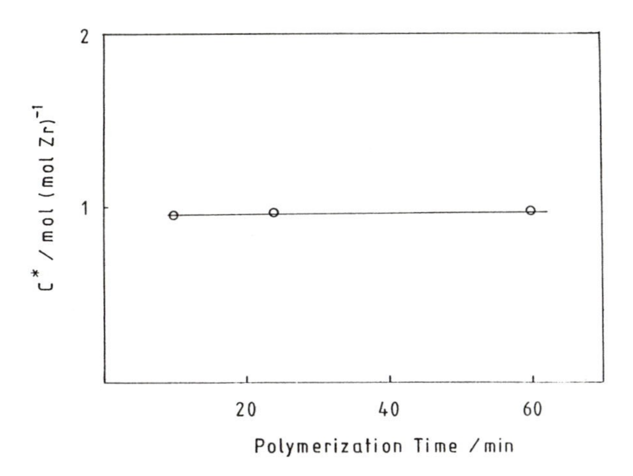

Figure 5. Variation of active centre concentration with duration of polymerization of ethylene using the catalyst system $(\eta^5\text{-MeC}_5H_4)_2ZrCl_2$/methylaluminoxane at 60 °C and 1 atm pressure. [Zr] = 0.024 mmol dm^{-3}; Al/Zr = 1000; toluene 0.30 dm^3; ethylene = 0.0353 mol dm^{-3}. Specific activity of ^{14}CO = 1.57 x 10^{10} dpm mol^{-1}; contact time = 4 min; mole ratio ^{14}CO : Zr = 23:1.

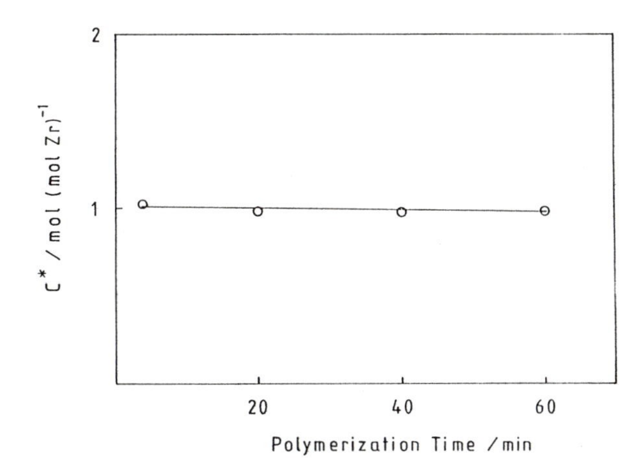

Figure 6. Variation of active centre concentration with duration of polymerization of ethylene using the catalyst system $(\eta^5\text{-t-BuC}_5H_4)_2ZrCl_2$/methylaluminoxane at 60 °C and 1 atm pressure. [Zr] = 0.024 mmol dm^{-3}; Al/Zr = 1000; toluene = 0.30 dm^3; ethylene = 0.0353 mol dm^{-3}. Specific activity of ^{14}CO = 1.57 x 10^{10} dpm mol^{-1}; contact time = 4 min; mole ratio ^{14}CO : Zr = 23:1.

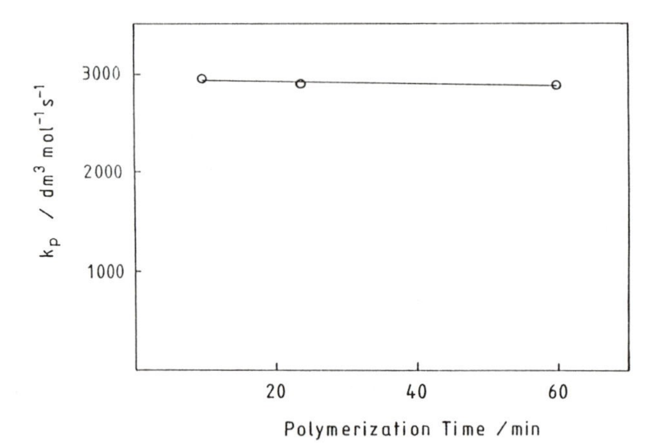

Figure 7. Variation of the propagation rate coefficient with duration of polymerization of ethylene using the catalyst system $(\eta^5\text{-MeC}_5H_4)_2ZrCl_2$methylaluminoxane.
Experimental details as for Figure 5.

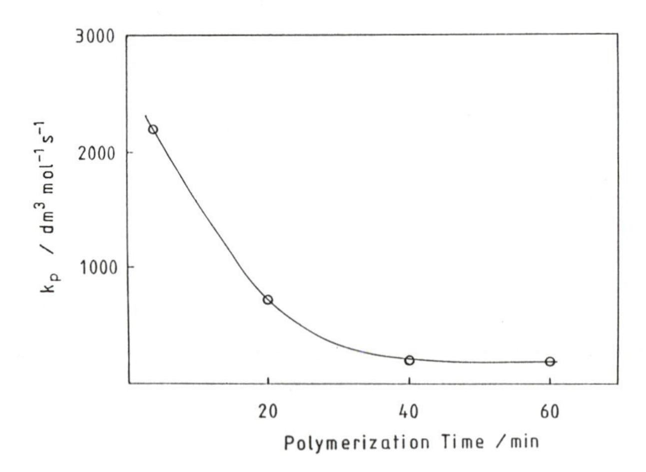

Figure 8. Variation of the propagation rate coefficient with duration of polymerization of ethylene using the catalyst system $(\eta^5\text{-t-BuC}_5H_4)_2ZrCl_2$/methylaluminoxane.
Experimental details as for Figure 6.

addition, a higher electron density at the zirconium atom may weaken the zirconium-carbon σ-bond to the growing polymer chain which could also facilitate insertion of the monomer, and hence lead to increased rate of polymerization.

TABLE II

Polymerization System $(RC_5H_4)_2ZrCl_2$/Methylaluminoxane/Ethylene: Variation of Active Centre Concentration and Propagation Rate Constant with Polymerization Time

R	t_p /min	R_p(add) /mol C_2H_4 (mol Zr)$^{-1}$ s^{-1}	C* /mol (mol Zr)$^{-1}$	k_p /dm^3 mol^{-1} s^{-1}
	10	89	1.07	23.6
H	24	89	0.93	27.2
	60	89	0.94	26.9
	10	99	0.95	29.4
Me	24	99	0.97	28.8
	60	99	0.98	28.5
	4	79	1.02	22.0
	20	25	0.98	7.2
Bu^t	40	6.9	0.98	2.0
	60	6.8	0.99	2.0

Contact time = 4 min; Specific activity of ^{14}CO = 1.57 x 10^{10} dpm mol^{-1};
R_p(add) is the rate of polymerization at the time of ^{14}CO addition;
t_p = polymerization time; Other conditions as in Table I.

Opposing these effects an increase in the steric bulk of the alkyl substituent may hinder the approach of the incoming monomer and decrease the rate of polymerization. Thus, steric crowding is expected to increase steadily from R = Me, to R = i-Pr. The observed rates of polymerization may then reflect a combination of the steric and the electronic effects and may then reach a maximum when R = Me. When R = t-Bu the steric factor now outweighs the electronic factor and the relative rate is less than when R = H (see Figure 3). This explanation agrees well with those advanced by Ewen(5). The same explanation is presumably true for the catalyst system $(\eta^5\text{-}C_5Me_5)_2ZrCl_2$-methylaluminoxane studied by Kaminsky(4), where the rate of polymerization was noted to be some orders of magnitude less than the unsubstituted cyclopentadienyl complex.

There is an interesting difference between the results obtained with $(\eta^5\text{-}RC_5H_4)_2ZrCl_2$ (R = H, Me) on the one hand, and those for

R = t-Bu on the other. In the former case, both C^* and k_p remain constant throughout the duration of the polymerization. Hence, the rate of polymerization remains steady over his period. In contrast, when R = t-Bu, although the value of C^* remains constant over the polymerization period the value of k_p decreases (see Figure 8). It can, therefore, be concluded that it is this decrease in the value of k_p which is responsible for the different kinetic rate-time profile observed for the η^5-$(t\text{-}BuC_5H_4)_2ZrCl_2$ catalyst.

In the model as proposed by Kaminsky(9) for the active centre, the zirconium atom has a sterically crowded environment. Thus, access of the monomer to the transition metal centre is expected to be sensitive to the size of the ligands attached to the transition metal. The presence of a t-Bu group could be expected to give rise to more severe steric crowding, evidently so much so that the k_p decreases with the extent of polymer formation. Steric considerations may also explain the decrease in activity which we, and others, have observed when ethylaluminoxane is used in place of methylaluminoxane.

Conclusions. The very high rates of ethylene polymerization which are obtained at 60 °C and 1 atmosphere pressure, using the homogeneous catalyst system $(\eta^5\text{-}RC_5H_4)_2ZrCl_2$-methylaluminoxane arise from the quantitative participation of the zirconium atoms in active centre formation, and the high values of propagation rate constants.

The variation in the rate of ethylene polymerization with variation in the nature of the R group on the cyclopentadienyl unit can be explained from the opposing effects of the electronic and the steric effects attributable to these substituent groups.

The values of C^* remain constant with polymerization time as do the derived values of k_p except when R = t-Bu. In this case, the catalyst system shows a kinetic rate-time profile which reflects a decrease in k_p with polymerization time, C^* remaining constant throughout the duration of polymerization.

Experimental. Bis-(η^5-cyclopentadienyl) or (η^5-alkylcyclopentadienyl)-zirconium (IV) dichloride(12-14) and methylaluminoxane(2,10,15) were prepared according to literature procedures. Radio-labelled carbon monoxide was supplied by ICI plc, Petrochemicals and Plastics Division, Billingham, U.K. Ethylene (CP Grade, Air Products) was dried by passage over 13X and 4A molecular sieves, and solvents were dried using standard procedures(16).

Polymerization and determination of active centre concentration. Details of the polymerization are as detailed previously(11). The procedure for active centre concentration determination has been described in another publication(7), a contact time of 4 min was used for this work.

References.

1. H. Sinn, W. Kaminsky, Adv. Organomet. Chem., **18**, 99 (1980).
2. W. Kaminsky, M. Miri, H. Sinn, R. Woldt, Makromol. Chem., Rapid Commun., **4**, 417 (1983).
3. J. Herwig, W. Kaminsky, Polymer Bull., **9**, 464 (1983).

4. W. Kaminsky, K. Kulper, S. Niedoba, Makromol. Chem., Macromol. Symp., **3**, 377 (1986).
5. J.A. Ewen in "Catalytic Polymerization of Olefins", Ed. T. Keii and K. Soga, Elsevier, Amsterdam, p.271, 1986.
6. B.L. Booth, G.C. Ofunne, P.J.T. Tait, J. Organomet. Chem., **315**, 143 (1986).
7. P.J.T. Tait, B.L. Booth, M.O. Jejelowo, Makromol. Chem., Rapid Commun., **9**, 393 (1988).
8. P.J.T. Tait in "Transition-Metal Catalyzed Polymerizations: Alkenes and Dienes", Ed. R.P. Quirk, Harwood Acad. Publ., New York, Vol. A, p.115, 1983.
9. W. Kaminsky in "Transition-Metal Catalyzed Polymerizations: Alkenes and Dienes", Ed. R.P. Quirk, Harwood Acad., New York, Vol. A, p.225, 1983.
10. J.A. Ewen, J. Am. Chem. Soc., **106**, 6355 (1984).
11. M. Abu-Eid, S. Davies and P.J.T. Tait, Polym. Prepr., Am. Chem. Soc., **24**, 114 (1983).
12. P.C. Wailes, R.S.P. Coutts, H. Weigold in "Organometallic Chemistry of Titanium, Zirconium and Hafnium", Academic Press, New York, 1974.
13. M.F. Lappert, C.J. Pickett, R.I. Riley, P.I.W. Yarrow, J. Chem. Soc., Dalton Trans., 805 (1981).
14. E. Samuel, Bull. Soc. Chim., France 3548 (1966).
15. G.A. Razuvaev, Yu.A. Sangalov, Yu.Ya. Nel'Kenbaum, K.S. Minsker, Bull. Acad. Sci. USSR, Div. Chem. Sci., Engl. Trans., **11**, 2434 (1975).
16. A.I. Vogel, "Textbook of Practical Organic Chemistry including Qualitative Organic Analysis", 4th Edn., p.264, Longmans, London - New York, 1978.

RECEIVED February 12, 1992

Chapter 7

Migratory Nickel(0)–Phosphorane Catalyst

α-Olefin Polymerization by 2,ω-Linkage

G. Fink, V. Möhring, A. Heinrichs, and Ch. Denger

Max–Planck Institut für Kohlenforschung, Kaiser Wilhelm Platz 1, 4330 Mülheim, Ruhr, Germany

The paper deals with the novel 2,ω-linkage of α-olefins with the catalyst system nickel (0) compound / bis(trimethylsilyl)aminobis-(trimethylsilylimino)phosphorane. When linear α-olefins are polymerized, the polymer contains only methyl branches, regularly spaced along the chain with a seperation corresponding to the chain length of the monomer. The monomer insertion is regioselective; only $C_{\omega} \rightarrow C_2$ coupling of the growing chain with the next monomer takes place. The nickel catalyst complex migrates along the polymer chain between two insertions. A kinetic model has been set up to describe the mutual superposition of the insertion reaction and the migration of the catalyst. In this model the migration steps are treated as preceding equilibria.
The use of optically active monomers gives no reaction of the pure enantiomeres, whereas the racemic mixture is well polymerized. This reaction leads to a strongly alternating "copolymer" of the (R) and (S) enantiomeres with regularly spaced chiral centers in the main chain.
Polymerization under high pressure increases the molecular weight from 6000 to 90000 g/mol.

Polymerization of ethylene with the homogeneous catalyst system nickel(0)compound / bis(trimethylsilyl)aminobis(trimethylsilylimino)phosphorane leads, according to *Keim et al.* (1) to short chain branched polymers. We have found that this system polymerizes a-olefins: surprisingly the structure of the product is consistent not with the usual 1,2-coupling of the monomers to give a comb-like branched product (Figure 1, above), but with a 2,ω-coupling (Figure 1, below)(2)(3).

Examples of nickel(0) compounds (4) / aminobis(imino)phosphorane (5) catalysts are shown in Figure 2. The catalyst components - preferably in equimolar ratio - are employed in situ in the pure liquid monomer or in aromatic solvents. The nature of the nickel(0) compound has no influence on the structure of the α-olefin polymer. But only the phosphoranes shown in Figure 2 form an active catalyst. Even phosphoranes in which only one silicon atom is substituted by a carbon atom give an inactive catalyst (6). On the other hand, the kinetic behaviour differs much

0097–6156/92/0496–0088$06.00/0

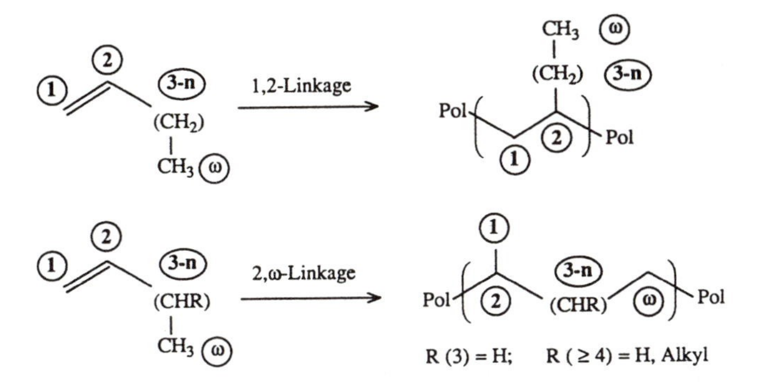

Figure 1: Polymerization of α-olefins under 1,2- and 2,ω-linkage

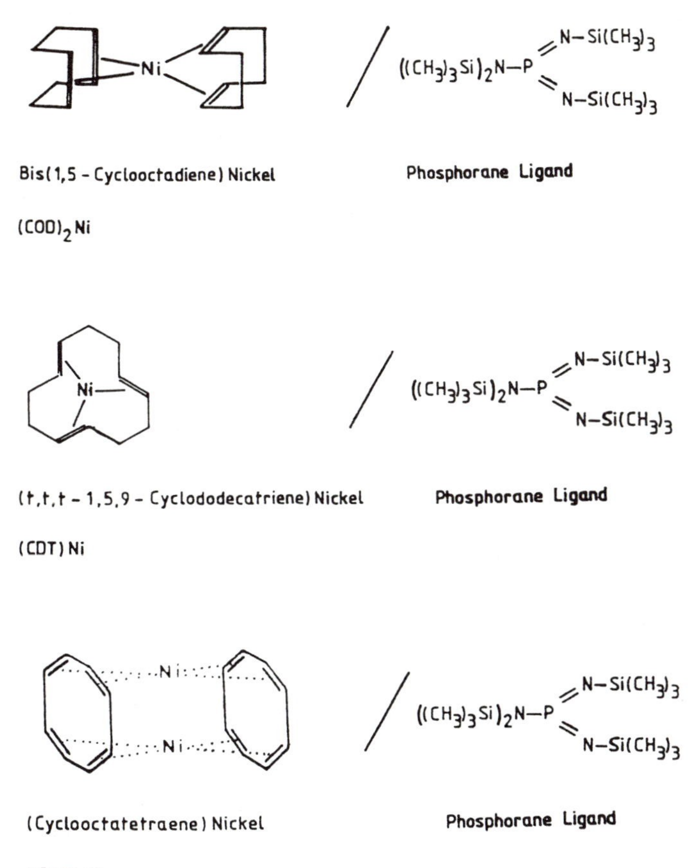

Figure 2: Nickel(0) / Phosphorane catalysts

with the nickel(0) compound used (7). These findings suggest that the rate-determing step in the formation of the active species is the release of the ligand from the nickel(0) compound.

It was proved that a 2,ω-linkage takes place by polymerizing 1-deutero-1-hexen. All the methyl branches in this polymer are bearing one deuterium.

How the 2,ω-linkage takes place formally is demonstrated again in Figure 3. In this way, linear α-olefins and singly branched α-olefins can be polymerized, but not α-olefins with quaternary C-Atoms in the chain (in other words: one hydrogen atom at every carbon atom is necessary) or olefins with vinylene or vinylidene groups yield polymers.

The structure of the formed poly-α-olefins is unusual. The polymer (see again Figure 1 or 3) contains only methyl branches, regularly spaced along the chain with a separation corresponding to the chain length of the monomer. Thus, in the polymer of a linear α-olefin with n ($-CH_2-$) groups the distance between two methyl branches is (n+1) ($-CH_2-$) groups. Thus, their structure is well defined and can be predetermined by selection of the appropriate α-olefin. Thus, for example, the structure of the polymer synthesized from 1-pentene corresponds to that of a strongly alternating copolymer of ethylene and propylene (see Figure 3).

The structure was confirmed using ^{13}C-NMR spectroscopy; this is demonstrated in Figure 4 for the example of poly-2,5-(1-pentene). The assignment of the signals was carried out with the help of the increment rules of *Lindemann and Adams* (8). All the signals to be expected for 2,ω-linked α-olefin polymers were found in the spectra in corresponding intensity ratios.

The details of the chain start and a resulting irregularity therefore in the structure were discussed earlier (3)(7).

Results and Discussion

Taking into account all the results, a scheme was developed with which the origin of the special structure of the poly-α-olefins can be explained. This scheme is shown in Figure 5 and its main points are:

- The monomer can only insert into a primary nickel-alkyl bond at the end of the growing chain;
- the insertion is regioselective, only C_ω - C_2 coupling of the growing chain with the next monomer takes place;
- the nickel catalyst "migrates" along the polymer chain between two insertions. During this "migration" transfer reactions can occur, but not insertions.

These processes are represented schematically in Figure 5 for the polymerization of 1-butene. Looking further on the transfer reaction from left to right in the scheme, we should note: if the transfer reaction occurs immediately after the insertion, then a vinylidene group will be formed. If the transfer reaction occurs during the migration of the nickel catalyst along the polymer chain, a vinylene group will be formed and finally, if the transfer reaction occurs at the end of the growing chain, a vinyl group will be formed. All these double bonds are

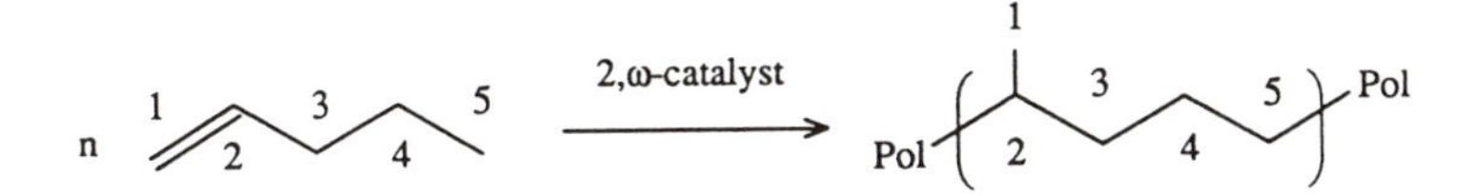

Figure 3: 2,ω-linkage of 1-pentene

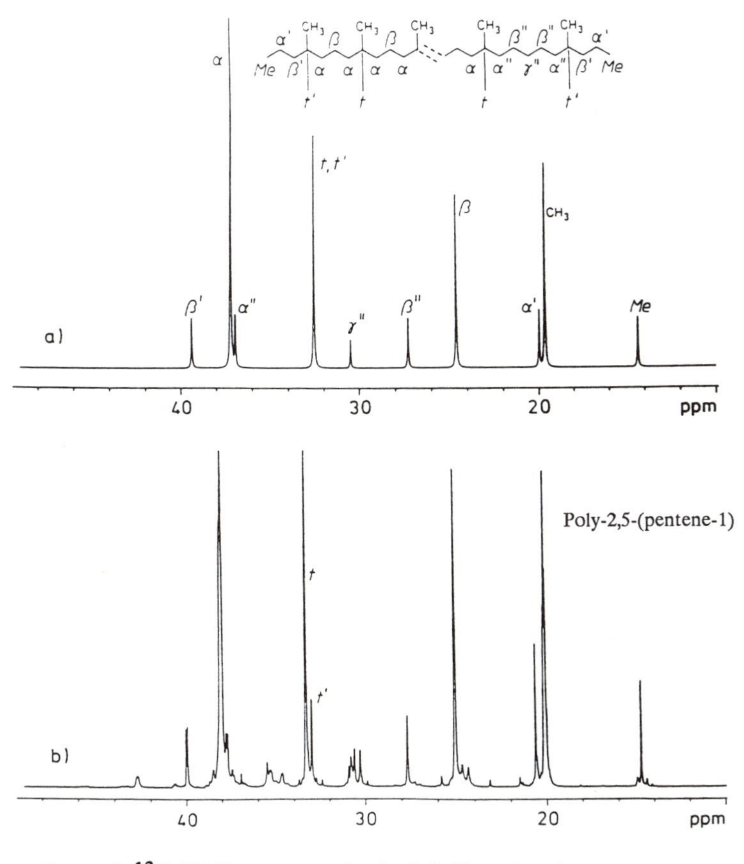

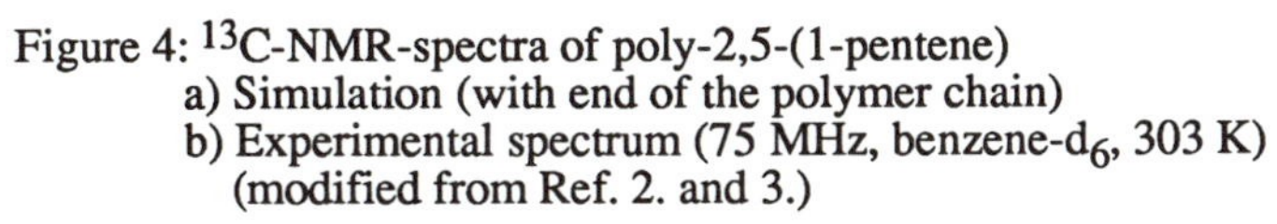

Figure 4: ^{13}C-NMR-spectra of poly-2,5-(1-pentene)
a) Simulation (with end of the polymer chain)
b) Experimental spectrum (75 MHz, benzene-d_6, 303 K) (modified from Ref. 2. and 3.)

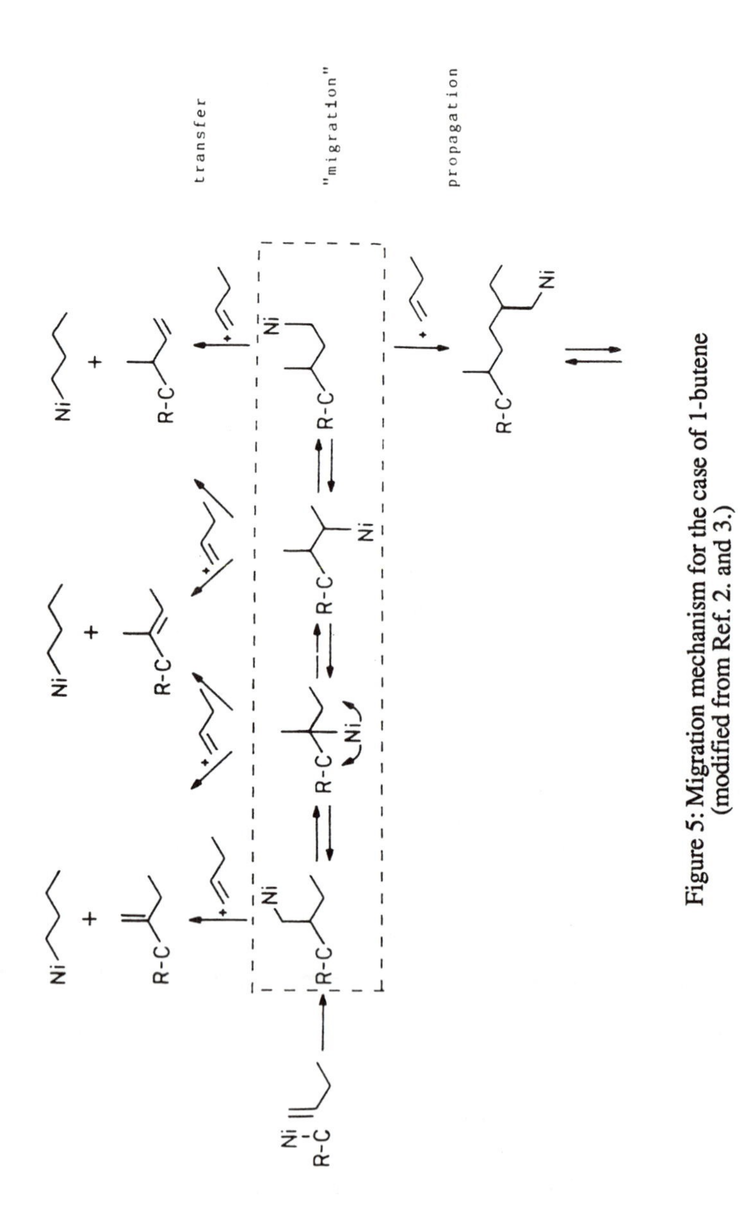

Figure 5: Migration mechanism for the case of 1-butene (modified from Ref. 2. and 3.)

found in the polymer. Furthermore we found that the ratios between these double bonds are depending on the monomer (see Figure 6). While the amount of vinyl and vinylidene groups are nearly constant, the number of vinylene groups per 100 C-atoms increases with the length of the monomer used. Using a longer α-olefin there have to be more migration steps. Therefore the possibilty for the formation of a vinylene group is much higher. The existence of double bonds at all positions of the chain in the corresponding amounts is one proof that the nickel catalyst really migrates through the chain. The double bonds marks the place where the catalyst "left" the chain via β-elimination.

Taking into account that one hydrogen at every carbon atom in the α-olefin is necessary for polymerization, the "migration in itself" can be explained in terms of an addition / elimination mechanism via alkylnickel / nickelhydride species with a 1,2-hydride shift. This is demonstrated in Figure 7. The result of the β-elimination as the first step according to this mechanism is a coordinated nickel hydride. Olefins with vinylene double bonds can not be polymerized. Therefore we must postulate that during migration only coordinated hydrides are formed. If once a free nickel hydride is generated it can only react with a new α-olefin, the result of this reaction is the start of a new chain or in other terms the transfer reaction to the monomer. It was possible to confirm the existence of a nickel hydride species by introducing deuteriumchloride into the polymerization flask. The hydride forms gaseous HD which could be isolated and analysed by means of mass spectrometry. The amount HD found is increased with the amount of catalyst used (Figure 8). But only 4 ± 1 % of the used nickel compound is found as HD. With this method we can not destinguish between coordinated and free nickel hydride. According to the mechanism, we should have nickel alkyl in the reaction flask as well. This nickel alkyl should react with the deuteriumchloride and should form polymer chains bearing a deuterium atom. Up to now it was not possible to find the deuterium atom in a polymer chain. This is probably due to problems with the low concentration. Because the analytical methods for molecules with a low moleculelar weight are much more sensitive, we isolated and analysed the unpolymerized monomer after deuterolysis. At the moment of deuterolysis there should be some newly started "polymer chains" constituted by one monomer only. These molecules are nickel alkyls as well, but they form deuterated alkanes when treated with DCl. Indeed it was possible to detect deuterated monomers with GC-MS-coupling in the liquid distilled from the reaction mixture after deuterolysis.

One problem of the 2,ω-polymerization is that the molecular weight of the product is always very low ($M_w \approx 6\ 000$ g/mol), whereas the molecular weight distribution is small ($M_w / M_n \approx 1.6$). The molecular weight remains constant no matter which α-olefin is employed (Figure 9). This can be explained by the migration mechanism, if it is assumed, that the migration steps proceed via equilibria (see later).

One of our experiments to increase the molecular weight, was by applying high pressure. As you can see in Figure 10, it was possible to increase the molecular weight of poly-2,6-(1-hexene) up to $M_w \approx 90\ 000$ g/mol by increasing the pressure up to 1 400 MPa. The reason for this is a kinetic pressure effect, which we examined in details. For this purpose we used the method of Buback (9). The polymerization was run in an optical high pressure cell. With this equipment it was possible to take FT-IR-spectra of the content of the autoclave during polymerization. In Figure 11 you can see the spectra of an experiment with 1-hexene at 300 MPa. The time between each spectra was two minutes. We used

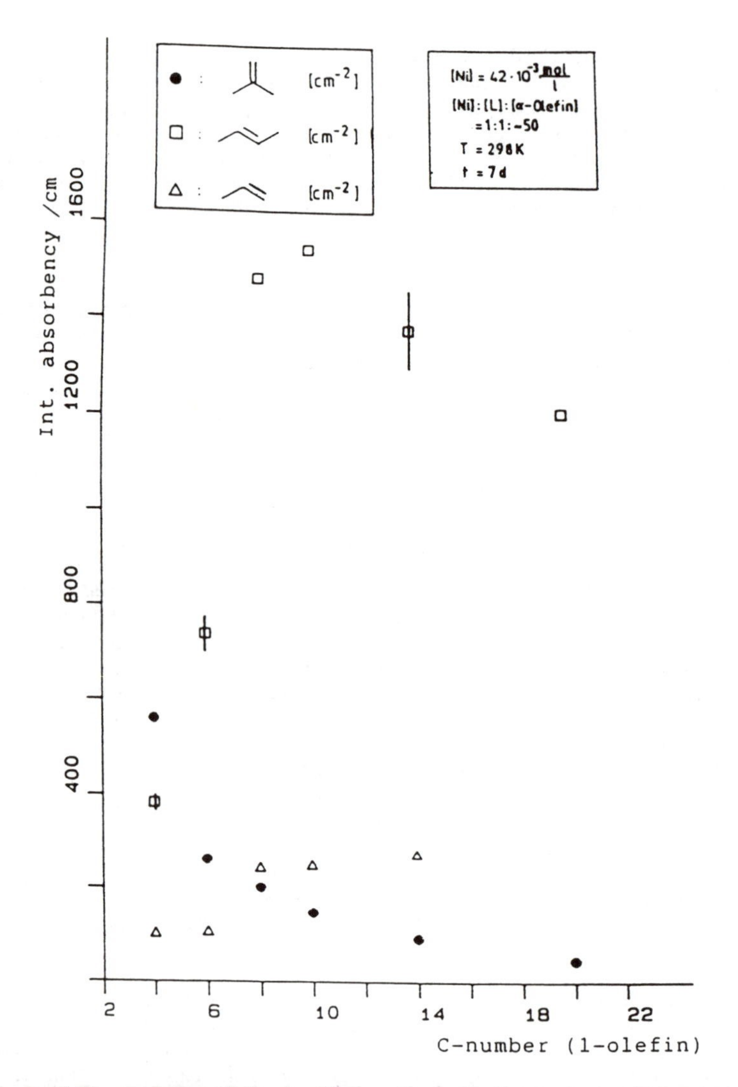

Figure 6: Dependance of the double bonds on the C-number of the α-olefin (modified from Ref. 3.)

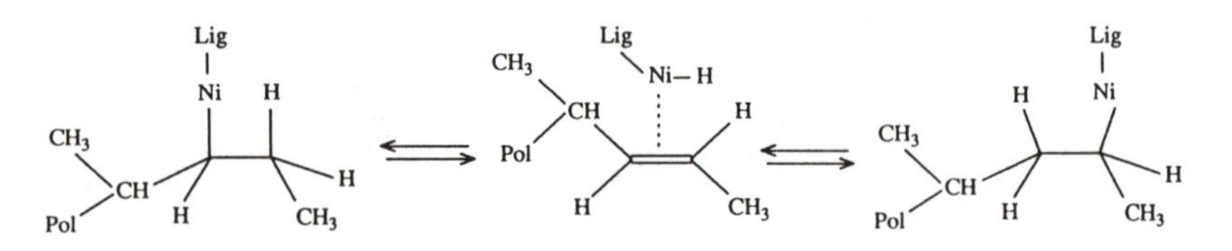

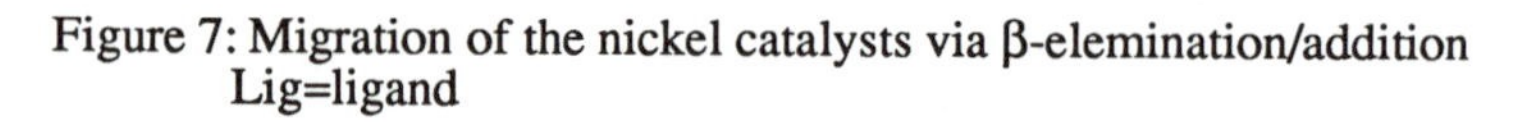
Figure 7: Migration of the nickel catalysts via β-elemination/addition Lig=ligand

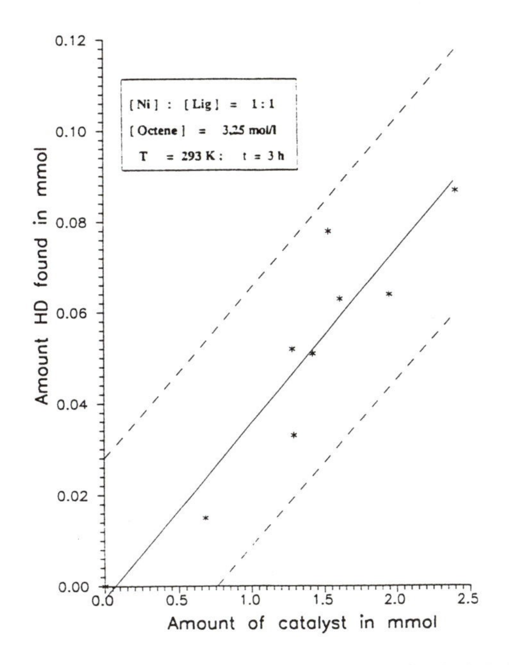

Figure 8: Amount of HD found versus the amount of nickel catalyst used

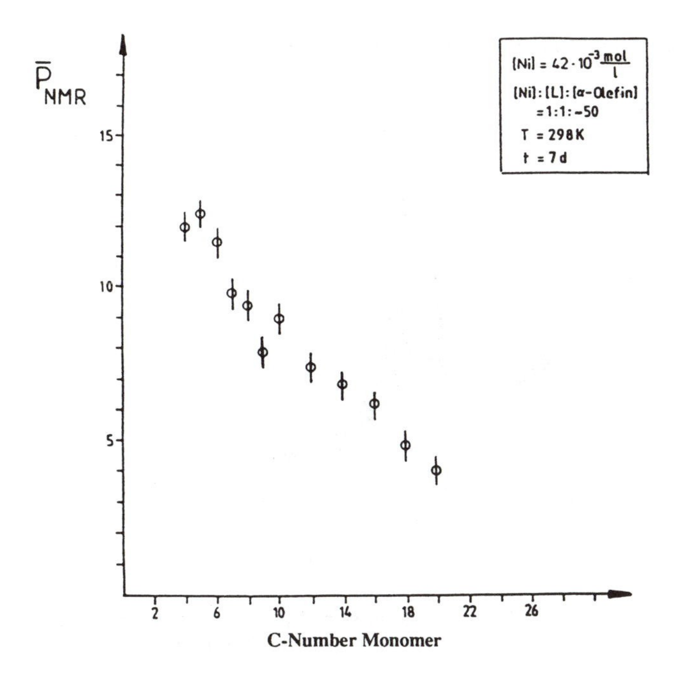

Figure 9: Degree of polymerization from NMR versus C number of the α-Olefin (modified from Ref. 3.)

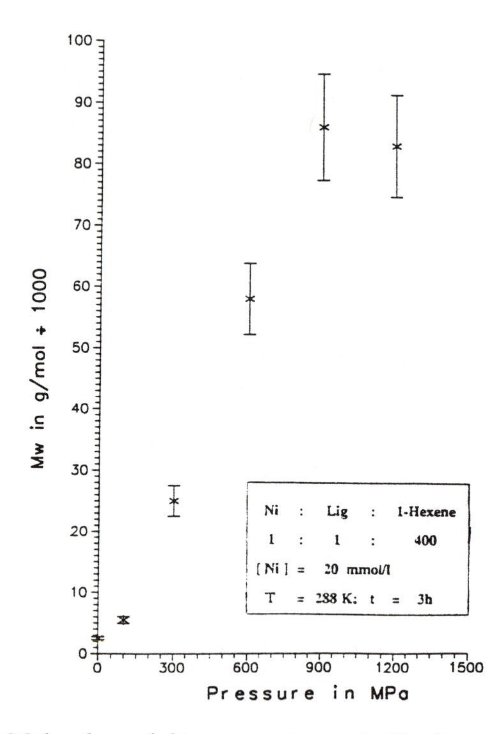

Figure 10:Molecular weight versus pressure inside the reaction vessel

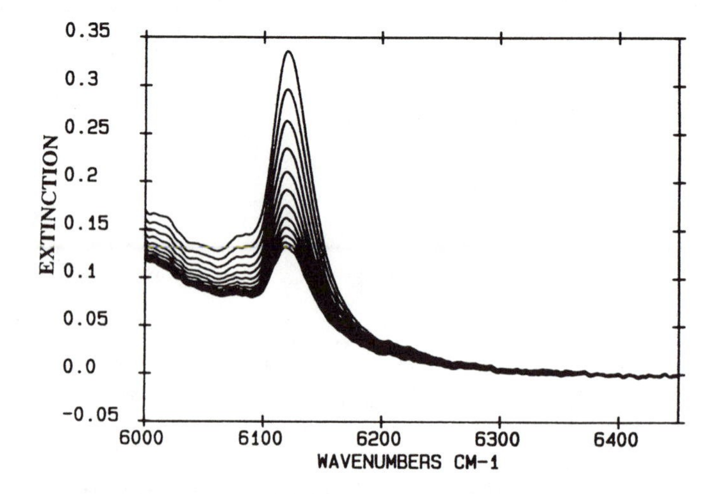

Figure 11:FT-IR-spectra of the polymerization of 1-hexene at 300 MPa; the time between each spectra was 2 minutes

the first overtone of the olefinic CH stretching mode at 6120 cm^{-1}. Integration yields, according to Lambert-Beer's Law, the concentration of the α-olefin. A linear relationship was found between the logarithm of the measured concentration of the α-olefin and the reaction time, which means that the reaction is first order concerning the monomer. Therefore, from the slope of this curve we got the rate constants of the polymerization. Doing this for a set of different pressures we could estimate the activation volume of the reaction (Figure 12). For the formation of a carbon-carbon single bond, we expect a value between -10 and -20 cm^3/mol. The value of - 17.8 ± 1 cm^3/mol lies in this area. This confirms that, for the case of 1-hexene, the insertion and not the migration is the rate determing step for the 2,ω-polymerization.

We evaluated the activation parameters for a series of different α-olefins (see Table 1). It seems to be surprising that the activation energy is decreased with the length of the monomer used. For the case of hexadecene we could not estimate any temperature dependence at all. The activation entropies show the same relation with the length of the α-olefin used. For the case of dodecene we evaluated an activation entropy of -220 J/(mol K). This is a rather small value for a bimolecular reaction.

Table 1: Activation parameter of the 2,ω-polymerization of α-olefins

	E_A [kJ/mol]	$\Delta H^{\neq}$ [kJ/mol] T = 288 K	$\Delta S^{\neq}$ [J/(mol K)] T = 288 K
Ethylene	63 ± 3	61	- 87
1-Hexene	51 ± 3	49	- 157
1-Octene	45 ± 3	43	- 180
1-Dodecene	33 ± 3	31	- 223

The reason for this phenomenon lies in the superposition of the insertion reaction and the migration of the catalyst. We have set up a kinetic model to describe the 2,ω-polymerization (Figure 13). The migration steps are treated as preceding equilibria. With the approximation that the equilibrium constant K is the same for all equilibria, we can set up the rate law for the whole reaction (Equation 1,2).

$$d[M] / dt = k_w [D] [M] \quad (1)$$

or

$$d[M] / dt = k_w \{K^n [A]\} [M] \quad (2)$$

The power n at the equilibrium constant in Equation 2 is given by the number of

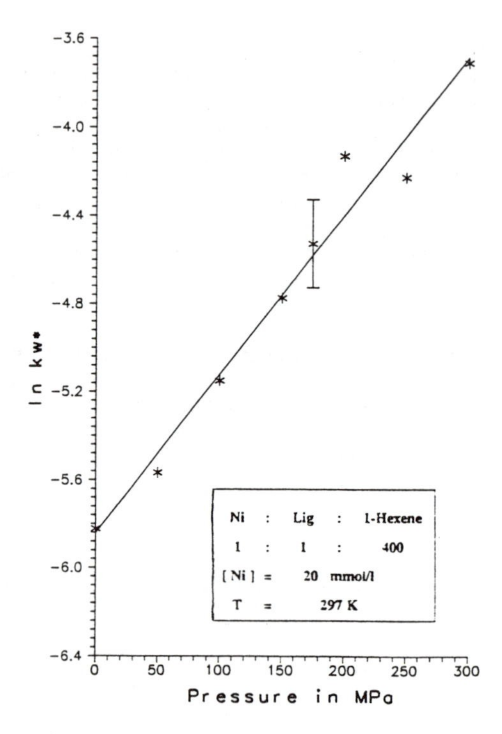

Figure 12: Rate-constant versus pressure for the 2,ω-polymerization of 1-hexene

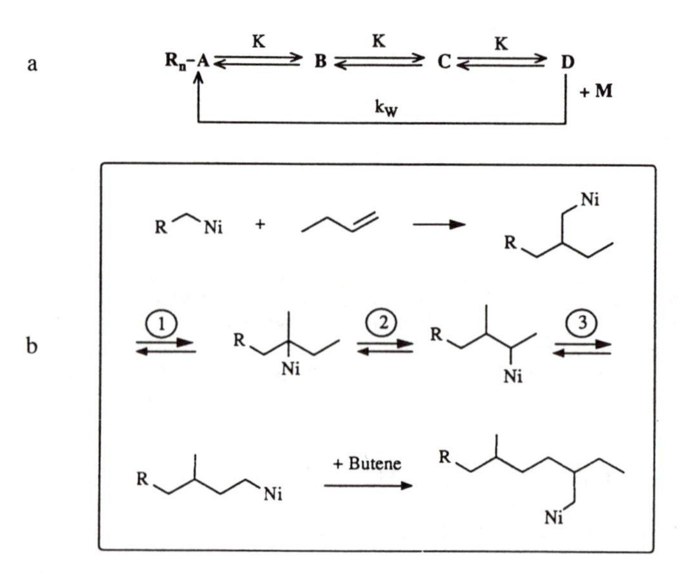

Figure 13: Kinetic model for the 2,ω-Polymerization
a) Kinetic scheme
b) Mechanism of the 2,ω-polymerization for the example of 1-butene

the preceding equilibria or, in other words, by the number of migration steps which have to be done by the catalyst between two insertions. According to this model we have a superposition of insertion and migration. One of these processes (the insertion) is endergonic and the other (the migration) is exergonic.

The only value which can be measured experimentally is the product of the rate constant for the insertion and the terms in arched brackets in Equation 2. Therefor the evaluation of the activation energy according to Arrhenius yields a value superposed of insertion and migration equilibria (see Equation 3).

$$\ln \{ k_w K^n \} = \ln A - (E_a / RT) + (n \Delta G / RT) \quad (3)$$

As you can see in Equation 3, the number of migration steps n is important for the overall activation energy. This value decides which process is dominating, the migration or the insertion. On the other hand, the number of migration steps between two insertions is closely related to the length of the α-olefin used. In Figure 13 the 2,ω-polymerization of 1-butene is shown. You can easily see that for this example there are three migration steps necessary between two insertions. For the general case, we need the number of carbon atoms of the α-olefin minus one migration step.

As we mentioned above, the length of an α-olefin is directly connected with the number of migration steps or the factor n in Equation 3. Therefore the overall activation paramters should depend in a linear manner on the length of the monomer; this is confirmed by the experiment as you can see in Figure 14. From the slope of the curve in Figure 14 the equilibrium constant was (K = 1.2) for one migration step. It is not surprising that the value is nearly one, because starting material and product are similar.

With the above discussed mechanism we can explain how the 2,ω-linkage happens and the unusual kinetic behaviour of the 2,ω-polymerization of α-olefins as well. For the polymerization of ethylene with the same catalyst system there is no migration necessary. Therefore it is clear that the activation energy and the activation entropy are in the expected range. But it is noteworthy, that they fit into the series of α-olefins (see Table 1).

Furthermore, the use of optically active olefins, e.g. 4-methyl-1-hexene, leads to very interesting insights into the complex polymerization mechanism and its stereochemistry. Surprisingly, the pure (R)- and (S)- enantiomers give no reaction, whereas the racemic mixture always is very well 2,ω-polymerized. The resulting polymer is optically inactive. This means that the catalyst needs both enantiomers to form the racemic polymer. Hence, we must conclude that the enantiomers are alternatively arranged along the polymer chain. In other words, a strongly alternating "copolymer" of the (R) and (S) enantiomers is formed. Hence, the initial chiral carbon-atom in the 4-methyl-1-hexene is not racemised by the migration of the nickel catalyst, i.e., the migration proceeds either by retention or inversion of the configuration of the chiral center.

The migration mechanism via nickelhydride as discussed above, leads for the case of the optically active monomer 4-methyl-1-hexene to the situation shown in Figure 7 before.

The crucial fact during the β-elimination is the formation of a nickel hydride

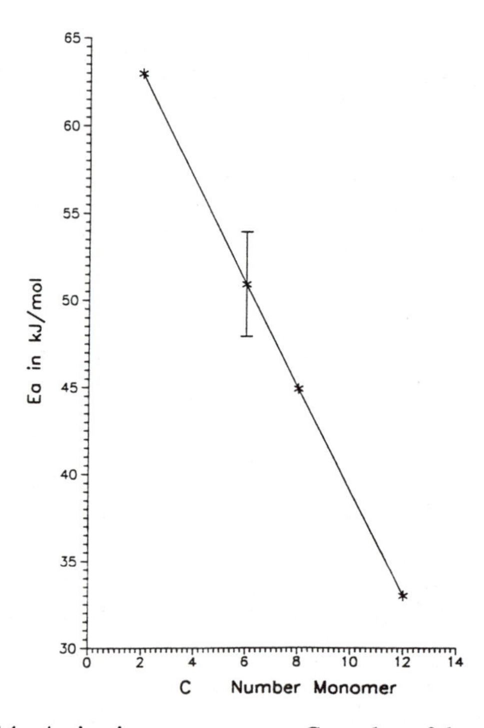

Figure 14: Activation energy versus C number of the α-olefin

which is coordinated to the double bond of the polymer chain and is not a free hydride. This makes sure that the stereochemical configuration is preserved. Addition leads either to the original molecule or to a molecule in which the catalyst has migrated one step further to the end of the polymer chain. Olefins with inner double bonds are not polymerized. Therefore it is clear that a free nickel hydride, as it is produced during the transfer reaction, can add only to α-olefins. A second argument is that there should be racemisation of chiral monomers if the migration would proceed via a free nickel hydride.

The 2,ω-linkage of 4-methyl-1-hexene leads according to Figure 15 to the structure in which the initial chiral center (*) in the Fischer-projection is fixed and the methyl group, resulting from the prochiral double bond, has a statistical position.

Because we need both enantiomers and there are two possible configurations for the new generated methyl group, we can construct four different units, the two racemic and the two meso forms (Figure 15).

Combining these units to a polymer chain with four monomers we get the nine possible structures in Figure 16.

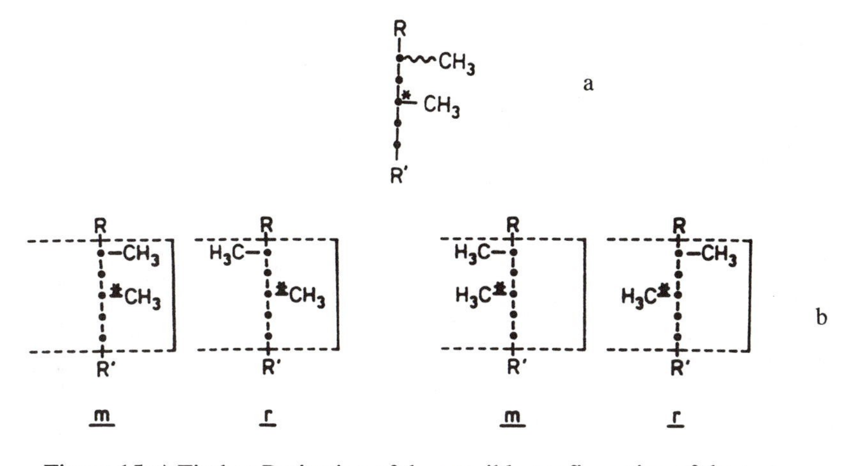

Figure 15:a) Fischer-Projection of the possible configuration of the two methyl groups for the general case
b) The four different units with the two racemic and the two meso positions of the methyl groups

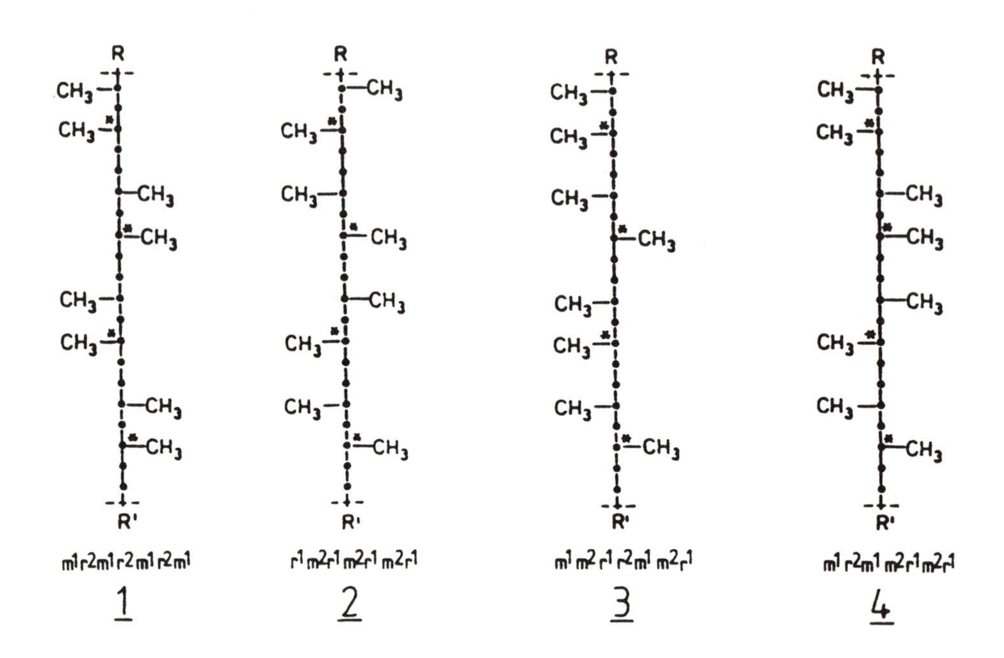

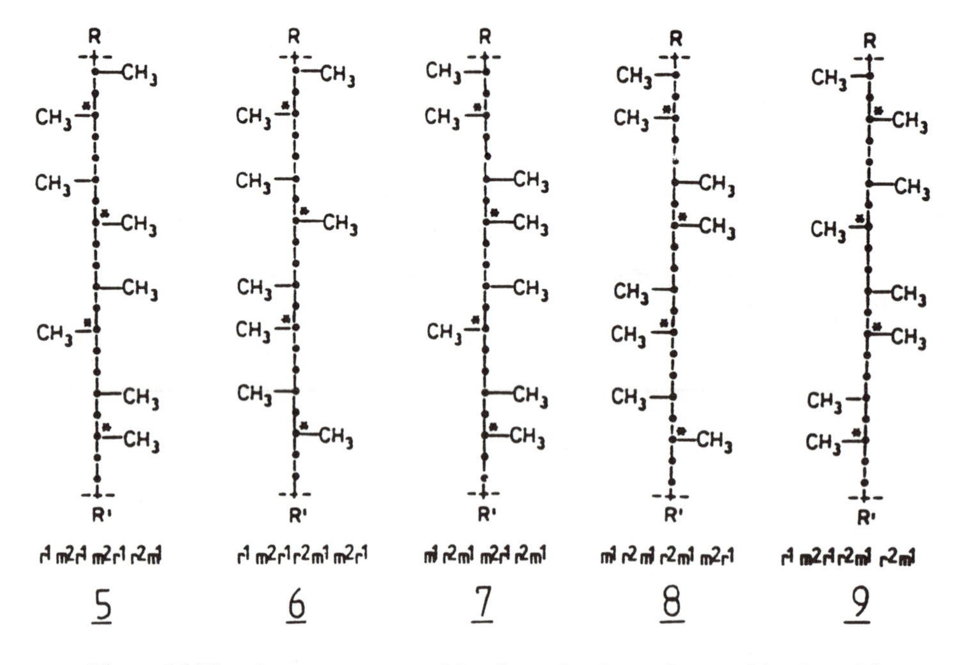

Figure 16:The nine structures resulting from the alternative combination of four (R) and (S) monomers

It was not possible up to now to decide which structure is formed exactly from the ^{13}C-NMR-spectra. But it seems to be more like structure 1 or 2 in Figure 16. What is important for this polymer is that in all structures we have a "copolymer" with regularly spaced chiral centers in the main chain.

The experimental details of this work are published elsewhere (6, 7, 10).

Literature Cited

1. Keim, W.; Appel, R.; Storeck, A.; Krüger, C.; Goddard, R.; *Angew. Chem.* **1981**, *93*, 91.
2. Möhring, V.M.; Fink, G.; *Angew. Chem* **1985,** *97*, 982; *Angew. Chem. Int. Ed. Engl.* **1985,** *24*, 1001.
3. Fink, G.; In *Recent Advances in Mechanistic and Synthetic Aspects of Polymerization*; Fontanille, M.; Guyot, A., Eds.; Nato ASI Series, Ser. C; Math. Phys. Sci. 215; Reidel Publ.: Dordrecht, 1987, 515-533.
4. Bogdanovic, B.; Kröner, M.; Wilke, G.; *Justus Liebigs Ann. Chem.* **1969,** *699*, 1
5. Scherer, O. J.; Kuhn, N.; *Chem. Ber.* **1974,** *107*, 2123.
6. A. Heinrichs, Dissertation, University of Duesseldorf 1990.
7. V.M. Möhring, Dissertation, University of Duesseldorf 1985.
8. Lindemann, L. P.; Adams, J. Q.; *Anal. Chem.* **1971,** *43*, 1245.
9. Buback, M.; *Z. Naturforsch.* **1984,** *39a*, 399.
10. Ch. Denger, Dissertation, University of Duesseldorf 1991.

RECEIVED November 15, 1991

Chapter 8

Structure and Morphology of Highly Stereoregular Syndiotactic Polypropylene Produced by Homogeneous Catalysts

Adam Galambos, Michael Wolkowicz, and Robert Zeigler

Himont Research and Development Center, 800 Greenbank Road, Wilmington, DE 19808

A highly stereoregular sample of syndiotactic polypropylene (s-PP) produced via homogeneous catalysis was examined using a variety of analytical techniques to determine how the crystallization properties compared to those of commercial isotactic polypropylene (i-PP) and to investigate the dependence of multiple crystalline forms and morphologies on thermal and mechanical processing conditions. The thermal properties suggested that with comparable thermal history, s-PP had a melting point about 10 °C less than that of i-PP, and had a slower spherulitic growth rate at a given crystallization temperature. Two crystal forms of s-PP were observed by solid-state NMR: An all-*trans* form promoted by cold drawing and the more typical "figure-8" helical form produced during quiescent crystallization. Multiple melting behavior of the "figure-8" crystalline form was dependent on the thermal processing conditions. Finally, a transition from a spherulitic to an axialitic growth habit of the "figure-8" crystalline form was observed to occur as the isothermal crystallization temperature was increased from 138 to 148 °C.

The thirty years following the groundbreaking work of Natta and coworkers on the polymerization of isotactic (*1,2*) and syndiotactic (*3*) polypropylene (i-PP and s-PP, respectively) have seen a marked divergence in the fortunes of the two polymers. Isotactic PP has enjoyed significant improvements in both catalyst technology (increasing manufacturers ability to control stereoregularity (*4,5*) and molecular weight (*6,7*)), and polymerization technology (greatly reducing production costs). The net result has been the emergence of

0097–6156/92/0496–0104$06.00/0

i-PP as a versatile engineering thermoplastic, with a US sales volume in excess of 8 billion pounds a year (*8*).

Syndiotactic PP, however, has enjoyed no such commercial success. The primary reason has been that, until quite recently, little improvement had been made in a catalyst technology which was only capable of producing a material with a relatively low molecular weight and poor stereoregularity (*9*), at a temperature (-78°C) decidedly unfriendly for commercial production.

In the last few years, developments in metallocene catalyst systems (often called homogeneous catalysts), particularly by Ewen and coworkers (*10*), have made possible the synthesis of highly stereoregular s-PP in good yield at conventional polymerization temperatures (25-70°C). The availability of s-PP at heretofore unacheivable levels of stereoregularity has spurred the recent renewed interest in the structure-property relationships for this polymer (*11,12*).

Previous investigations of the solid-state structure of s-PP have proposed the existence of both multiple crystal forms and multiple crystalline morphologies. Three crystal forms have been proposed for s-PP - two consist of chains which adopt a $(t_2g_2)_2$, or "figure-8" helical conformation (*13-15*) and differ only in the pattern of chain chirality incorporated along the *b* crystallographic axis, while the third is made up of polymer chains in an all- *trans* form, which is promoted by cold drawing (at 0°C) of s-PP quenched rapidly from the melt state (*11,16*). Recent work by Lovinger and coworkers (*14,15*) has corrected the erroneous original indexing of the "figure-8" unit cell (*11*) - the indexing errors were the result of crystallographic reflections from small amounts of crystalline isotactic PP present in all early s-PP samples being mistakenly assigned to the diffraction pattern of s-PP. Solid-state ^{13}C NMR experiments substantiated the existence of the helical chain conformation in the crystalline state (*17*), while this report presents NMR data on the planar zig-zag crystalline conformation.

Multiple crystalline morphologies have been proposed to explain two prominent endotherms observed in DSC experiments (*18*) - the lower melting endotherm was theorized to be the result of the incorporation of chain defects in the crystalline lamellae, while the higher melting form was thought to consist of a more perfect crystalline structure from which chain defects were excluded.

This study employs thermal analysis, wide-angle x-ray diffraction (WAXD), small-angle x-ray scattering (SAXS), solid state ^{13}C NMR, and optical and electron (SEM) microscopy to examine the relationship between solid state structure and thermal history for a highly stereoregular s-PP material.

Experimental

The samples used in this study were provided by M. Galimberti of HIMONT Italia S.R.L. and were prepared according to the following procedure. Polymerization of the s-PP was performed in a jacketed 2 liter Buchi glass autoclave equipped with mechanical stirrer and thermometer. Isopropyl (cyclo-pentadienyl-1-fluorenyl) Zr(IV) dichloride was synthesized according to the method of Ewen (*10*). A solution of 0.4 vol % methylalumoxane (Schering, 30 wt % in toluene) in toluene was charged to the autoclave in a propylene atmosphere. After pressurization with propylene to 4 atm. (at 11°C), catalyst (8 mg. in 10 ml of toluene/0.7 ml methylalumoxane) was injected. Polymerization was quenched by injecting 5 ml of methanol and the polymer recovered by precipitation with methanol/HCl.

Differential scanning calorimetry experiments were performed using a Perkin-Elmer Model DSC-7 equipped with a Model 7700 computer. Data were acquired at a scan rate of 20°C/min except where noted.

Wide-angle x-ray diffraction patterns were acquired on a Phillips diffractometer in the transmission mode using Cu Kα radiation. The detector was a scintillation counter equipped with a curved crystal focalizer; data was acquired by step scanning in 0.1° 2q increments. Small-angle x-ray scattering data was acquired using the rotating anode source with pinhole collimation and a 2-D multiwire position sensitive detector at the National Institute of Standards and Technology (NIST). Spectra were rotationally averaged and corrected for background scattering before analysis.

The solution-state ^{13}C NMR spectra were run at 75.4 MHz on a Varian UNITY-300 NMR spectrometer. The samples were run in 10 mm tubes as 10% solutions in tetrachloroethane-d_2 at 130 °C. 2000 transients were acquired for each solution-state NMR spectrum. The solid-state ^{13}C NMR spectra were run at 67.9 MHz on a Chemagnetics CMX-270 NMR spectrometer. The samples were spun at the magic angle at 5 kHz. Cross-polarization with a 1 ms contact period was used, with rf power levels set so that γH_2 was $4x10^5$ s^{-1} for both the ^{13}C and proton channels. 500 transients were acquired for each solid-state NMR spectrum.

Spherulite growth rate studies were performed using as polymerized s-PP powder in a Linkman TH600 programmable hot stage. Samples were heated to 225 °C, held in the melt for 10 minutes, then cooled at 50 °C/min to the chosen crystallization temperature and allowed to crystallize isothermally to completion.

SEM photomicrographs of the quiescently crystallized material were acquired on a Hitachi S-800 Field Emission instrument using an accelerating voltage of 5 KV. Samples were prepared by etching using a 0.7 % $KMn0_4$ in 2:1 sulphuric/phosphoric acid solution.

Results

Molecular Characterization. The s-PP used for solid-state studies had a higher degree of stereoregularity than any s-PP discussed in prior morphological studies. The fraction of syndiotactic pentads as determined by solution ^{13}C NMR was 0.95, which corresponds to an average of one error in the syndioregular structure every 126 propylene monomer units. The measured intrinsic viscosity was 1.6 dl/g at 135°C in decahydronaphthaline.

Both hexane extraction and temperature rising elution fractionation (TREF) in xylene were used in an attempt to separate the sample into fractions of varying stereoregularity. ^{13}C NMR, DSC, and GPC of fractions recovered from both techniques indicated that separation was based on molecular weight, with no significant differences in tacticity. This behavior is different from that seen with typical isotactic polypropylene (i-PP), reflecting the expected differences in performance of homogeneous and heterogeneous catalysts.

Solid State Structure. The solid state structure of samples with five distinctly different processing histories were examined by DSC, WAXD, SAXS, and ^{13}C NMR. The samples were:

I. as-polymerized powder
II. a 0.5 mm film cooled from the melt to room temperature at 20°C/min
III. a 0.5 mm film rapidly quenched in water from a 230°C melt, stored at room temperature.
IV. above film annealed for 8 hours at 150°C.
V. III above cold drawn at 0°C.
VI. a 0.5 mm film rapidly quenched in water from a 230°C melt, stored below 0 °C.

Thermal Behavior. DSC 1st heat data acquired at a scan rate of 20°C/min for the first four samples listed above are displayed in Figure 1, and values of onset and peak melting temperature, heat of fusion, and sample density for all five samples are compiled in Table I. The as-polymerized powder (I) appears to be the most crystalline of the materials studied by DSC, as it has the highest heat of fusion. There are two distinct endotherms in the melting behavior of the melt crystallized samples, with peak temperatures typically around 145 and 152°C. The lower melting form is promoted by slow cooling from the melt (II), while the higher temperature form is prominent in the quickly cooled sample (III). Figure 3 illustrates the systematic transition between the two peaks after thermal conditioning at cooling rates between 5 and 80 °C/min. Annealing between 140 and 150°C produces a single endothermic transition about 10°C above the annealing temperature (Sample IV).

At room temperature, solid-state ^{13}C NMR indicates that the all-*trans* crystalline form is the only ordered crystalline structure present in Sample V. Upon heating in the DSC (Figure 2) above room temperature this form converts to the "figure-8" crystal structure, with some evidence of conversion (via a shift in DSC baseline) starting at around 40 °C, followed by a more rapid conversion between 80 and 100°C, as evidenced by the exotherm observed in the DSC trace in that temperature range. The rapid conversion between s-PP crystal forms

Table I. Thermal and Morphological Properties of Various s-PP Specimens

Sample Number	1st Melt onset (°C)	peak (°C)	ΔHf (J/g)	Density (g/cc)	SAXS Long Spacing corr.[1] (nm)	uncorr.[1] (nm)
I	138.7	149.5	79	-	-	-
II	139.9	146.2	51	0.888	12.3	14.6
III	138.0	150.1	39	0.876	-	-
IV	155.4	159.2	65	0.897	20.3	23.3

[1]Corrected (corr.) and uncorrected (uncorr.) refer to the Lorentz correction, which involves multiplying observed intensities by the square of the scattering vector

$$(q = \frac{4\pi}{\lambda}\sin\Theta).$$

Reported periodicities are the Bragg's Law spacings based on the maxima of the resultant plots of intensity vs. scattering angle. The spacing,

$$d = \frac{1}{2\sin\Theta} = \frac{2\pi}{q_{max}}$$

where λ is the wavelength of the radiation used (λ = 0.154 nm in this case) and Θ is one-half the observed scattering angle at which the maximum intensity is observed.

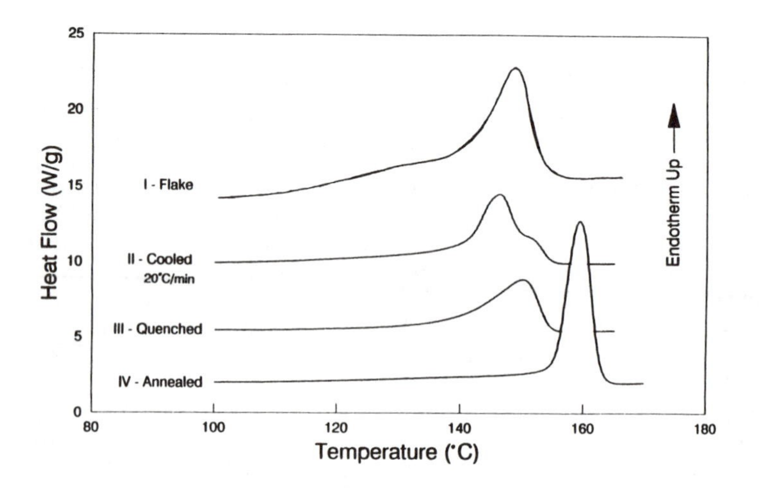

Figure 1. DSC thermograms of four s-PP specimens: I - as polymerized reactor flake, II - Film cooled from melt at 20 °C/min, III - film water quenched from melt, IV - Film from sample III annealed under vacuum for 8 hrs at 150 °C. The thermograms were acquired at a heating rate of 20 °C/min.

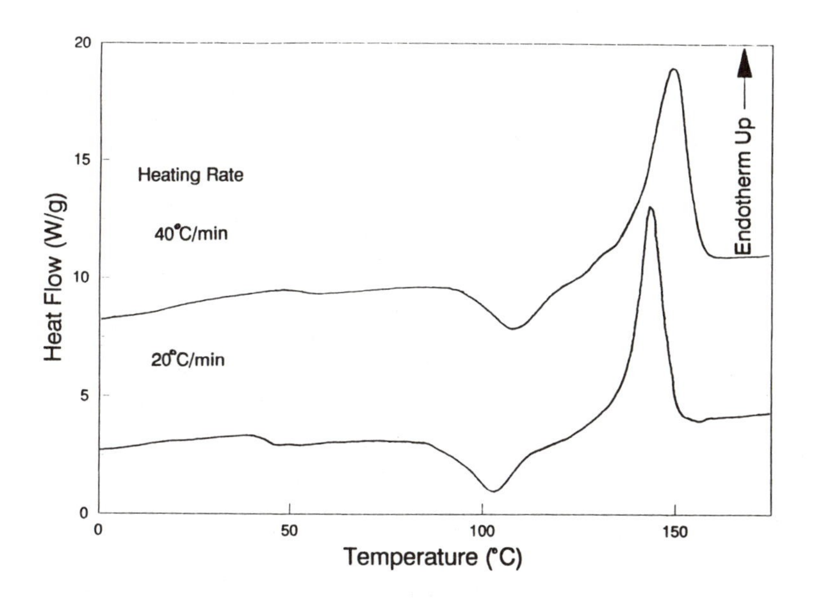

Figure 2. DSC thermograms of s-PP specimens in the all-*trans* crystalline form. Data acquired at heating rates of 20 and 40 °C/min.

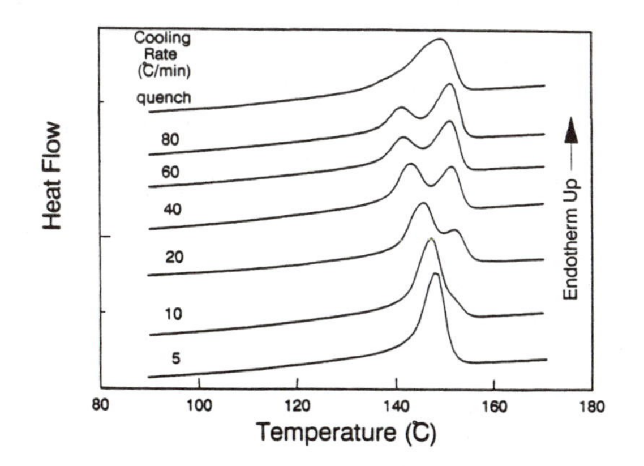

Figure 3. DSC thermograms of s-PP specimens crystallized using varying cooling rates. Cooling rates: 5, 10, 20, 40, 60, 80 ° C/min, plus rapid quench from melt in ice water. Data acquired at a heating rate of 20 ° C/min.

in that temperature range has been documented by previous researchers (*12,16*).

Wide- and Small-Angle X-Ray Scattering (WAXS and SAXS) Results. WAXD (Figure 4) and SAXS (Figure 5) spectra were recorded for the two samples with the greatest temperature difference in melting point - the specimen cooled at 20°C/min from the melt (II - peak T_m at 146°C), and the annealed sample (IV - peak T_m at 159°C). The WAXD patterns of the two samples were similar, with the exception of a distinct peak at a Bragg spacing of 0.47 nm which was present in the annealed, high melting point sample, and absent in the lower melting point specimen. This particular peak does not correspond to any reflection from the "figure-8" crystalline structure as indexed by Lovinger and coworkers (*14,15*) - rather, this is the most intense reflection from the diffraction pattern for the "all-*trans*" crystalline form as documented by Chatani, et al (*12*). The intensity of this peak relative to those of the "figure-8" form indicates that the all-*trans* crystal form is present in this sample at a level no greater than a few percent. Peak intensities were greater and peak widths were narrower for the annealed material, which is consistent with the higher degree of crystallinity observed in this sample by DSC and density measurements, and the larger crystallite size predicted by small-angle x-ray scattering (SAXS) results (see Table I).

Solid-State ^{13}C NMR Results. The solid-state ^{13}C CP-MAS NMR spectra of samples I-IV are shown in Figure 6. The methine carbons resonate near 26.5 ppm and several methyl carbons appear near 20.9 ppm. Four separate methylene resonances are observed; those at 47.8 ppm and 39.2 ppm were described by Bunn et al. (*17*), who assigned them to those carbons on the exterior and interior portions of the "figure-8" helices of crystalline syndiotactic polypropylene. If it is assumed that the chemical shifts of the methylene carbons in polypropylene are determined by local conformations (*19*), then the resonance at 49.1 ppm represents methylene carbons in an "all-*trans* configuration", such as that described by Natta *et. al.* (*16*). The peak at 43.7 ppm is very broad and probably represents polymer with a distribution of local conformations such as would be found in a non-crystalline polymer.

Two experiments were carried out to verify these assignments of the methylene resonances of s-PP. The first experiment confirmed the assignment of the 49.1 ppm resonance by running a solid-state NMR spectrum of a sample of "all-*trans*" s-PP that was prepared by stretching the sample by 100% at 0°C, as described by Natta et. al. (*16*). This sample (sample V) is shown in Figure 7 and shows that the major methylene resonance in all-*trans* s-PP resonates at 50.1 ppm, verifying the assignment of the 49.1 ppm resonance to s-PP chains in an all-*trans* conformation. The broad methyl resonance at 19.5 ppm can also be associated all-*trans* s-PP. The width of the resonances in

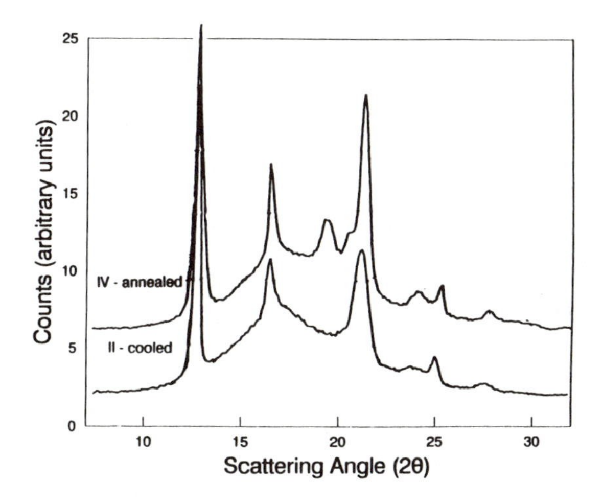

Figure 4. Wide-Angle X-Ray Diffraction Spectra of s-PP Samples II and IV.

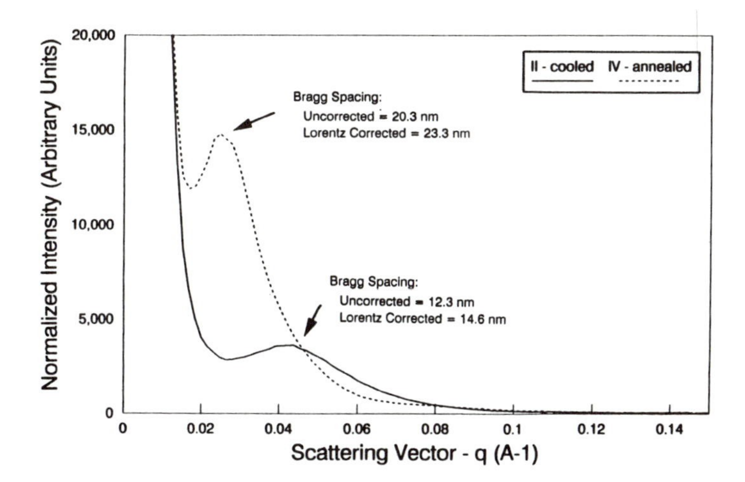

Figure 5. Small-Angle X-Ray Scattering Spectra of s-PP Samples II and IV.

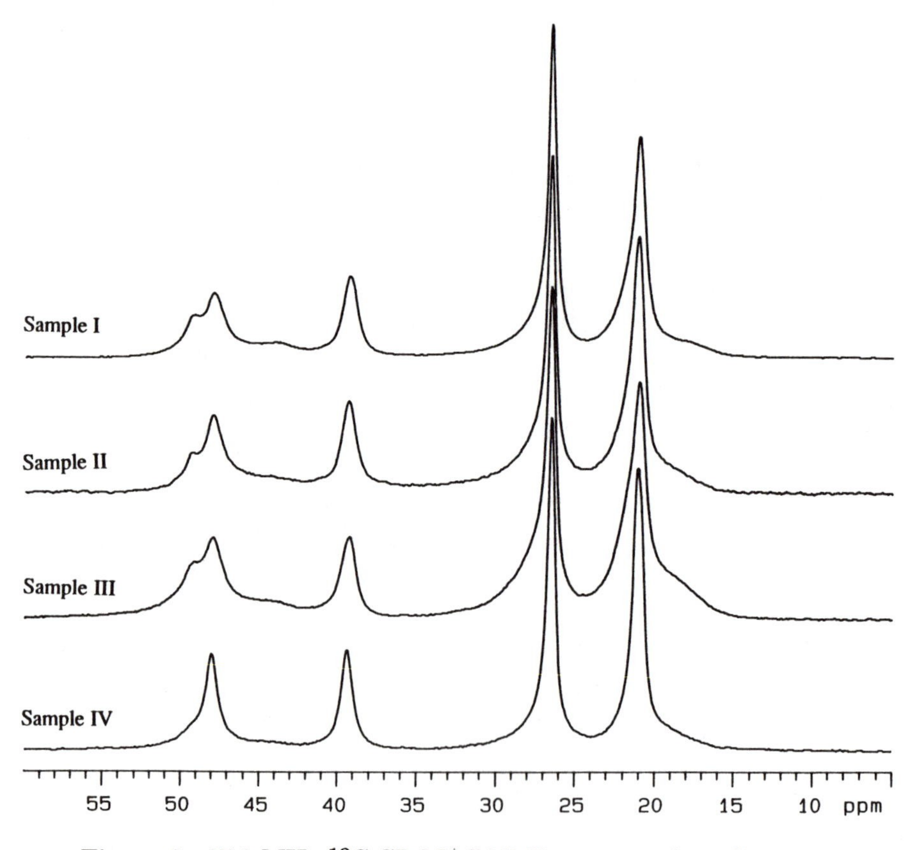

Figure 6. 67.9 MHz ^{13}C CP-MAS NMR spectra of syndiotactic polypropylene samples.

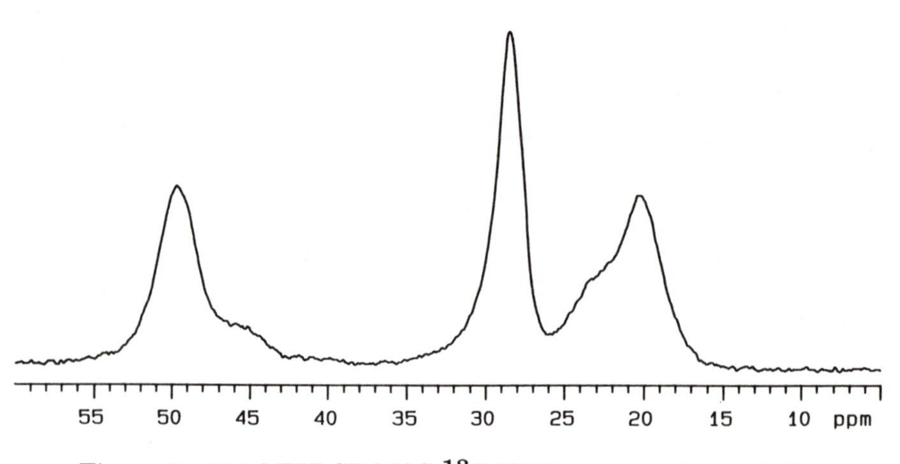

Figure 7. 67.9 MHZ CP-MAS ^{13}C NMR spectra of sample V.

sample V suggest that the crystalline regions in this sample are less ordered than those of sample IV. A comparison of the solid-state NMR spectra to X-ray diffraction data are planned. This comparison should provide greater insight into both the microstructures and larger-scale structures of this polymer.

The second experiment to verify the assignments of the methylene resonances was to compare the spectra of s-PP that was ice-quenched from the melt (sample VI) and run at - 20 °C (below T_g) to that run at room temperature. These spectra are shown in Figure 8 and show that the intensity of the resonance at 43.7 ppm is affected by the change in sample temperature, while the intensities of the other resonances are not. This variation in intensity is due to the relative efficiency of the cross-polarization (CP) NMR pulse sequence for static molecular structures, with CP efficiency decreasing with increasing molecular motion (*20*). As the sample temperature rises through T_g, the molecular motion of the non-crystalline component of the polymer will increase and the intensity of the peak representing the non-crystalline polymer will decrease. Therefore a resonance that decreases in intensity as the sample temperature goes above T_g can be associated with non-crystalline structures in the s-PP. The resonance at 43.7 ppm clearly represents non-crystalline s-PP.

A splitting of the methyl resonance at 21.0 ppm is observed for sample IV, the most highly annealed syndiotactic polypropylene film shown in Figure 6. This splitting is an indicator that the methyl groups exist in two separate environments in the crystalline polymer. It appears that this methyl splitting could result from inter-helix interactions, since a very similar splitting in i-PP was assigned to helices with opposing "direction of coiling" of the crystalline helices (*21*).

Optical and Scanning Electron Microscopy Results. Measurements of isolated spherulite dimensions by optical microscopy during isothermal crystallization at 138, 145, and 148°C indicated that the rate of crystalline growth was linear with time at approximately 0.11, 0.034, and 0.019 μm/min, respectively. While these rates are about 10 times higher than previous measurements on less stereoregular s-PP (*22*), they are still about one tenth the reported growth rates for i-PP in that temperature range, both in the literature (*23-25*), and from our measurements as shown in Figure 9.

SEM photomicrographs of etched s-PP crystallized at 145 °C (Figure 10) show a spherulitic morphology with clearly visible lamellae having periodicity in the 25-50 nm size range. This in agreement with the periodicity measured by SAXS on Sample IV, which had been annealed for 8 hours at 150 °C.

Discussion

Estimation of Thermodynamic Parameters. Several s-PP thermodynamic properties were estimated from the combined SAXS,

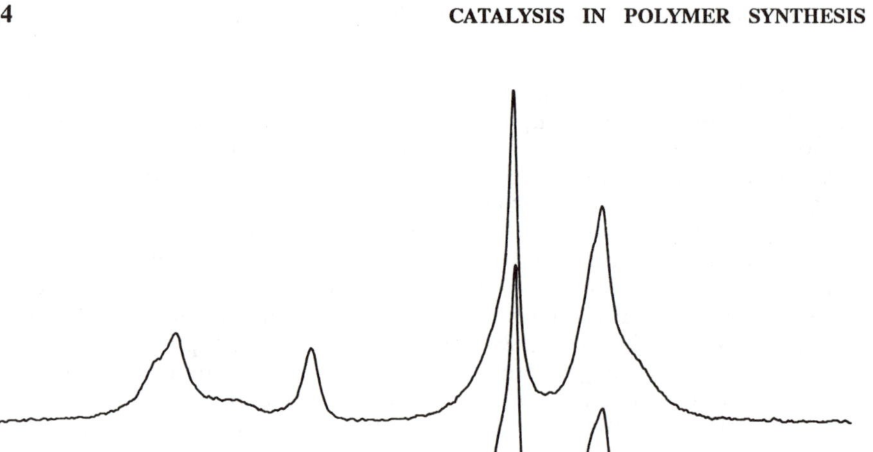

Figure 8. 67.9 MHZ CP-MAS ^{13}C NMR spectra of sample VI.
Bottom: Spectrum run at a sample temperature of 0 °C.
Top: Spectrum run at a sample temperature of 22 °C.

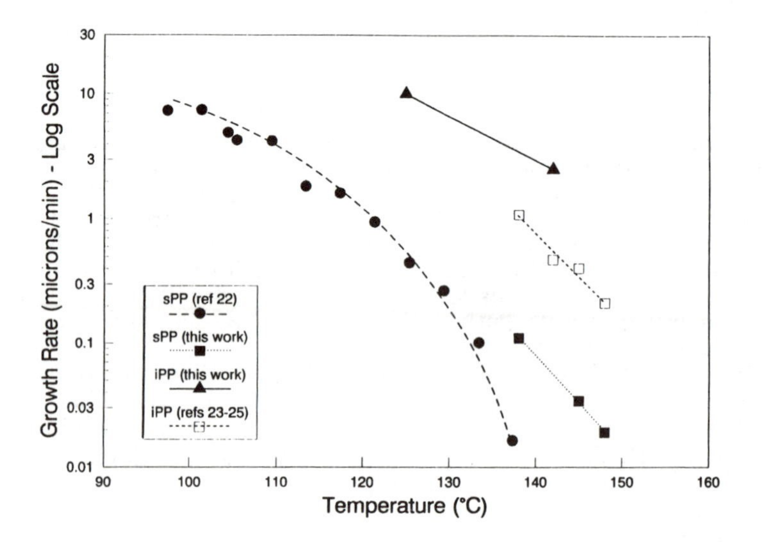

Figure 9. Spherulitic Growth rate of s-PP - comparison of values from this study to literature values for s-PP and experimental and literature values for i-PP.

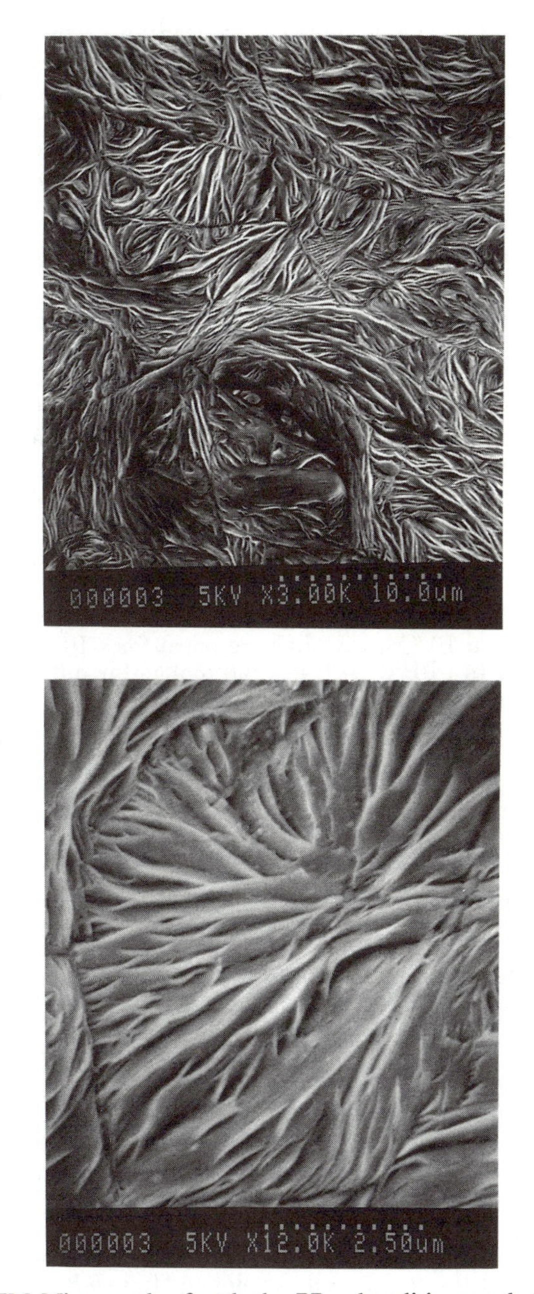

Figure 10. SEM Micrograph of etched s-PP spherulitic morphology. a. 2160X. b. 8640X.

density, and DSC experiments of this study. Extrapolating heat of fusion values as a function of density results in a prediction of 105 J/g for the DH_f of a 100% crystalline sample, in good agreement with a literature determination of 105.5 J/g (*26*), and well below the value of 165 J/g determined for i-PP[27]. An extrapolation of the melting point to infinite crystal thickness from DSC and SAXS data predicted a melting point for this sample of 171°C, which is 10°C higher than the highest value reported for less stereoregular materials available to previous researchers (*26*).

No attempt was made from the limited data of this study to estimate the equilbrium T_m^{∞} for s-PP. Previous researchers working with less stereoregular s-PP have estimated this temperature as high as 220 °C (*22*). More recent work by Balbontin and coworkers (*21*), who provided s-PP for our study, have calculated T_m^{o} values as high as 186 °C for the most stereoregular sample, and have predicted a T_m^{∞} near 220 °C by extrapolating melting point data from samples with varying stereoregularity to 100% stereoregularity. Literature values for the T_m^{∞} of i-PP are generally in the range of 180-190°C (*29*).

In a practical sense, our work seems to indicate that the equilibrium T_m^{∞} of s-PP is actually lower than the T_m^{∞} of i-PP, since we have observed both an order of magnitude slower crystallization kinetics and about a 10°C lower melting point for s-PP when compared directly to i-PP with the same molecular defect level (rrrr = mmmm pentad fraction by solution NMR).

Multiple Crystal Forms of s-PP. This study has identified thermal/processing conditions which produce multiple s-PP crystalline forms:

1. The planar "all-*trans*" crystal form was promoted by drawing a quenched polymer film at 0 °C, as observed by previous researchers (*11,16*). Solid-state NMR measurements (Figure 7) indicate that there are not detectable levels of the "figure-8" crystal form in the stretched sample.

2. The "figure-8" crystal form is a common form for s-PP in the absence of applied stress. The "all-*trans*" form was observed to convert to the "figure-8" form at temperatures above 40 °C in a DSC experiment. Rapid conversion, as indicated by a pronounced exotherm during the DSC scan, was observed to occur near 100 °C, again in agreement with work by previous workers (*11,16*).

3. Room temperature WAXD of a sample annealed for 8 hours at 150 °C indicated that although a well developed "figure-8" crystalline structure existed in the sample film, a small amount (< 5%) of the "all-*trans*" crystalline form was also present. This observation was consistent with solid-state NMR measurements, which also detected a

small fraction of the "all-*trans*" form in the sample. Since the "all-*trans*" form of s-PP has been observed to convert to the "figure-8" form at elevated temperatures, it might be that the small fraction of all-*trans* crystal formed upon cooling from the annealing temperature to room temperature.

Previous work by Lovinger and coworkers (*14,15*) describe a transition between 3-dimensional (spherulitic) and 2-dimensional (axialitic,rectangular single crystal) structures in thin melt crystallized s-PP films as the crystallization temperature increases. They note spherulitic growth at 70 °C, axialitic growth at 90 °C, and lathlike structures growing at 105 °C. In this study, we have observed a similar trend toward increasingly 3-dimensional growth patterns at lower crystallization temperatures, but because of the increasing stereoregularity of the available s-PP, and the fact that we were using thicker sample films, the 3-D-to-2-D transition temperature is significantly higher. At crystallization temperatures as high as 138 °C the crystalline morphology is primarily spherulitic (see Figure 11), while at higher temperatures, such as 148 °C (Figure 12), the crystallization habit is primarily axialitic.

Marchetti and Martuscelli (*18*) noted two endotherms in the DSC behavior of dried, solution crystallized single crystal aggregates of s-PP. These endotherms were also evident upon second heating, so they represented structure that developed during melt crystallization, as well as during solution crystallization. The investigators attributed the two peaks to the melting of "figure-8" crystals with and without stereo-defects incorporated in the crystal structure. They offered no direct evidence for this theory, however.

Lovinger and coworkers (*14,15*) presented evidence for two crystallographic modifications for s-PP made up of chains in the "figure-8" helical conformation. The forms had identical chain packing along the *a* crystallographic axis, but could have a spectrum of packing schemes along the *b* axis, ranging from regularly antichiral at high crystallization temperatures, to a random distribution of chiral packing under conditions of rapid crystallization. In this sense, "disorder" unrelated to errors in stereoregularity could be incorporated in the crystal structure. The authors did not relate the crystalline melting point to variations in crystal structure.

We observed a systematic change in the relative area of low and high temperature DSC endotherms as a function of cooling rate (Figure 3). Slower cooling rates (less than 40 °C/min) produced a larger lower melting temperature endotherm (145 °C), while fast cooling (greater than 40 °C/min) produced a larger higher melting temperature endotherm (152 °C). A rapidly quenched film had lower crystallinity, as indicated by lower density and lower enthalpy change upon melting, and a peak melting temperature intermediate, but closer to the higher of the two previously observed endotherms (150 °C).

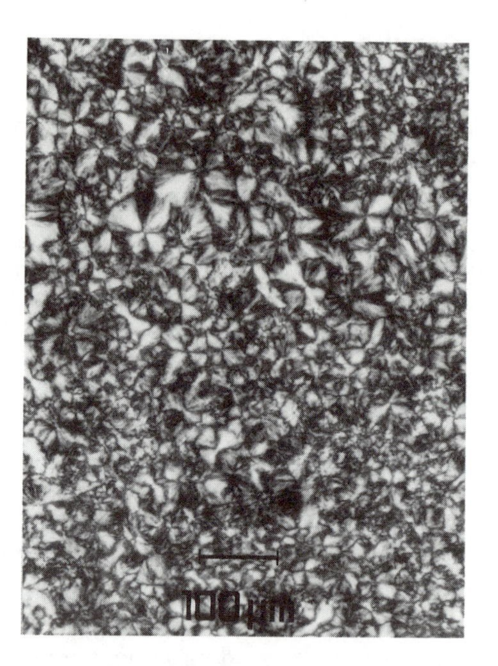

Figure 11. Optical micrograph of melt crystallized s-PP (T_c = 138) °C. 100.5X.

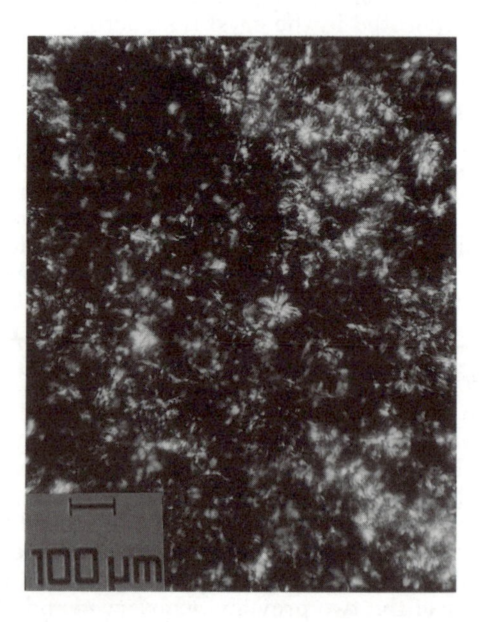

Figure 12. Optical micrograph of melt crystallized s-PP (T_c = 148) °C. 65X.

Theoretically, if the different melting endotherms were related to variations in the "figure-8" crystal structure, WAXD should be able to detect the differences, as Lovinger (*14,15*) indicated distinct changes in the diffraction pattern with changes in ordering along the *b* crystal axis. The WAXD data we present in this work - one spectra each from the low melting slow cooled film and the high melting annealed film (see Figure 4) - does not indicate marked differences in the "figure-8" crystal structures resulting from these two thermal histories. Additional diffraction peaks in the annealed sample were attributed to a small fraction of the "all-*trans*" crystal form in an earlier section of this report.

Acknowledgements

The authors would like to express appreciation to HIMONT, Incorporated, for supporting this work, and the following individuals who were instrumental in the acquisition of the data presented: N. P. Pyrros (Hercules, Incorporated), J. D. Barnes (NIST), and J. Browne, J.P. Faline, T. Orr, and D. Morgan (HIMONT, Incorporated). The s-PP samples were provided by M. Galimberti of HIMONT Italia S.R.L. We would also like to thank A.J. Lovinger of AT&T Bell Labs for his valuable advice and insight during our discussions.

Literature Cited

1. Natta, G.;Pino, P; Mazzanti, G. US Patent 3,112,300 (1963), and prior Italian Patent (1954).
2. Natta, G.; Pasquon, I.; Corradini, P.; Peraldo, M.; Pegoraro, M.; Zambelli,A. *Rend. Acc. Naz. Lincei,* **1960,** 28, 541.
3. Natta, G.; Pasquon, P.; Zambelli, A. *J. Am. Chem. Soc.,* **1962,** 84, 1488.
4. Boor, J. *Macromolecular Rev.,* **1967,** 2, 115-268.
5. Barbe, P.C.; Cecchin, G.; Noristi, L. *Advances in Polym. Sci,* **1987,** 1-81.
6. Vandenberg, E.J. US Patent 3,051,690, (assigned to Hercules Powder Co.), Aug. 28, 1962.
7. Bua, E.; Luciani, L. Italian Patent 554,013, (assigned to Monticatini), Jan. 5 , 1957.
8. *Modern Plastics,* Jan 1991 issue, p. 113, McGraw-Hill publishing, NY,NY.
9. Zambelli, A.; Natta, G.; Pasquon, I.; Signorini, R. *J. Polym. Sci, Pt. C,* **1967,** 16, 2485.
10. Ewen, J.A.; Jones, R.L.; Razavi, A.; Ferrara, J.D. *J. Am. Chem. Soc.,* **1988,** 110, 6255.
11. Chatani, Y.; Maruyama, H.; Noguchi, K.; Asanuma, T.; Shiomura, T. *J. of Polymer Sci., Pt. C: Letters,* **1990,** 28, 393-398.
12. Balbontin, G.; Dainelli, D.; Galimberti, M.; Paganetto, G. submitted to *Mak. Chem.*
13. Corradini, P.; Natta, G.; Ganis, P.; Temussi, P.A. *J. Polym. Sci., Pt. C,* **1967,** 16, 2177.

14. Lotz, B.; Lovinger, A.J.; Cais, R.E. *Macromolecules,* **1988,** 21, 2375-2382.
15. Lovinger, A.J.; Davis, D.D.; Lotz, B. *Macromolecules,* **1991,** 24, 552-560.
16. Natta, G.; Peraldo, M.; Allegra, G.*Mak. Chem.,* **1964,** 75, 215.
17. A. Bunn, A.; Cudby, M.E.A.; Harris, R.K.; Packer, K.J.; Say, B.T. *J.C.S. Chem. Comm.,* **1981,** p. 15.
18. Marchetti, A.; Martuscelli, E. *J. Polym. Sci., Polym. Phys. Ed.,* **1974,** 12, 1649.
19. Tonelli, A.E. "NMR Spectroscopy and Polymer Microstructure", VCH Publishers, New York, (1989).
20. Komoroski, R. A. "High-Resolution NMR Spectroscopy of Synthetic Polymers in Bulk", VCH Publishers, Deerfield Beach, FL.
21. Bunn, A.; Cudby, M.E.A.; Harris, R.K.; Packer, K.J.; Say, B.T. *Polymer,* **1982,** 23, 694.
22. Miller, R.L.; Seeley, E. G. *J. of Poly. Sci: Polym Phys. Ed.,* **1982,** 20, 2291.
23. Falkai, B.V.; Stuart, H.A. *Kolloid Z.,* **1959,** 162, 138.
24. Binsbergen, F.L.; DeLange, B.G.M. *Polymer,* **1970,** 11, 309.
25. Goldfarb, L. *Makromol. Chem.,* **1978,** 179, 2297.
26. Gee, D. R.; Melia, T. P. *Polymer,* **1969,** 10, 239.
27. Gaur, U.; Wunderlich, B. *J. Phys. Chem. Ref. Data,* **1981,** 10(4), 1051.
28. Boor, J. Jr.; Youngman, E.A. *J. Polym. Sci., Pt. B.,* **196,** 577.
29. Clark, E. J.; Hoffman, J.D. *Macromolecules,* **1984,** 17, 878.

RECEIVED February 3, 1992

Chapter 9

Dicyclopentadiene Polymerization Using Well-Characterized Tungsten Phenoxide Complexes

Andrew Bell

Research Center, Hercules Advanced Materials and Systems Company, Hercules Incorporated, 1313 North Market Street, Wilmington, DE 19894–0001

Pure $WOCl_{4-x}(OAr)_x$ complexes (x = 1, 2, 3 and 4; and OAr = phenoxide) were prepared in essentially quantitative yield by reacting tungsten oxytetrachloride ($WOCl_4$) with the requisite amount of phenol or metal phenoxide. The ring-opening metathesis polymerization (ROMP) of dicyclopentadiene (DCPD) by $WOCl_3(OAr)$, $WOCl_2(OAr)_2$, and $WOCl(OAr)_3$, in combination with trialkyltin hydride (R_3SnH) was assessed. The tungsten catalysts produced are capable of bulk-polymerizing DCPD to very high polymer yields. The polymerization ability of a particular procatalyst was correlated with the reduction potential ($W(VI) \rightarrow W(V)$) of the complex and the charge on the oxygen of the di- or tri-substituted phenoxide ion. The effects of changing DCPD:W, Sn:W, and rate moderator:W ratios on residual monomer levels were studied.

Transition metal-catalyzed ring-opening metathesis polymerization (ROMP) of cyclic olefins is an important application of the olefin metathesis reaction (Scheme 1) [*1*]. Polynorbornene, polyoctenamer, and polydicyclopentadiene are currently produced on an industrial scale using this process [*2*].

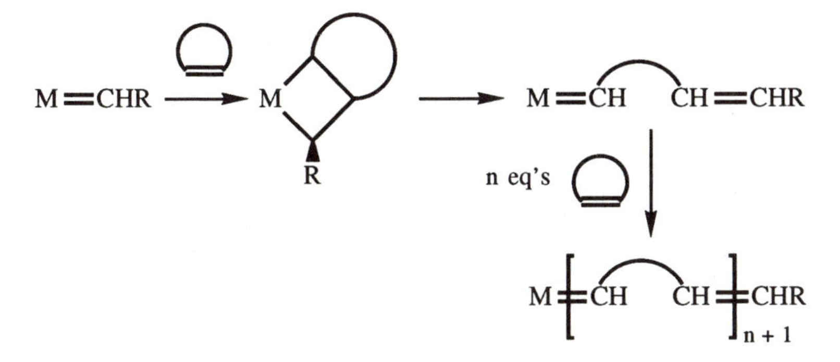

Scheme 1. Mechanism for ROMP of Cycloolefins

0097–6156/92/0496–0121$06.00/0

Dicyclopentadiene (DCPD) polymerizes quite readily using metathesis catalysts, but the nature of the product is highly dependent on specific details of catalyst structure and polymerization conditions. Known products range from soluble or only partially soluble thermoplastics [*3*] to an insoluble, brittle polymer [*4*]. Ring-opening metathesis of the very strained norbornene ring with concomitant co-metathesis of a limited number of less-strained cyclopentene rings results in the highly cross-linked thermoset network depicted in Scheme 2. Recently, it has been found that such a highly crosslinked thermoset resin has properties which make it useful as an engineering plastic [*5, 6*]. METTON® liquid molding resin is a Hercules Incorporated trademark for a proprietary blend of polydicyclopentadiene formed by the ring-opening metathesis polymerization of DCPD.

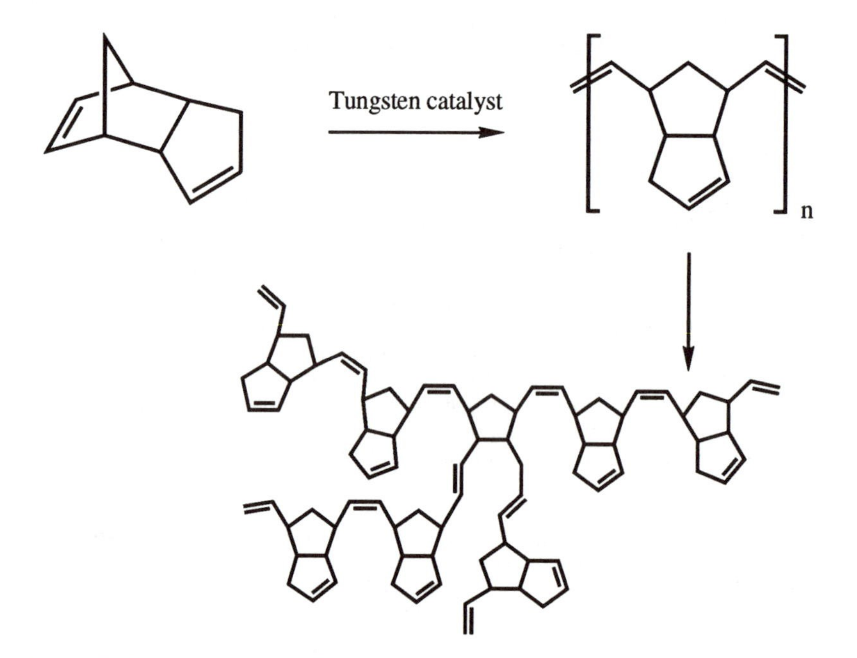

Scheme 2. Formation of PolyDCPD Utilizing a Tungsten Catalyst (Depicted in a Stepwise Fashion for Clarity)

The DCPD polymerization reaction has characteristics which make it readily adaptable to reaction injection molding (RIM) [*7*], resin transfer molding (RTM), pour molding, and other liquid-molding processes. The active metathesis catalyst is formed when two separate reagents, a procatalyst component and an activator component, are combined. The procatalyst is defined as a metal complex which, when combined with a suitable activator, generates the ring-opening metathesis polymerization catalyst. This makes it possible to use separate solutions of the individual components in DCPD monomer to generate an active catalyst when the components are combined in the mixhead prior to injection into a mold. (See Scheme 3.)

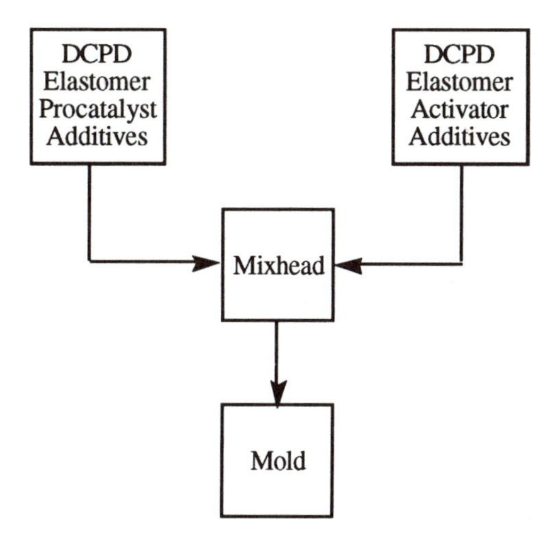

Scheme 3. Schematic for METTON polyDCPD RIM System

In our continuing efforts to develop new catalyst systems, we have focused our attention on tungsten chemistry supported by bulky phenoxide ligation. Of particular interest are a number of tungsten complexes with phenoxide ligands which have been utilized as olefin metathesis catalyst precursors, e.g., tungsten hexaphenoxide ($W(OC_6H_5)_6$) [*8,9*], tungsten oxytetraphenoxide ($WO(OC_6H_5)_4$) [*10*], dichlorotetrakis(phenoxide)tungsten ($WCl_2(OAr)_4$) and related complexes [*8-13*]. Most recently, Basset and co-workers have completed a study [*14*] whereby treatment of tungsten hexachloride (WCl_6) with various substituted phenols in toluene or carbon tetrachloride leads to chlorophenoxide complexes $WCl_{6-x}(OAr)_x$, where $x = 2$ to 4. These complexes are also quite active catalysts for the ring-opening metathesis polymerization of strained olefins once activated by an appropriate alkylating agent, e.g., Et_2AlCl [*15*].

In similar fashion, polyDCPD has been produced [*16, 17*] using $WCl_5(OAr)$, $WCl_4(OAr)_2$, $WCl_3(OAr)_3$, and $WCl_2(OAr)_4$ procatalysts activated by non-alkylating tin hydride reagents, i.e., tri-*n*-butyltin hydride (n-Bu_3SnH) or triphenyltin hydride (Ph_3SnH).

We are now in a position to describe a new catalyst system for the bulk polymerization of DCPD utilizing tungsten phenoxide-based procatalysts, derived from tungsten oxytetrachloride ($WOCl_4$) and various 2,6-disubstituted phenols, in combination with trialkyltin hydrides or triaryltin hydrides.

Experimental

General Procedures and Starting Materials. All manipulations were performed anaerobically in nitrogen-sparged pop bottles or under an argon atmosphere (Vacuum Atmospheres Dri-Lab) or using Schlenk techniques. All liquid transfers were made by either syringe or cannulae to maintain an inert atmosphere. Sublimed tungsten oxytetrachloride ($WOCl_4$) was obtained from H. C. Starck (Berlin, Germany) and used as received for the synthesis of $WOCl_{4-x}(OAr)_x$ procatalysts. 2,6-dibromophenol, 2,6-dichlorophenol, 2,6-

diisopropylphenol, 2,6-dimethoxyphenol, 2,6-dimethylphenol, 2,6-dimethylphenol, and 2,6-diphenylphenol were purchased from Aldrich Chemical Company and used as received. 2,4-dichloro-6-methylphenol was obtained from American Tokyo Kasei, Inc. and used as received. Elemental analyses were performed by Galbraith Laboratories, Inc. Knoxville, TN 37950.

Cyclic Voltammetry. Dichloromethane (CH_2Cl_2, HPLC grade, Burdick and Jackson) was dried over 4A molecular sieves (Aldrich Chemical Company) and sparged with dry nitrogen before use. The electrolyte, tetra-*n*-butylammonium hexafluorophosphate $[N(n\text{-}Bu)_4][PF_6]$ (TBAH), was prepared by the metathesis reaction of potassium hexafluorophosphate and tetra-*n*-butylammonium iodide. The TBAH obtained was recrystallized three times from ethanol/water and dried at 80 °C for 24 hours. When performing measurements using $WOCl_{4-x}(OAr)_x$, an extra drying step was introduced to remove any H_2O brought into the system by the electrolyte. The 0.2M TBAH-CH_2Cl_2 solutions were prepared in the dry box and placed over 4A molecular sieves for a minimum of 24 hours prior to use.

Cyclic voltammograms (CV) were performed using a BioAnalytical Systems Inc., Model CV-1B instrument in conjunction with a Hewlett-Packard 7045A X-Y recorder. A three-compartment (H) cell (≈10 mL volume of main compartment) was used with a platinum disk working electrode, platinum gauze auxiliary electrode, and a silver wire quasi-reference electrode (AgQRE) to which all potentials refer. The potential of the working electrode was checked by the addition of a small amount of ferricinium hexafluorophosphate ($[Cp_2Fe](PF_6)$) into the analysis chamber. The $Cp_2Fe^{0/+}$ couple for $[Cp_2Fe](PF_6)$ is at +0.45 V versus AgQRE. Alternatively, decamethylferrocene $[(Me_5Cp)_2Fe]$ could be substituted for $[Cp_2Fe](PF_6)$ when analyte and redox couples were coincident. Cyclic voltammograms were obtained in the dry box in dichloromethane containing 0.2 M TBAH as supporting electrolyte. Analyte concentrations were approximately 1×10^{-3} M under these conditions.

Polymerization Studies. All manipulations were performed anaerobically in nitrogen-sparged pop bottles or under an argon atmosphere (Vacuum Atmospheres Dri-Lab) or using Schlenk techniques. All liquid transfers were made by either syringe or cannulae. Tri-*n*-butyltin hydride (packaged in Sure/Seal™ bottle) was purchased from Aldrich Chemical Company and stored refrigerated (~0 °C). Dicyclopentadiene (DCPD) of ≥98% purity was used to prepare all catalyst and activator polymerization stock solutions.

Polymerizations were conducted in nitrogen-sparged test tubes by adding together the procatalyst/DCPD and activator/DCPD components (2.5 mL of each), mixing on a vortex mixer, and then inserting the tube into a heated block (~32 °C) or oil bath (80 °C). Residual monomer levels were determined by gas chromatography after extracting the unpolymerized dicyclopentadiene by swelling the polymer sample in toluene.

Tungsten Phenoxide Procatalysts. Pure $WOCl_{4-x}(OAr)_x$ (x = 1, 2, 3 and 4; and OAr = 2,6-disubstituted phenoxide, e.g., OC_6H_3-2,6-Me_2) complexes were prepared in essentially quantitative yield by allowing $WOCl_4$ to react with the requisite amount of phenol or lithium phenoxide in hydrocarbon or etheral solvents. The procedures used were typical of those described earlier [*12, 13, 14(a), 18- 27*].

$WOCl_3(OC_6H_3$-2,6-$Cl_2)$ To a quantity of $WOCl_4$ (3.79 g; 11.1 mmol) stirred in cyclopentane (50 mL) was added dropwise a solution of 2,6-dichlorophenol (HOC_6H_3-2,6-Cl_2) (1.18 g; 11.1 mmol) in cyclopentane (25 mL). The dropwise addition of the phenolic solution was accomplished over a period of 30 minutes. During the phenol addition, the solution changed from orange to deep red and purple crystals precipitated from the reaction solution. The reaction mixture was stirred at room temperature for two hours. After this time, the reaction mixture was filtered to remove the dark red-purple crystalline product $WOCl_3(OC_6H_3$-2,6-$Cl_2)$, and the collected solids were washed with pentane (10 mL), and dried in vacuo; yield = 4.20 g (81%). The filtrate was evaporated to dryness under reduced pressure and was shown by electrochemistry to be of the same identity as the isolated solids. Anal. Calcd. for $C_6H_3Cl_5OW$: C, 15.38%; H, 0.64%; Cl, 37.86%. Found: C, 15.21%; H, 0.67%; Cl, 37.00%.

$WOCl_2(OC_6H_3$-2,6-$Cl_2)_2$ A quantity of $WOCl_3(OC_6H_3$-2,6-$Cl_2)$ (3.50 g; 7.48 mmol) was dissolved in a minimum of diethyl ether (~50 mL). To this solution was added a saturated solution of lithium 2,6-dichlorophenoxide ($LiOC_6H_3$-2,6-Cl_2) (1.28 g; 7.48 mmol) in diethyl ether (~10 mL). The phenolic solution was slowly added dropwise. Almost instantly, the deposition of red crystals occurred. The reaction mixture was stirred at room temperature with a slow nitrogen purge over the ether solution for 150 minutes. After an additional hour, the reaction mixture was filtered to remove a dark red, crystalline product $WOCl_2(OC_6H_3$-2,6-$Cl_2)_2$. The solid was then washed with a small volume of dried pentane (5 mL) and dried under vacuo (3.90 g). This material was dissolved in dichloromethane (~25 mL) and filtered to remove the lithium chloride by-product. Evaporation of the filtrate under reduced pressure led to pure $WOCl_2(OC_6H_3$-2,6-$Cl_2)_2$ in 72% yield (3.21 g). Anal. Calcd. for $C_{12}H_6Cl_6OW$: C, 24.21%; H, 1.01%; Cl, 35.77%. Found: C, 23.86%; H, 1.08%; Cl, 36.07%.

Theoretical. The calculations were carried out using the standard AM1 procedure as implemented in the MOPAC 5.0 (molecular orbital package) program [*28*]. Prior to calculating atomic charges for the various phenols and phenoxide anions, the geometries of the species were calculated by minimizing the energy with respect to all geometrical variables.

Results and Discussion

Procatalyst Synthesis. The synthesis of $WCl_x(OAr)_{6-x}$ complexes (x = 0, 1, 2, 3, 4, and 6) has been reported by a number of authors [*14, 18-23*]. In general, treatment of tungsten hexachloride (WCl_6) with the stoichiometric amount of phenol in toluene or carbon tetrachloride solvents leads to chlorophenoxide and phenoxide complexes of the following stoichiometries: $WCl_5(OAr)$, $WCl_4(OAr)_2$, $WCl_3(OAr)_3$, $WCl_2(OAr)_4$, and $W(OAr)_6$.

There is, however, a distinct lack of information available on the preparation of phenoxide complexes derived from tungsten oxytetrachloride ($WOCl_4$). Previous work by Funk and Mohaupt [*24*] showed that mono- and bis-phenoxide complexes could be prepared from $WOCl_4$ and salicylic acid, salicylaldehyde or methyl salicylate. By employing quite harsh reaction conditions, Prasad and co-workers [*25*] prepared a number of tetrakis phenoxide complexes, i.e., $WO(OAr)_4$, by reacting $WOCl_4$ with cresols and nitro phenols. For the purpose of comparison to tungsten(IV) phenoxides, Schrock et al. [*26*] prepared the tetrakis(2,6-dimethylphenoxide) oxytungsten complex, $WO(OC_6H_3$-2,6-$Me_2)_4$, through chloride substitution with lithium 2,6-dimethylphenoxide.

The use of phenoxides and alkoxides as ligands in the preparation chemistry of transition metal and organometallic complexes is now commonplace. The particular advantages associated with phenoxide ligands are as follows: (i) the tungsten center is more readily solubilized by attachment of the phenoxide, (ii) bulky ligands at the tungsten center thwart catalyst decomposition pathways involving the combination of metal centers through bridging ligands, (iii) the ligands themselves are able to control the sterics at the reaction site, and (iv) the electronic character of the reaction site can be altered by the substituents on the phenoxide ligand. Thus, an indication of the electronic and steric demands at the metal center during the metathesis polymerization reaction can be investigated.

It should be noted that in the synthesis of phenoxide derivatives of $WOCl_4$-based complexes is not as simple as might first be supposed. One complication reported by Funk [*19*] and Mortimer [*20*] is the facile conversion of tungsten oxytetrachloride ($WOCl_4$) to the tungsten hexaphenoxide ($W(OAr)_6$) complex in excess phenol. Further, reaction of the W=O bond apparently also occurs in the preparation of $WCl(OC_6H_5)_5$ from $WOCl_4$ and phenol [*20*].

We find that it is now possible to prepare phenoxide complexes of $WOCl_4$ in a stepwise fashion by using phenols or their metal salts (e.g., NaOAr or LiOAr) in hydrocarbon or ether solvents [*27*]. Scheme 4 shows the synthetic routes to such $WOCl_{4-x}(OAr)_x$ complexes. The monophenoxide complexes can be directly prepared from $WOCl_4$ and one equivalent of phenol. Further, the *bis*-phenoxide complexes may be synthesized by direct addition of two equivalents of phenol to $WOCl_4$ or, depending on the phenol substitutents, by using the monophenoxide tungsten complex and one equivalent of a lithium phenoxide. We find it is best not to use phenols as the reagents of choice for the whole reaction scheme as they are not discriminating enough in their reactivity. Chloride substitution to generate *tris*- and *tetra*-phenoxide oxytungsten complexes, i.e., $WOCl(OAr)_3$, and $WO(OAr)_4$, is most cleanly accomplished using lithium phenoxide ligand reactions with the intermediate $WOCl_2(OAr)_2$ complex. The $WOCl_3(OAr)$, $WOCl_2(OAr)_2$, $WOCl(OAr)_3$, and $WO(OAr)_4$ complexes are isolated as crystalline solids in essentially quantitative yield and are readily dissolved in pure DCPD.

Electrochemical Characterization of Procatalysts. Recently, we reported the electrochemical behavior (cyclic voltammetry and coulometry) of tungsten(VI) phenoxide complexes, $WCl_{6-x}(OAr)_x$, where $x = 0, 1, 2, 3, 4,$ and 6 [*29*]. Rothwell [*22, 23*] and Schrock [*21*] have also discussed the electrochemistry of $WCl_5(OAr)$, $WCl_2(OAr)_4$, and $W(OAr)_6$ complexes. In general, the cyclic voltammetric characterization of tungsten(VI) phenoxide compounds appears to be quite straightforward. The cyclic voltammetric experiment shows that the stepwise replacement of chloride by phenoxide in $WCl_{6-x}(OAr)_x$ complexes leads to species that become harder to reduce by about 500 mV on each subsequent substitution.

It is known that the reduction potentials for the $WCl_{6-x}(OAr)_x$ complexes correspond with the degree of phenoxide substitution. More important, however, is the fact that the reduction potential indicates the electronic character of the metal center. Schrock's study showed that the redox potentials of some tungsten(VI) phenoxide species follow trends that one would predict on the basis of the electron-withdrawing ability of the phenoxide ligands [*21*]. Indeed, it is found by analyzing olefin metathesis data that more active catalysts are formed when the most electropositive metal centers are activated toward metathesis [*12, 15*].

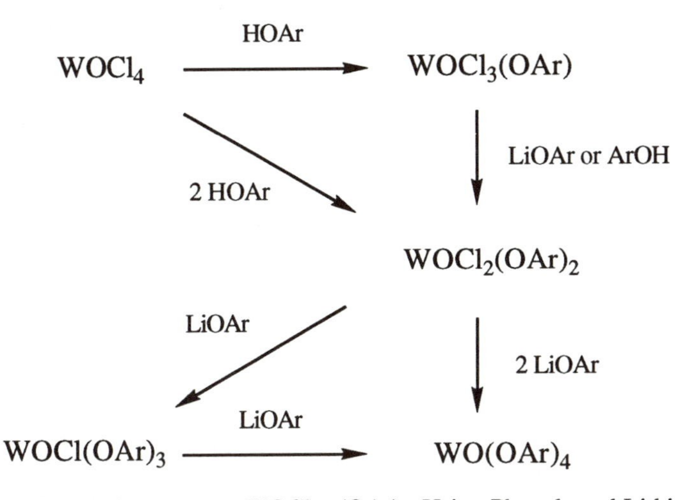

Scheme 4. Synthetic Routes to $WOCl_{4-x}(OAr)_x$ Using Phenols and Lithium Phenoxides.

In a similar fashion, we find that the W(VI)→W(V) reduction potentials of $WOCl_{4-x}(OAr)_x$ complexes may be altered by varying both the degree of substitution and the substituents on the phenoxide ligands. Subtle changes in the electrochemistry of the $WOCl_{4-x}(OAr)_x$ complexes are associated with changes in the amount of phenoxide π-bonding to the tungsten center. The π-bonding is manipulated by the use of various electron-withdrawing or releasing groups attached to the phenol. For any given procatalyst, the reduction potential is related to the nature of the phenoxide 2,6-substituents. The charge on the free phenoxide (or free phenol oxygen) decreases in the following order: *i*-Pr > CH_3 > CH_3O ≈ Ph > Cl > Br. The order corresponds roughly to the decrease in the pK_a of the corresponding phenol. In this study, we have calculated the electronic charges on a number of 2,6-disubstituted or 2,4,6-trisubstituted phenols and correlated them to the reduction potential of the $WOCl_{4-x}(OAr)_x$ (x = 1, 2, 3, or 4) complexes prepared.

The relationship between the reduction potentials of tungsten phenoxide complexes and charges on the phenoxide oxygen is given in Figure 1. One could equally well correlate the sum of the Hammett parameters or the charge on the phenol used in the procatalyst preparation in place of phenoxide oxygen charge and achieve a similar relationship. Four series of reduction potentials are shown and these correspond to various $WOCl_{4-x}(OAr)_x$ (x = 1, 2, 3, or 4) complexes. Monophenoxide complexes ($WOCl_3(OAr)$) have the most positive reduction potentials and so are the easiest to reduce. The following seven phenol ligands were used to prepare the tungsten phenoxide complexes: 2,6-Me_2; 2,6-*i*-Pr_2; 2,6-$(OMe)_2$; 2,6-Ph_2; 2,4-Cl_2-6-Me; 2,6-Cl_2; and 2,6-Br_2. As anticipated, the compounds which are easiest to reduce are those with a phenoxide substituted with the best electron-withdrawing groups (lowest electronic charge), e.g., Br and Cl.

Polymerization Activity. A number of systems have been described in the literature in which the phenol substituents dictate the activity of the catalyst employed [*12-15*]. Dodd [*12*] showed that for *trans*-$WCl_2(OAr)_4$ procatalysts based on WCl_6, the activity of the catalysts increased with the electron-withdrawing

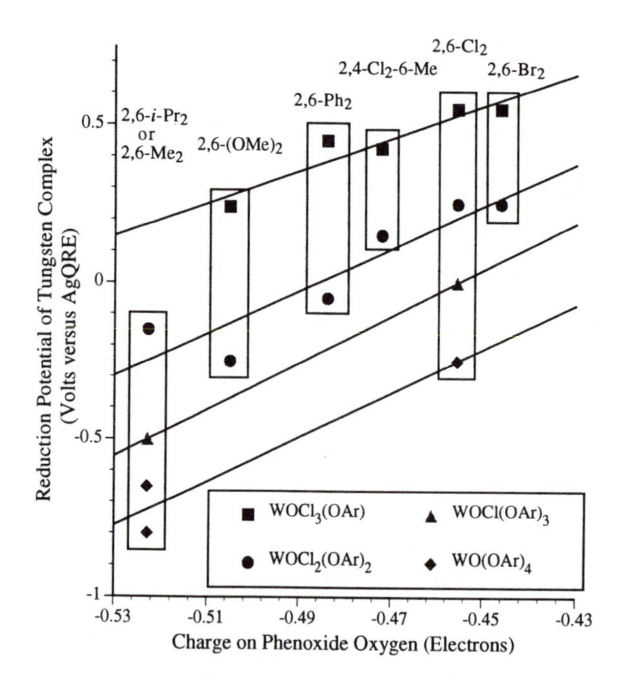

Figure 1. Correlation Between Reduction Potential of $WOCl_{4-x}(OAr)_x$ Complexes and Charge at the Phenoxide Oxygen

ability of the para substitutents. As expected, a linear Hammett relationship was observed.

In a similar fashion, Basset isolated some uni-component catalysts from the reaction of various $WCl_4(OAr)_2$ complexes with alkylating agents [*14, 15*]. By employing pent-2-ene as the olefinic reagent, Basset determined that catalyst activity increased with the electron-withdrawing ability of the phenol substitutents [*14(b), (c)*]. Selectivity was seen to be in reverse order.

Recent work completed using WCl_6-derived phenoxide complexes has shown that trialkyltin hydrides and triaryltin hydrides can be used to polymerize DCPD in bulk in high yield [*16,17*]. We are now in a position to discuss the polymerization chemistry based on ROMP catalysts which are formed by the activation of the prepared $WOCl_{4-x}(OAr)_x$ complexes using alkyltin hydride or aryltin hydride activators [*27*].

The polymerization ability of $WOCl_3(OAr)$, $WOCl_2(OAr)_2$, and $WOCl(OAr)_3$ in combination with trialkyltin hydrides or triaryltin hydrides has been assessed [*27*]. Both $WOCl_3(OAr)$ and $WOCl(OAr)_3$ complexes can be activated readily by trialkyltin hydride or triaryltin hydride to polymerize pure DCPD in bulk [*27*]. In discussing the intricacies of the polymerization reactions employing $WOCl_{4-x}(OAr)_x$ complexes, however, we will illustrate them by using the $WOCl_2(OAr)_2$ series. These complexes are quite easy to prepare, the full series of substituted phenols are available, the complexes are stable in DCPD, and a wide range of reduction potentials have been measured for them.

The degree of monomer conversion can be related directly to the reduction potential of the particular $WOCl_2(OAr)_2$ complex employed (i.e., electronic charge on the oxygen of the phenoxide) (see Figure 1). It is apparent that the activity of a particular catalyst can be empirically related to the electron-withdrawing characteristics of the phenol substituents, e.g., Hammett constants. Dodd [*12, 13*] and Basset [*14, 15*] report that metal electrophilicity (or olefin-philicity) is important in the olefin metathesis reaction. For the series of $WOCl_{4-x}(OAr)_x$ procatalysts prepared in this work, we are now able to gauge this electrophilicity by measuring the reduction potentials associated with each procatalyst. Electron-withdrawing phenoxides, such as 2,6-dichlorophenoxide and 2,6-dibromophenoxide, ensure that the DCPD polymerization reaction with the corresponding W-phenoxide complex occurs almost completely to give polymer possessing a residual DCPD level of 0.1 wt%. When $WOCl_2(OC_6H_3\text{-}2,6\text{-}i\text{-}Pr_2)_2$, prepared using the less-acidic 2,6-diisopropylphenol, is activated by a trialkyltin hydride a slightly less active catalyst is generated, since the residual monomer level in the polyDCPD is closer to 0.4 wt%.

In order to understand the range of reactivity which may arise under molding conditions the effect of varying reactant stiochiometries on residual monomer level were systematically studied. The effect of the following reaction variables were investigated: (i) **monomer** to **catalyst** ratio, (ii) **activator** to **catalyst** ratio, and (iii) **rate moderator** to **catalyst** ratio, and the outcomes outlined below.

Based on the electron-withdrawing abilities of the phenol substituents, the $WOCl_2(OAr)_2$ complexes ligated by 2,6-diisopropylphenoxide, and 2,6-dichlorophenoxide moieties were selected as the extremes of the series of phenols investigated. Using these procatalysts, the effect of **DCPD:W** reactant ratio on residual monomer was studied [*27(b)*]. It is quite apparent that two distinct reactivity profiles exist (Figure 2(a)). Although both procatalysts polymerize DCPD to very high conversion even at the highest monomer to tungsten ratio, the procatalyst based on 2,6-dichlorophenol afforded slightly lower residual monomer levels.

In some catalyst systems, over-reduction of the procatalyst metal center, e.g., $WCl_2(OAr)_4$, by an activator, e.g., Et_2AlCl, leads to loss in catalyst activity [*12*]. The examples shown here are based on the 2,6-diisopropylphenol and 2,6-dichlorophenol derived procatalysts in combination with trialkyltin hydride (Figure 2(b)). Overall, there is little affect of increasing the trialkyltin activator to tungsten ratio on residual monomer levels even at high **R_3SnH:W** reactant ratios [*27(b)*]. The slight difference in residual monomer levels is attributed to the substitution at the phenol.

We find that the onset of gelation and curing can be adjusted by altering the Lewis base to tungsten reactant ratio (**rate moderator:W**) [*27, 30, 31*]. Suitable rate moderators can be selected from the compounds typically used: ethers, phosphines, phosphites, phosphites, phosphonites, phosphinites, pyridines, pyrazines, and many other Lewis bases. Once again, we find that the polymerization reaction is essentially immune to reactant stoichiometry. As previously observed, the activated 2,6-dichlorophenol procatalyst is more effective at polymerizing DCPD than the corresponding 2,6-diisopropylphenol procatalyst.

Polymer Properties. A comparison of the properties found for two polyDCPD samples generated using different reaction conditions is shown in Table I. Under the molding conditions employed, tensile, flexural, and impact property values for METTON 1537 [*32*] and a sample generated by using $WOCl_2(OAr)_2$/*n*-Bu_3SnH combination are essentially equivalent. These polymer properties, again, reflect the excellent balance of impact, modulus, and toughness which are characteristic of the polyDCPD material.

Table I. Poly(dicyclopentadiene) Properties

PROPERTY	METTON 1537	POLYMER FROM $WOCl_2(OAr)_2$
Tensile		
Strength	6,200 psi	5,900 psi
Modulus	260,000 psi	280,000 psi
Flexural		
Strength	10,400 psi	9,800 psi
Modulus	284,000 psi	278,000 psi
Notched Izod (23 °C)	8.0 ftlb/in	9.1 ftlb/in
Heat Deflection Temperature (264 psi)	103 °C	120 °C
Glass Transition Temperature (T_g)	137 °C	158 °C

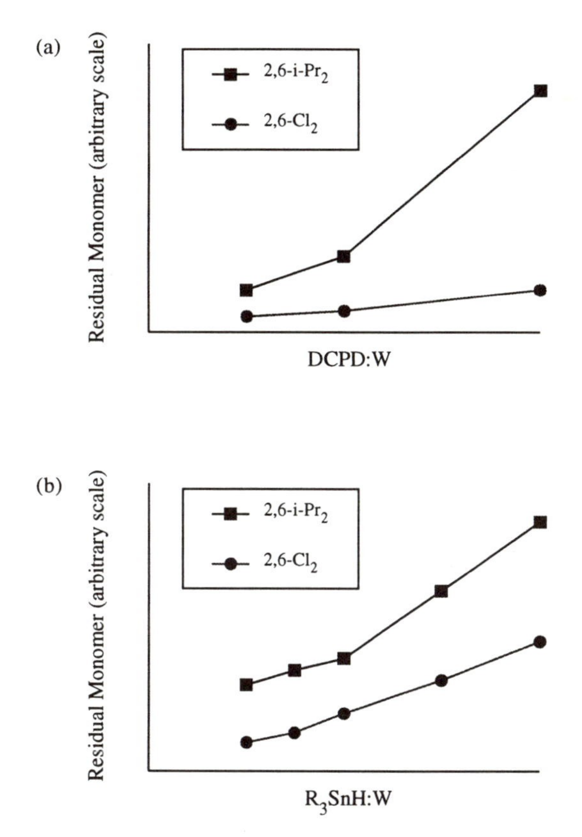

Figure 2. Effect of Reactant Stoichiometry on Residual Monomer for $WOCl_2(OAr)_2$ Complexes: (a) DCPD:W and (b) R_3SnH:W

Concluding Remarks

In summary, the stepwise synthesis of an entire family of pure tungsten oxytetrachloride-based phenol procatalysts, i.e., $WOCl_{4-x}(OAr)_x$, complexes, has been described. The cyclic voltammetric technique was demonstrated as being the analytical tool of choice for distinguishing between $WOCl_{4-x}(OAr)_x$ complexes and for discerning the electronic influences of a variety of 2,6-disubstituted or 2,4,6-trisubstituted phenol ligands. The polymerization ability of a particular $WOCl_{4-x}(OAr)_x$ procatalyst can be correlated with the charge on the oxygen of the 2,6-disubstituted phenol or 2,6-disubstituted phenoxide ion, and reduction potential.

Pure $WOCl_{4-x}(OAr)_x$ procatalysts, activated by trialkyltin or triaryltin hydrides, polymerize DCPD in bulk to very high polymer yields. Further, these new catalyst systems were found to be quite insensitive to reactant stoichiometry (monomer/procatalyst/activator/moderator).

Acknowledgements

Ms. Robyn J. Hanson prepared the procatalysts compounds described and performed the DCPD polymerization experiments. Atomic charge calculations were performed by Dr. Denise R. Fitzgibbon using the MOPAC semi-empirical molecular orbital package program.

Literature Cited

1. Reviews on Olefin Metathesis: (a) Dragutan, V.; Balaban, A. T.; Dimonie, M. *Olefin Metathesis and Ring-Opening Polymerization of Cycloolefins;* Wiley Interscience, Chichester, 1985. (b) Ivin, K. J. *Olefin Metathesis*, Academic Press, London, 1983. (c) Grubbs, R. H. In *Comprehensive Organometallic Chemistry*; Wilkinson, G. Ed.; Pergamon Press Ltd, Oxford, 1982; Vol. 8, pp 499-551. (d) Amass, A. J. In *Comprehensive Polymer Science*; Allen, G. Ed.; Pergamon Press Ltd, Oxford, 1989; Vol. 4, Ch. 6, pp 109-134. (e) Feast, W. J. In *Comprehensive Polymer Science*; Allen, G. Ed.; Pergamon Press Ltd, Oxford, 1989; Vol. 4, Ch. 7, pp 135-142.
2. Streck, R. *J. Mol. Cat.* **1988**, *4,* 305.
3. Dall'Asta, G.; Montrono, G.; Mannetti, Z.; Tosi, C. *Makromol. Chem.* **1969**, *130,* 153.
4. Devlin, P. A.; Lutz, E. F.; Pattan, R. J. U.S. Patent 3,627,739 **1971**.
5. Klosiewicz, D. W. U.S. Patent 4,400,340 **1983**.
6. Breslow, D. S. *CHEMTECH* **1990**, *20,* 540-544.
7. Macosko, C. W. *RIM - Fundamentals of Reaction Injection Molding*, Hanser Publishers, New York, 1989.
8. Natta, G.; Dall'asta, G.; Mazzanti, G. Ital. Pat. 733,857 **1964**.
9. Das, K. Neth. Appl. 7,014,132 **1972**.
10. Koche, H. Ger. Offen. 2,024,835 **1970**.
11. Höcker, H.; Musch, R.; Jones, F. R.; Luederwald, I. *Angew. Chem., Int. Ed. Engl.* **1973**, *12*, 430.
12. Dodd, H. T.; Rutt, K. J. *J. Mol. Catal.* **1982**, *15,* 103.
13. Dodd, H. T.; Rutt, K. J. *J. Mol. Catal.* **1985**, *28*, 33.
14. (a) Quignard, F.; Leconte, M.; Basset, J. -M.; Hsu, L. Y.; Alexander, J. J.; Shore, S. G. *Inorg. Chem.* **1987**, *26*, 4272. (b) Quignard, F.; Leconte, M.; Basset, J. -M. *J. Mol. Catal.* **1986**, *36*, 13. (c) Quignard, F.; Leconte, M.; Basset, J. -M. *J. Chem. Soc., Chem. Comm.* **1985**, 1816.

15. Basset, J. -M.; Leconte, M.; Ollivier, J.; Quignard, F. U.S. Patent. 4,550,216 **1985**.
16. (a) Sjardijn, W.; Kramer, A. H. U.S. Patent 4,729,976 **1986**. (b) Sjardijn, W.; Kramer, A. H. U.S. Patent 4,810,762 **1987**.
17. Bell, A. U.S. Patent 5,019,544 **1991**.
18. For a general review of techniques available for the synthesis of metal phenoxide complexes see: Chisholm, M. H.; Rothwell, I. P. In *Comprehensive Coordination Chemistry*, Wilkinson, G.; Gillard, R.; McCleverty, J. A. Eds.; Pergamon Press, Oxford, 1987. Chapter 15.3, Volume 2. (b) Bradley, D.C.; Malhotra, K. C.; Gaur, D. P. In *Metal Alkoxides*; Academic Press, New York, 1978. (c) Malhotra, K. C.; Martin, R. L. *J. Organometal. Chem.* **1982**, *239*, 159-187.
19. Funk, V. H.; Weiss, W.; Mohaupt, G. *Z. Anorg. Allg. Chem.* **1960**, *304,* 239-240.
20. Mortimer, P. I.; Strong, M. I. *Aust. J. Chem.* **1965**, *18*, 1579.
21. Kolodziej, R. M.; Schrock, R. R.; Dewan, J. C. *Inorg. Chem.* **1989**, *28*, 1243-1248.
22. (a) Kerschner, J. L.; Fanwick, P. E.; Rothwell, I. P.; Huffman, J. C. *Inorg. Chem.* **1989**, *29,* 780-786. (b) Kerschner, J. L.; Torres, E. M.; Fanwick, P. E.; Rothwell, I. P.; Huffman, J. C., *Organometallics* **1989**, *8*, 1424-1431.
23. Beshouri, S. M.; Rothwell, I. P. *Inorg. Chem.* **1986**, *25,* 1962-1964.
24. Funk, V. H.; Mohaupt, G. *Z. Anorg. Allg. Chem.* **1962**, *315*, 204-212.
25. Krishnaiah, K. S. R.; Prasad, S. *J. Indian Chem. Soc.* **1961**, *38*, 400-402.
26. Listemann, M. L.; Schrock, R. R.; Dewan, J. C.; Kolodziej, R. M. *Inorg. Chem.* **1988**, *27*, 264-271.
27. (a) Bell, A. *Polymer Preprints (PMSE Division)* **1991**, *64*, 102-103. (b) Bell, A. U.S. Patent 5,082,544 **1992**.
28. Available from the Quantum Chemistry Program Exchange (QCPE), Program No. 455 (MOPAC (Version 5.0): a Semiempirical Molecular Orbital Program), Creative Arts Building, Indiana University, Bloomington, Indiana 47405.
29. Bell, A.; DeWulf, D. W. "Cyclic Voltammetric Studies of Tungsten(VI) Phenoxide Complexes," presented at 199th American Chemical Society Meeting, Boston, MA, April 22-27, 1990.
30. Nelson, L. L. U.S. Patent 4,727,125 **1988**.
31. (a) Matlack, A. S. U.S. Patent 4,933,402 **1990**. (b) Matlack, A. S. U.S. Patent 4,883,849 **1989**.
32. (a) Geer, R. P. "Poly(dicyclopentadiene): A New RIM Thermoset," paper presented at Society of Plastic Engineers, NATEC, Detroit, Michigan, September 21, 1983. (b) See also METTON® Liquid Molding Resin Product Information, Hercules Incorporated, Product Information, P. O. Box 1071, Hercules Plaza, Wilmington, DE 19899-9973.

RECEIVED January 24, 1992

Ring-Opening Coordination Polymerization of Epoxides

HIGH-MOLECULAR WEIGHT POLYMERS were not available from a wide variety of epoxides until the late 1950s. At that time, a number of investigators developed new coordination catalysts derived from the organometallic compounds of Al, Zn, and Mg. One of the most versatile types was that discovered by Vandenberg and was based on alkylaluminoxanes alone and after modification with a chelating agent as acetylacetone. The alkylaluminoxanes alone behave either as coordination catalysts or as superior cationic catalysts, depending on the monomer. These catalysts led to the most important commercialization in the area (i.e., to the epichlorohydrin and propylene oxide elastomers). A number of other unique catalysts have been developed for epoxides. Two are included in this book. The chapter by Teyssié et al. in this section covers the mechanism aspects of the soluble polynuclear μ-oxometal alkoxide aggregates, usually based on Al and Zn, discovered by Osgan and Teyssié, which do not require residual organometal bonds as do many other catalysts. The chapter by Inoue in the Anionic Polymerization section reviews the very important metalloporphyrin catalysts based on Al and Zn. The exact mechanism of this catalyst is unknown. Although we have assigned it to the Anionic Polymerization area, it is still possible that coordination polymerization is involved. Three of the chapters in this section (Wicks and Tirrell, Cheng, and Parris and Marchessault) cover the reactions and characterization of high-molecular-weight polyepoxides prepared with Vandenberg-type catalysts. The Lau chapter presents an interesting cure study on more conventional amine–epoxy resins.

Chapter 10

Structural Effects and Reactivity of Polymer-Bound Functional Groups

Kinetics of Nucleophilic Displacement Reactions on Chlorinated Polyethers

Douglas A. Wicks[1] and David A. Tirrell

Department of Polymer Science and Engineering, University of Massachusetts, Amherst, MA 01003

The reactions of chlorinated polyethers with tetrabutylammonium benzoate in N,N-dimethylacetamide have been examined. In all cases, the reactions are described by simple second-order kinetics at low conversion but exhibit retardation at extents of reaction greater than 60%. The magnitude of the retardation is a function of the length of the side chain that connects the polymer backbone and the reactive site, and the effect nearly vanishes for the C_4 side chain. Suppression of the reaction rate is proposed to arise from steric inhibition of attack by the relatively bulky benzoate ion.

We have reported a series of studies of the synthesis and reactions of halogenated polyethers *(1-6)*. The objectives of this work have been: i). preparation of polyether elastomers of enhanced reactivity toward nucleophiles, and ii). elucidation of the determinants of reactivity in such systems. The conclusions drawn from these studies have been the following. First, extension of the chloroalkyl side chain from C_1 to C_4 in polymers of structure **1** enhances about sixfold the rate of chloride displacement by benzoate anion *(4)*. Second, use of bromide – rather than

$$\left[CH_2 - \underset{\underset{(CH_2)_n - Cl}{|}}{CH} - O \right] \qquad \mathbf{1}$$

a: n = 1 c: n = 3
b: n = 2 d: n = 4

chloride – as leaving group leads to rate increases of a factor of about sixty *(5)*. And third, removal of the ether oxygen β to the leaving group in polyepichlorohydrin (by "insertion" of a backbone methylene group as in **2**) results in a doubling of the reaction rate *(6)*.

[1]Current address: Polymer Research, Mobay Corporation, Pittsburgh, PA 15205

0097–6156/92/0496–0136$06.00/0

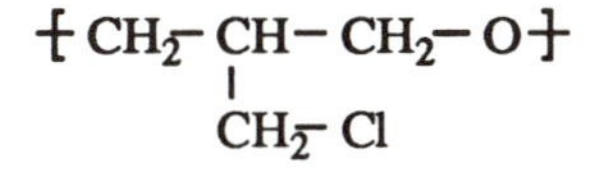

$$\left[CH_2-\underset{\displaystyle CH_2-Cl}{CH}-CH_2-O \right] \qquad \mathbf{2}$$

In all of this work, we have used the benzoate anion, in the form of its tetrabutylammonium salt, as the attacking nucleophile (Equation 1).

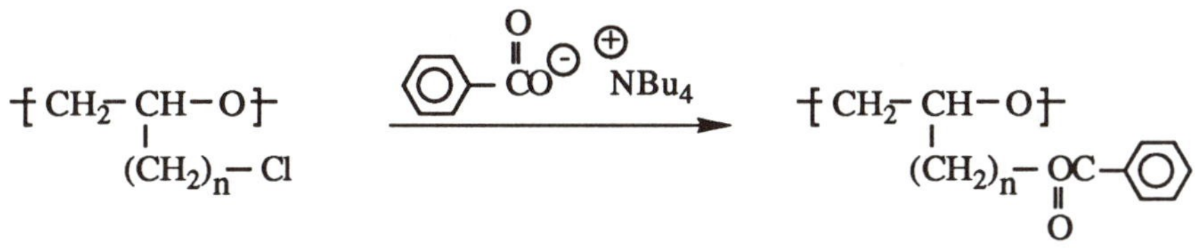

$$\left[CH_2-\underset{\displaystyle (CH_2)_n-Cl}{CH}-O \right] \xrightarrow{C_6H_5COO^{\ominus}\ \overset{\oplus}{N}Bu_4} \left[CH_2-\underset{\displaystyle (CH_2)_n-OC(=O)-C_6H_5}{CH}-O \right]$$

Equation 1

Without exception, these reactions are well described by second-order kinetics at low conversion, but show negative deviation from second-order rate laws as the reaction proceeds *(5)*. In the present paper, we examine this displacement reaction more thoroughly, with particular emphasis on the magnitude and origins of the rate suppression observed at high conversion.

Experimental

Materials. Poly(epichlorohydrin) (PECH, **1a**) was purchased from Aldrich Chemical Co. Oxirane monomers were prepared as described previously *(1)*. [CAUTION: The preparations of (3-chloropropyl)oxirane and (4-chlorobutyl)oxirane described in ref. 1 involve sodium reductions of tetrahydrofurfuryl chloride and (2-chloromethyl)tetrahydropyran, respectively. These reactions can become very vigorous. We experienced one serious explosion and fire when the reaction mixture overflowed the vessel and splashed into the cooling bath. The ice-water bath recommended by Brooks and Snyder *(7)* should be avoided.] Triethylaluminum (25% in toluene), tetrabutylammonium hydroxide (40% in water) and N,N-dimethylacetamide (DMAc) were used as received from Aldrich. 1-Chloropentane (CP, fractionally distilled, b.p. 108°C) and 2,4-pentanedione (AcAc, distilled from CaO, b.p. 125-36°C) were also obtained from Aldrich. Poly(3-chlorooxetane) (P3CO) *(3)*, poly[2-(chloromethyl)oxetane] (P2CMO) *(6)* and poly[3-(chloromethyl)oxetane] (P3CMO) *(6)* were prepared as described previously.

Preparations.

Initiator *(8,9)*. To a dry N_2-purged Schlenk tube equipped with a magnetic stir bar was added 10 mL (19 mmol) of 25% triethylaluminum in toluene. The tube was cooled in an ice bath and 1.96 mL (19 mmol) of AcAc was added slowly with vigorous stirring. Distilled water (0.17 mL, 9.5 mmol) was added in similar fashion. The initiator solution was aged for at least 10 hr at room temperature under a positive N_2 atmosphere, protected from light, and used with 48 hr of preparation. All initiator solutions prepared in this way appeared to be homogeneous.

PCEO. (2-Chloroethyl)oxirane (17.0 g, 160 mmol) was purged with N_2 in a septum capped polymerization ampule and cooled to 0°C. To this was added 5.0 mL (8.0 mmol Al) of initiator solution. The ampule was heated in an oil bath at 80°C for 24 hr. Two precipitations from CH_2Cl_2 into methanol yielded 6.3 g (59 mmol equivalents, 37%) of PCEO. η_{inh} (30°C, 0.5 g/dL $CHCl_3$): 2.3 dL/g. GPC ($CHCl_3$): M_n-237,000; M_w-688,000. ^1H NMR (200 MHz, $CDCl_3$) δ: 3.5-3.8 (5H, broad multiplet); 2.0 (2H, broad multiplet). ^{13}C NMR (75 MHz, CD_2Cl_2) δ: 76.8, 75.6, 42.1, 35.7.

PCPO. (3-Chloropropyl)oxirane (8.0 g, 66 mmol) was purged with N_2 in a septum capped polymerization ampule and cooled to 0°C. To this was added 2.0 mL (3.2 mmol Al) of initiator solution. The ampule was heated in an oil bath at 80°C for 24 hr. Two precipitations from CH_2Cl_2 into methanol yielded 6.0 g (50 mmol equivalents, 75%) of PCPO. η_{inh} (30°C, 0.5 g/dL $CHCl_3$): 1.9 dL/g. GPC ($CHCl_3$): M_n-198,000; M_w-1,250,000. ^{13}C NMR (75 MHz, C_6D_6) δ: 79.2, 74.1, 45.3, 29.6, 29.2.

PCBO. (4-Chlorobutyl)oxirane (20.0 g, 150 mmol) was purged with N_2 in a septum capped polymerization ampule and cooled to 0°C. To this was added 4.5 mL (7.2 mmol Al) of initiator solution. The ampule was heated in an oil bath at 80°C for 24 hr. Two precipitations from CH_2Cl_2 into methanol yielded 11.6 g (87 mmol equivalents, 58%) of PCPO. η_{inh} (30°C, 0.5 g/dL $CHCl_3$): 4.6 dL/g. GPC ($CHCl_3$): M_n-260,000; M_w-1,110,000. ^1H NMR (200 MHz, $CDCl_3$) δ: 3.2-3.8 (5H, broad multiplet); 2.9 (2H, broad); 2.5 (4H, broad). ^{13}C NMR (75 MHz, C_6D_6) δ: 79.4, 73.1, 45.0, 33.4, 31.6, 23.1.

Tetrabutylammonium benzoate (TBAB). Benzoic acid (245 g, 2 mol) was added to 1300 g of aqueous tetra-n-butylammonium hydroxide (2 mol Bu_4NOH) in an open flask at room temperature. After the disappearance of the solid benzoic acid the solution was extracted twice with 1L $CHCl_3$. After drying over $MgSO_4$ removal of the solvent left an off-white solid which was triturated with 1 L of anhydrous ether and then dissolved in 500 mL anhydrous ethanol. The ethanol solution was heated to reflux under N_2 at which time an equal volume of hexanes was slowly added. Cooling of the solution to -10°C resulted in crystallization. After filtration and vacuum drying at room temperature over P_2O_5, 385 g (10.5 mol, 51 %) of TBAB was isolated. ^1H NMR (200 MHz, D_2O) δ: 8.09 (2H, multiplet); 7.28 (3H, quartet); 3.31 (8H, multiplet); 1.61 (8H, multiplet); 1.43 (8H, multiplet); 0.980 (12H, triplet). ^{13}C NMR (75 MHz, $CDCl_3$) δ: 170.6, 140.0, 129.3, 128.6, 126.9, 58.2, 23.7, 19.4, 13.4.

Measurements.

Routine measurements. ^1H NMR spectra were obtained using Varian XL-200 and XL-300 spectrometers with observe frequencies of 200 and 300 MHz, respectively. ^{13}C NMR were obtained on the same spectrometers operating at 50 and 75 MHz, respectively. Chemical shifts (δ) are reported in parts per million (ppm) downfield from trimethylsilane (TMS). For samples that contained no TMS, shifts are referenced to the published shifts of resonances from the lock solvent. Variable temperature measurements were made utilizing the standard Varian XL temperature control system, which was calibrated to ±0.1°C using the known change in relative peak frequencies with temperature of ethylene glycol standards *(10)*. ^{13}C spin-lattice relaxation time (T_1) measurements were performed on the Varian XL-300 (75 MHz) using the standard inversion-recovery pulse sequence of 180° pulse-delay time array (τ)-90° pulse-observe. Choice of delay times and actual T_1 calculations were performed using standard commands (DOT1 and T1) *(11)* contained within the Varian VXR version 4.1 software. Error limits given for T_1 values are those determined by the T_1 program and are based on the goodness of fit of the calculated values of peak height at each τ value and those observed. Nuclear Overhauser enhancements (NOE) were determined by comparison of the peak heights from ^{13}C NMR spectra obtained using continuous broadband ^1H decoupling with those obtained using a gated decoupling pulse sequence in which broadband ^1H decoupling was only used during the acquisition time. For all spectra used in NOE determinations, the delay time between the end of the data acquisition time and the next pulse was a minimum of 5 times T_1 for the peaks of interest.

Infrared spectra were obtained on either a Perkin Elmer 283 or a Perkin Elmer 1320 infrared spectrometer. All spectra were referenced to the 1601 cm^{-1} band of a thin polystyrene film. Purity of all monomers was checked to be >99% before attempting polymerization using a Varian Aerograph Model 920 gas chromatograph equipped with a 12 ft aluminum (1/4" ID) SE-30 (15% on Chromosorb) column and thermal conductivity detector.

Gel permeation chromatographic measurements were made using Waters instruments, with chloroform as solvent at a flow rate of 1.0 mL/min with two linear ultra-styragel columns. Number average molecular weights (M_n) and weight average molecular weights (M_w) were calculated using a Waters 160 Data Module based on calibration curves derived from narrow-dispersity polystyrene standards obtained from Polysciences. Inherent viscosities (η_{inh}) for polymer solutions were measured using a Cannon-Fenske #35 viscometer, with a solvent elution time of 128 sec for $CHCl_3$ at 30°C.

Kinetics of substitution by benzoate ion. In a typical run, PECH (137.9 mg, 1.5 mmol repeating units) was dissolved in 3.47 g N_2-purged DMAc over a 36 hr period with continuous stirring at 50°C. A solution of 545 mg (1.5 mmol) TBAB in 1.37 g DMAc was added via syringe and the reaction mixture was then transferred to a 10 mm NMR tube. The tube was placed in the probe of the Varian XL-300 spectrometer, which was pre-equilibrated to 50°C. ^{13}C NMR spectra were recorded periodically by using a 50° pulse width, a pulse delay of 0.5-2.0 sec, and an acquisition time of 0.9-1.0 sec. Conversions were determined from spectra averaged over 200-400 transients (P2CMO required 1500 transients). All reactant solutions and NMR tubes were pre-equilibrated at the reaction temperature for a minimum of 30 min. Reactant concentrations and reaction temperatures are listed in detail in Table I.

Results and Discussion

Kinetic Method. The kinetics of substitution of chloride by benzoate ion were readily determined by integration of ^{13}C NMR spectra recorded under quantitative conditions. In particular, determination of the relative intensities of signals arising from chlorinated and benzoyloxylated carbon sites gave estimates of reaction conversion that were within 3% of conversions measured from ^1H NMR spectra on isolated samples of partially reacted chains. Care was taken to ensure full relaxation of the nuclei of interest (pulse delay >2.5 T_1) and signal intensities were corrected by measured nuclear Overhauser enhancement factors.

Kinetic Results. Figure 1 shows plots of conversion vs. time for a series of reactions of PCEO with TBAB at temperatures ranging from 5°C to 50°C. These results are typical of those obtained on all of the chlorinated polyethers examined in this work, in that the conversion data can be fit by a simple second-order rate law (solid lines in Figure 1) only up to about 60% conversion. At higher extents of reaction, negative deviation from second-order kinetics is observed, as reported previously *(5)*.

Table I. Substitution Reactions in DMAc

Reaction	*Substrate*	*T* (°C)	$[CH_2Cl]_0$ (mol/L)	$[TBAB]_0$ (mol/L)	*k* ($M^{-1}s^{-1}$)
1	PECH	30	0.252	0.252	$(4.9\pm0.5)\times10^{-5}$
2	PECH	40	0.250	0.250	$(1.4\pm0.1)\times10^{-4}$
3	PECH	50	0.253	0.253	$(3.6\pm0.4)\times10^{-4}$
4	PECH	60	0.257	0.255	$(6.1\pm0.1)\times10^{-4}$
5	PECH	70	0.250	0.250	$(1.2\pm0.2)\times10^{-3}$
6	PCEO	5	0.220	0.220	$(6.1\pm0.8)\times10^{-3}$
7	PCEO	25	0.220	0.220	$(3.5\pm0.3)\times10^{-4}$
8	PCEO	40	0.219	0.219	$(1.3\pm0.1)\times10^{-3}$
9	PCEO	50	0.220	0.220	$(2.5\pm0.2)\times10^{-3}$
10	PCEO	40	0.220	0.275	$(1.4\pm0.1)\times10^{-3}$
11	PCEO	40	0.219	0.109	$(1.3\pm0.2)\times10^{-3}$
12	PCEO	40	0.146	0.220	$(1.4\pm0.2)\times10^{-3}$
13	PCEO	40	0.175	0.222	$(1.4\pm0.2)\times10^{-3}$
14[a]	PCEO	40	0.176	0.089	$(1.3\pm0.1)\times10^{-3}$
15	PCPO	30	0.195	0.200	$(8.6\pm0.8)\times10^{-4}$
16	PCPO	40	0.194	0.198	$(1.9\pm0.1)\times10^{-3}$
17	PCPO	50	0.193	0.197	$(3.5\pm0.6)\times10^{-3}$
18	PCBO	40	0.279	0.300	$(1.8\pm0.1)\times10^{-3}$
19	PCBO	40	0.174	0.174	$(1.7\pm0.2)\times10^{-3}$
20	P3CMO	50	0.246	0.246	$(8.2\pm0.8)\times10^{-4}$
21	P2CMO	50	0.159	0.159	$(4.0\pm0.3)\times10^{-4}$
22	P3CO	90	0.290	0.290	9.4×10^{-6}
23	CP	50	0.164	0.170	$(8.1\pm0.2)\times10^{-3}$
24	CP	40	0.163	0.163	$(4.3\pm0.2)\times10^{-3}$
25	CP	35	0.164	0.154	$(2.7\pm0.1)\times10^{-3}$
26	CP	30	0.219	0.228	$(1.9\pm0.2)\times10^{-3}$

[a] Included 0.137 g (0.46 mol) tetrabutylammonium chloride (TBACl) $[TBACl]_0 = 0.089$ mol/L.

Figure 2 shows second-order kinetic plots of the low conversion data for PCEO. Rate constants determined from the slopes of these plots (cf. Table I) were used to estimate Arrhenius activation parameters for the substitution reaction. Both the activation energy (E_a = 14.9 kcal/mol) and the pre-exponential factor (log A = 7.4 sec^{-1}) were found to be similar to those determined for the model compound 1-chloropentane (E_a = 14.3 kcal/mol, log A = 7.6 sec^{-1}). The insensitivity of the measured rate constant to variation in the initial concentrations of substrate and TBAB established that the reaction is first-order in each of the reactants.

Table I summarizes the initial rate constants for substitution on each of the halide substrates examined, and Figure 3 illustrates the quality

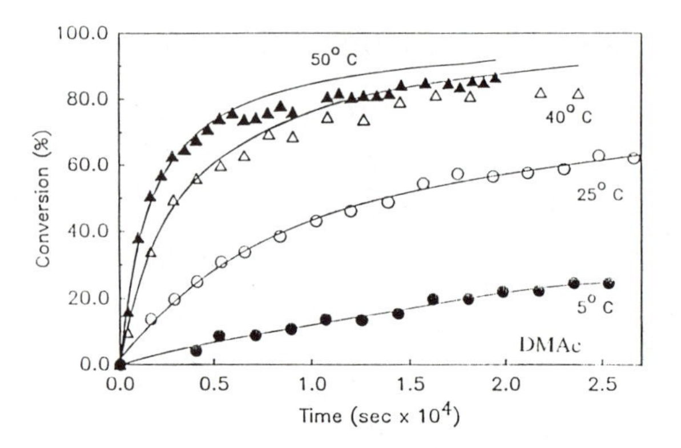

Figure 1. Kinetics of benzoate ion substitution on PCEO. Points are measured conversions; solid lines are conversions calculated by using a single second-order rate constant determined at low conversion.

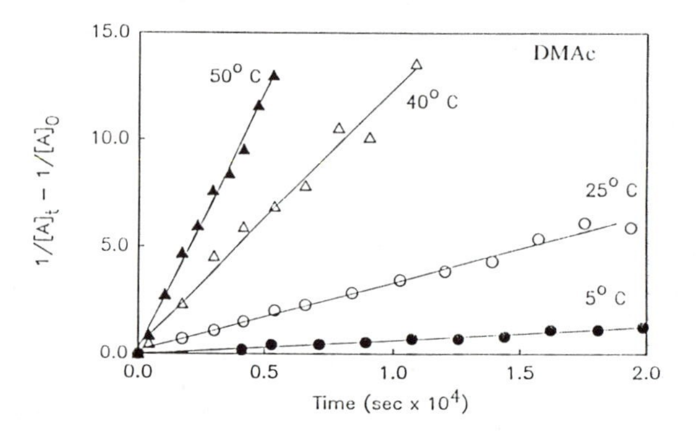

Figure 2. Kinetics of substitution on PCEO; second-order plots of low conversion data.

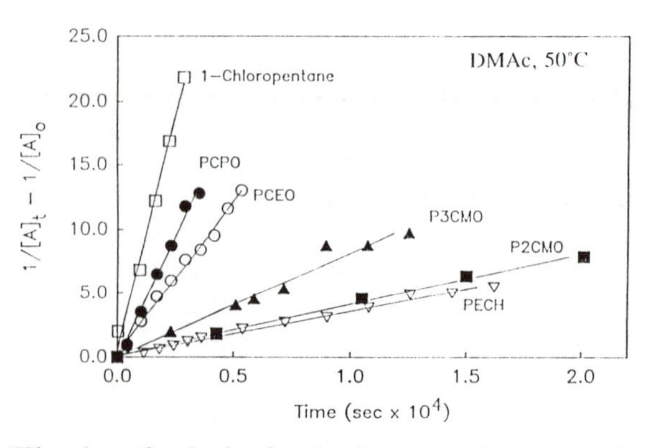

Figure 3. Kinetics of substitution by benzoate ion; second-order plots of low conversion data.

of the second-order fits to the low-conversion data. These results confirm the conclusions of our earlier work: extension of the side chain from C_1 to C_4 enhances the substitution rate approximately tenfold, and insertion of the backbone methylene group in P3CMO doubles the reaction rate as compared to PECH. The latter result appears to arise from screening of the inhibitory electronic effect of the β-ether oxygen atom in PECH, since the isomeric poly(oxetane) P2CMO (which retains the β-chloroether structure) is virtually identical to PECH in reactivity.

Two additional results deserve comment. First, the model compound 1-chloropentane reacts about threefold faster than PCPO or PCBO at 40°C. This is a bit surprising; the fact that these two polymers are equally reactive suggests that the chloromethyl groups have been removed from any local effects of the chain backbone. We propose instead that the reduced reactivity of the polymers results from a "solvent effect," i.e., from a change in local solvent composition near the chain. Displacement of DMAc by polyether chain units should affect the solvation, and thus the concentration and reactivity, of the attacking nucleophile in the vicinity of the polymer-bound chloride. No such effect would be anticipated in the reaction of 1-chloropentane with TBAB *(12)*.

The second point concerns the reactivity of P3CO (3), an isomer of PECH bearing a secondary chloride. P3CO is for practical purposes inert

$$\left[CH_2-\underset{\displaystyle Cl}{\underset{|}{CH}}-CH_2-O \right] \qquad 3$$

toward substitution by TBAB at 50°C; reliable rate data could not be obtained in a reasonable time interval at any temperature below 90°C. At 90°C a rate constant of $9.4 \times 10^{-6}\ M^{-1}s^{-1}$ was determined. Extrapolation of PECH reactivities measured at five temperatures between 30°C and 70°C leads to an estimate of $k = 4 \times 10^{-3}\ M^{-1}s^{-1}$ at 90°C. The reactivity of P3CO is thus ca. 400-fold lower than that of PECH, as a result of the combined effects of the secondary chloride *(13)* and the flanking β-ether oxygens of P3CO.

Deviation from Second-Order Kinetics. As discussed above, each of the reactions examined in this work shows negative deviation from second-order kinetics at conversions above 60%. In order to discuss this effect in a quantitative fashion, we have adopted the treatment of Boucher *(14)*, in which one allows the rate constant for reaction at a polymer-bound functional group to vary with the state of reaction of neighboring sites. As shown in Equation 2, the reaction is then described in terms of three rate constants, k_0, k_1 and k_2, for displacement at sites flanked by 0,

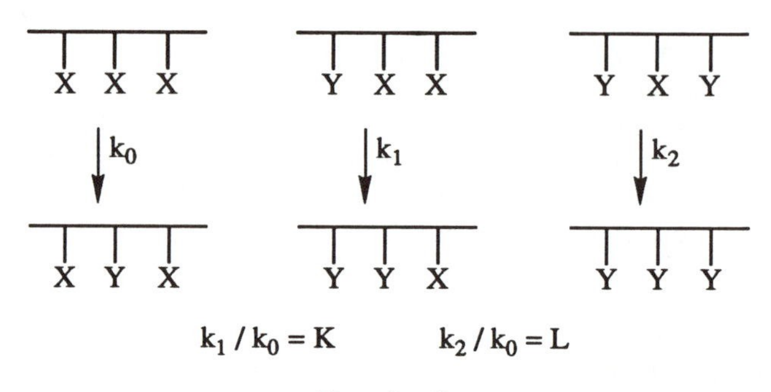

Equation 2

1 or 2 reacted neighbors, respectively. The initial rate constant k_0 is determined directly from low conversion data, as described above, and the overall conversion is then calculated in terms of two ratios of rate constants: $K = k_1 / k_0$ and $L = k_2 / k_0$. This treatment does not take into account the directionality of the polyether backbone or the consequent inequivalence of the relations between a functional group and each of its neighbors, but it seems unlikely that much will be gained by introducing this additional complication.

Figure 4 shows conversion vs. time data for the reaction of PCEO with TBAB in DMAc at 40°C. The upper curve represents the extent of reaction calculated assuming simple second-order kinetics with a single rate constant determined at low conversion. The high conversion data are not fit well. The lower curve in Figure 4 was calculated by using the same value of k_0, but with $k_1 = 1.02\ k_0$ and $k_2 = 0.47\ k_0$. The fit is clearly improved.

The goodness-of-fit was measured in terms of the sum (e) of the squares of the differences between the calculated and observed conversions over the full course of the reaction. Figure 5 shows a typical contour plot of e as a function of the rate constant ratios K and L. In all such plots, there is a well-defined minimum in e, but also a clear correlation between the consequences of changing each of the rate constant ratios; i.e., an increase in K can be offset in some measure by a reduction in L, and vice versa. Nevertheless, the data can be fit in reasonable fashion only within a restricted region of K, L space.

We were also concerned about the sensitivity of K and L to the value assumed for k_0. Table II lists the results of fitting trials with k_0 varying by plus-or-minus one standard deviation from the value calculated from the low conversion data. Variation in k_0 has the expected consequences for K and L, but the sensitivity is not large; indeed, L is nearly unaffected by modest changes in k_0. The goodness-of-fit is in all cases compromised when k_0 deviates from its measured value. The rate

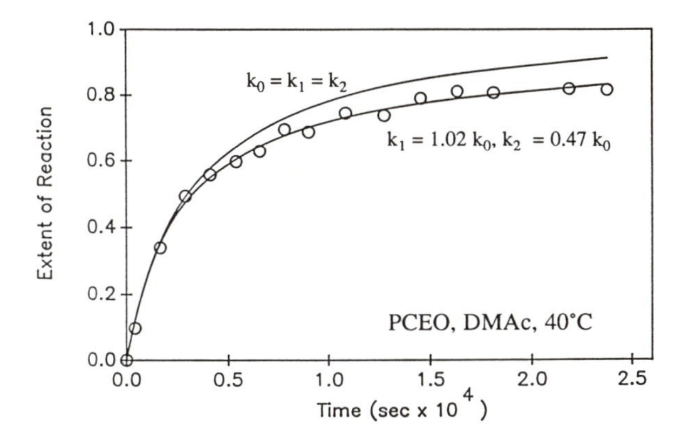

Figure 4. Kinetic curves for TBAB substitution on PCEO at 40°C; relative rate constants as shown.

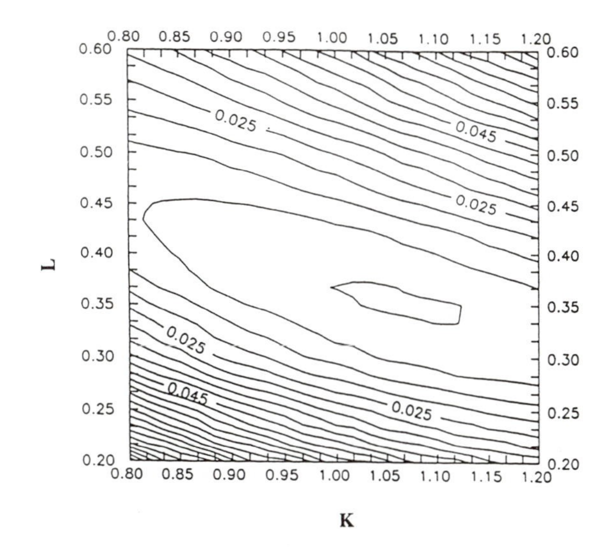

Figure 5. Contour plot of the error of calculated conversion vs. actual conversion as a function of K and L for the reaction of PCEO with TBAB in DMAc at 50°C.

constant ratios K and L thus appear to be useful as descriptors of the extent of deviation from simple second-order kinetics.

Table III lists the values of K and L determined for the reactions of the poly[ω-chloroalkyl)oxirane]s **1a-d** with TBAB in DMAc. Although we do not attach strict quantitative significance to these parameters, it seems clear that K is relatively constant for this series of polymers while L increases systematically with increasing side chain length. For the C_4 side chain (i.e., for PCBO), K and L are nearly unity, suggesting that neighboring group effects are minimal. For PECH, on the other hand, significant retardation is observed at high conversion, consistent with the notion that reaction at sites directly linked to the chain backbone should be sensitive to the state of reaction of neighboring functional groups.

Table II. Calculated K and L Values for the Reaction of TBAB with PCPO in DMAc at 40°C for Different Estimates of Initial Rate Constant

Estimate of Initial k[a]	*K*	*L*	*e*[b]
k_0	1.02	0.72	0.00482
$k_0 + 0.5\,\sigma$	0.92	0.72	0.00630
$k_0 + \sigma$	0.85	0.72	0.00667
$k_0 - 0.5\,\sigma$	1.07	0.73	0.00573
$k_0 - \sigma$	1.20	0.72	0.00557

[a] k_0 ($1.9 \times 10^{-3}\ M^{-1}s^{-1}$) is the average value derived from the slopes of origin-to-point lines for measurements at low conversion. The value σ is one standard deviation in the slope ($8.6 \times 10^{-5}\ M^{-1}s^{-1}$).
[b] e is the sum of squares of the differences between calculated and observed conversions.

Table III. Calculated K and L Values for Some Reactions of TBAB with Poly[(ω-chloroalkyl)oxirane]s in DMAc

Polymer	*Temperature (°C)*	*K*	*L*
PECH	70	1.20	0.33
PCEO	50	1.05	0.37
PCEO	40	1.02	0.47
PCPO	40	1.02	0.72
PCBO	40	0.99	0.93

In our earlier work *(5)* on the bromide analogues of polymers **1a-d**, we proposed that retardation of these reactions arises from local steric effects that suppress the rate of attack by the relatively bulky benzoate ion. In support of this proposal, we noted a systematic decrease in the ^{13}C T_1 measured over the course of the reaction. Global conformational effects were assigned a lesser role, since the inherent viscosities of the reactant and product forms of these polymers were nearly identical.

We now find that polymers **1a-d** behave in similar fashion. The inherent viscosity of each polymer increases slightly over the course of the reaction, while each measured T_1 decreases. Table IV summarizes the T_1 data for PCEO in DMAc. T_1 for the backbone methine carbon is halved over the course of the reaction, while the side-chain termini (of either reacted or unreacted chain units) suffer reductions of ca. 30% in T_1. As before, we propose that the substitution reaction should be insensitive to the rate of segmental diffusion, owing to the relatively high activation energy (ca. 15 kcal/mole) of the S_N2 process. On the other hand, suppression of T_1 may signal an increase in steric crowding at the side chain terminus, which may pose a barrier to nucleophilic attack.

Table IV. Spin-Lattice Relaxation Times[a] of Partially Benzoyloxylated PCEO

Conversion %	*T_1(CH)[b]* *sec*	*T_1(CH_2Cl)[c]* *sec*	*T_1(CH_2Benz)[d]* *sec*
0	0.523	0.618	–
11	0.453	0.564	–
19	0.452	0.572	0.320
28	0.396	0.542	0.293
37	0.369	0.519	0.272
41	0.377	0.516	0.260
45	0.335	0.497	0.247
53	0.337	0.457	0.254
59	0.333	0.469	0.253
83	0.293	0.421	0.239
100	0.256	–	0.220

[a]Error limits in T_1 measurements less than ±10%.

[b]T_1 of backbone methine signal; represents an average of reacted and unreacted repeating units.

[c]T_1 of chloromethyl carbon.

[d]T_1 of benzoyloxymethyl carbon.

Acknowledgments

This work was supported by a Presidential Young Investigator Award of the National Science Foundation. NMR spectra were recorded in the University of Massachusetts NMR Laboratory, which is supported in part by the NSF Materials Research Laboratory at the University. We thank Dr. Brian P. Devlin for the computer programs used to evaluate the kinetic parameters K, L and e.

Literature Cited

1. Shih, J.S.; Brandt, J.F.; Zussman, M.P.; Tirrell, D.A. *J. Polym. Sci. Polym. Chem. Ed.* **1982**, 20, 2839.
2. Shih, J.S.; Tirrell, D.A. *J. Polym. Sci. Polym. Chem. Ed.* **1984**, 22, 781.
3. Wicks, D.A.; Tirrell, D.A. *New Polymeric Mat.* **1987**, 1, 13.
4. Zussman, M.P.; Shih, J.S.; Wicks, D.A.; Tirrell, D.A. *Am. Chem. Soc. Symp. Ser.* **1988**, 364, 60.
5. Wicks, D.A.; Tirrell, D.A. *Makromol. Chem.* **1989**, 190, 2019.
6. Wicks, D.A.; Tirrell, D.A. *J. Polym. Sci. Pt. A: Polym. Chem.* **1990**, 28, 573.
7. Brooks, L.A.; Snyder, H.R. *Organic Syntheses, Coll. Vol. 3*, Wiley, New York, 1955, p. 698.
8. Vandenberg, E.J. *J. Polym. Sci.* **1960**, 47, 486.
9. Vandenberg, E.J. *J. Polym. Sci. A-1* **1969**, 7, 525.
10. XL-300 Systems Operator Manual, Varian Instruments.
11. VXR Software Operators Manual, Varian Instruments.
12. We thank one of the referees for useful comments on this point.
13. Streitwieser, A. *Solvolytic Displacement Reactions*, McGraw-Hill, New York, 1962.
14. Boucher, E.A. *Prog. Polym. Sci.* **1978**, 6, 63.

RECEIVED January 8, 1992

Chapter 11

Mechanism of Oxirane Coordination Polymerization of Soluble Polynuclear μ-Oxometal Alkoxide Aggregates

Ph. Condé, L. Hocks, Ph. Teyssié[1], and R. Warin

Laboratory of Macromolecular Chemistry and Organic Catalysis, University of Liège, B6, Sart–Tilman, 4000 Liège, Belgium

A detailed NMR study of the behaviour of soluble polynuclear μ-oxometalalkoxides has been performed in view of their high activity and stereoselectivity in oxirane polymerization. The results indicate a rather rigid, globular coordination aggregate structure ("tecto-alkoxides"), undergoing very slow exchanges. When correlated with other kinetic data, they support a general insertion-coordination mechanism, producing lower molecular weight atactic chains on the outermost "freer" Al atoms, and a stereoregular high polymer at higher rate on a few hindered sites located deeper in the aggregate.

The design of catalysts displaying a very high activity for the stereoselective polymerization of oxiranes to high molecular weight polyethers often relies upon activation of otherwise inactive metal derivatives, by reaction with compounds able to promote the formation of μ-bridged complexes *(1-3)* and in particular μ-oxo ones obtained by controlled hydrolysis or alcoholysis of alkyl-zinc or aluminum derivatives. That very fact, together with different kinetic and structural data, prompted E. Vandenberg to express the hypothesis that such a propagation should take place by a coordinative flip-flop mechanism (see Scheme 1) on specific sites of a polynuclear aggregate *(4)*.
Along the same lines of thought, well-defined initiators were synthesized, i.e. polynuclear bimetallic μ-oxoalkoxides $[(RO)_2Al\text{-}O\text{-}Zn\text{-}O\text{-}Al(OR)_2]_n$ with a mean degree of coordinative association n varying from 2 to 8 *(5-7)* as usual in most alkoxides. They proved to rank among the most active catalysts (typical half-polymerization time

[1]Corresponding author

0097–6156/92/0496–0149$06.00/0

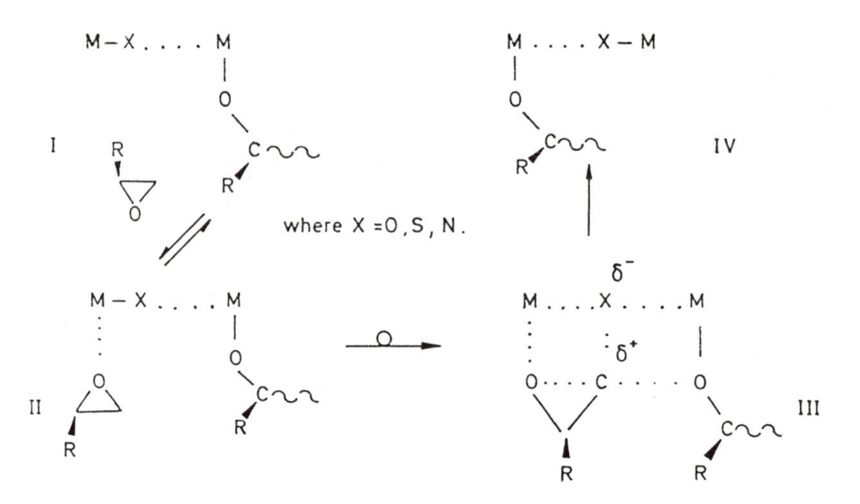

Scheme 1. Coordinative Insertion Mechanism for Oxiranes Polymerization

of 5 minutes at 50°C in 1 M solutions for R = n-butyl) for the polymerization of methyloxirane (MO) into a high M.W. (up to 10^6), partially isotactic (from 4 to 80 %) polyether *(8)* (PMO). Considering their well-controlled structure and their perfect solubility in several types of organic media, it appeared worthwhile to perform a detailed structural study of that family of compounds *(9)*. This paper will summarize the data obtained during a current NMR study, and will concentrate on their relation with other experimental data as well as their relevance to the aggregate coordination mechanism outlined above.

Experimental

The complex alkoxides were synthesized *(5,8,10)* in decalin under nitrogen atmosphere by a direct condensation between zinc acetate and aluminum isopropoxide in a 1 to 2 molar ratio, all reagents being carefully dried before use. Temperature was progressively raised up to boiling point and the evolved isopropylacetate continuously distilled off in order to completely displace the equilibrium. The end of the reaction was monitored by g.l.c. (disappearance of isopropylacetate). Decalin was then replaced by heptane or aromatics through distillation-redissolution cycles. The glassy dry residue was totally soluble in all of these media, yielding yellowish solutions. Complex alkoxides containing different R groups were obtained by quantitatively exchanging the isopropyl derivative with another alcohol in excess, i.e. n-butyl, continuously displacing the equilibrium by distillation and monitoring the course of the reaction by g.l.c. All of the reagents and solvents had to be thoroughly dried, using conventional methods.

The composition of the complexes obtained was quantitatively characterized by the Zn/Al content-ratio, as measured by complexometry, and the OR/Al content-ratio obtained from g.l.c. of alcohols after hydrolysis. Both fitted the expected values as well as elemental analysis data. The mean degree of coordinative association of these alkoxides was determined by cryoscopic measurements in benzene or cyclohexane, with a precision of ± 2 %. NMR spectra were recorded from -70 up to 125°C, mainly on solutions in d_8-toluene, using TMS or HMDS as a reference using a ^{13}C22,63 MHZ Brucker 90 spectrometer equipped with a variable temperature control. Before sealing, the NMR tubes were prepared by introducing, through rubber septums, the complex and monomer solutions under rigorously dry nitrogen atmosphere, using stainless steel capillaries and syringes.

Results and Discussion

Structure of the complex aggregate. Although some of the complex alkoxides belonging to this family exhibit rather simple ^{13}C-NMR spectra with narrow resonance bands *(8)* when they are not (or little) associated, the spectra of the $(RO)_4Al_2O_2Zn$ complexes are very different, both when R is n-butyl or iso-propyl (n in C_6H_6 being 6 or 4 respectively, and also somewhat dependent on the complex concentration). In these cases, one still observes two different resonances corresponding to coordinatively bridged and "free" OR groups, in a ratio of about 1 to 3, but these bands are broad, partially overlapping, and are not well resolved ever when corresponding to "free" OR's (see Table I).

In fact, they appear as non-symmetrical envelopes, including a certain number of badly resolved bands

Table I : Mean Values for ^{13}C Chemical Shifts in Bimetallic μ-oxo Alkoxides (d_8-Toluene, vs. TMS, 25°C)

	$OC_\alpha H_2$ bridged	$OC_\alpha H_2$ free	$C_\beta H_2$ "bridged"	$C_\beta H_2$ free	$C_\gamma H_2$	CH_3
$(\underline{n}\text{-BuO})_4Al_2O_2Zn$	66.4 (±3)	64.0	38.4	35.7	19.6	14.5
	OCH bridged	OCH "free"	CH_3 bridged	CH_3 free		
$(\underline{iso}\text{-PrO})_4Al_2O_2Zn$	67.7	63.9 ; 65.2	28.7	27.9		

corresponding to an equal number of different situations of a given nucleus in the aggregate, that number itself depending on the value of $\bar{n}$. Such a situation obviously arises from a considerable "rigidity" of the aggregate, as confirmed by the fact that on the NMR time-scale, no rapid exchange of OR groups was ever observed when changing temperature (up to 125°C) or when adding monomer or ligands; a noticeable exception being alcohols which are known to extensively dissociate these aggregates and indeed promote a better resolution of the spectra. That rigidity is strikingly illustrated by the broadness of the band corresponding to the β-carbon of the n-butyl group, indicating that only the extremity of these alkyl groups is free from the influence of the coordinative architecture of the complex aggregate. Also, no indication ever appeared of a scrambling process yielding ZnOR groups.

These aggregates most probably display a more or less globular shape, since $\bar{n}$ values are usually moderate ($\leq$ 8) and close to integers (whatever the nature of the R groups and the metals); that also explains their amazing solubility (actually, a total miscibility when R = n-Bu) in hydrocarbon solvents, due to the fact that a "mineral" core of metal ions linked by μ-oxo-bridges is surrounded by a continuous layer of highly lipophilic alkyl groups (the corresponding methyl and ethyl derivatives being indeed insoluble). Once built up from random coordinative interactions, they become rather rigid except for the carbons at the top of that external lipophilic layer. The rigidity might be explained by analogy (necessarily limited) with the structure of spinelles, such as in Gahnite, where zinc atoms are in a tetrahedral and aluminum atoms in an octahedral arrangement. In our case however, exchange of OR groups, although very slow, must exist, so preventing the establishment of a perfectly regular structure, in agreement with the fact that the complexes could not be crystallized. Supported by structural studies *(11)* of an alkylzinc alkoxide tetramer displaying a cubane-like structure, a representation of these aggregates should thus imply a rigid core of oxo-zinc groupings surrounded by oxo-aluminum alkoxide entities located in different coordination environments, as tentatively sketched in Figure 1. It is worthwhile to mention here that the ^{27}Al NMR spectra also exhibit broad, complex resonances suggesting at least 2 different environments for the Al atoms (n-butyl derivative), i.e. a dominant 5-coordinated one (ca. 45 ppm) and a 6-coordinated one (ca. 10 ppm) in a ratio of roughly 5 to 1 *(12)*.

The existence of such a core in which electron delocalization is possible is also supported by the fact that the corresponding $(nBuO)_4Al_2O_2Mo^{II}$ solutions exhibit a very dark-brown color and absorbs throughout the visible spectra and close to the far infrared.

In summary, these entities for which we coined the general denomination of "tecto-alkoxides" (by analogy

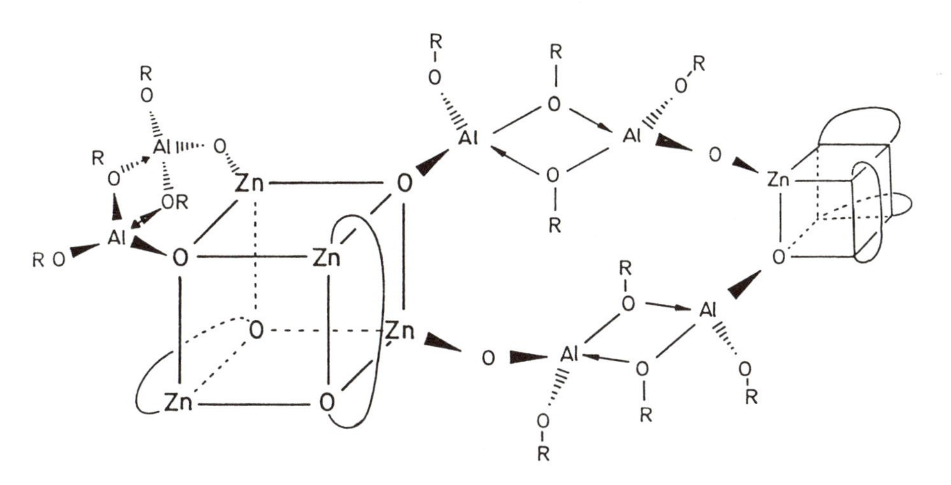

Figure 1. Schematic Representation of the Main Features of an $(RO)_4Al_2O_2Zn$ Aggregate (n = 8)

with tacto-silicates), may behave as soluble "heterogenous" catalysts, the lipophilic "surface" of which may offer particular sites of different configurations, displaying specific activation properties. In particular, it might very well explain the experimental fact that, despite their unambiguously defined structure, they produce in simultaneous competitive reactions and by the same insertion mechanism into the Al-OR bond, linear random oligomers (D.P. ca. 6 to 24), atactic PMO (M_n above 10^5) and partially isotactic PMO (M_n ca. 10^6), their variable ratio depending on the actual steric environment of these Al-OR bonds (R being an initial alkyl group or a growing chain).

Fate of the monomer. In order to check the coordinative nature of the process and possibly the in-situ environment of the monomer, an attempt was made to also observe these phenomena by ^{13}C-NMR spectroscopy. Since with most monomers the stationary concentration of the coordinated monomer is usually very low versus free monomer and growing chains (due to fast polymerization), an extremely "slow" substrate was selected for that purpose, i.e. tert-butyloxirane (tBO), and it was indeed checked that no significant amount of polymer was ever formed under the conditions of an NMR run.

Table II summarizes the observed ^{13}C chemical shifts of the "monomer" when adding tBO to $(nBuO)_4Al_2Zn_2O$ in base-mole ratios from 0 to 2.

A number of rather striking observations emerge from an analysis of these experiments:

1) the coordination of the oxirane on the complex is extremely strong. Up to a 1:1 molar ratio of oxirane

Table II : ^{13}C Chemical Shifts (ppm) of tBO, Alone or in the Presence of $(nBuO)_4Al_2O_2Zn$ (in d_8-Toluene at 25°C, ref. HMDS)

tBO	CH	CH_2	C	CH_3
alone	59.6	43.4	30.8	25.8
1 to 2 equiv. per Zn(*)	77.7	71.4	32.5	26.8
	"	"	"	27.4

(*) : for tBO/Zn = 2, signals of free tBO also appear)

to initiator, a quantitative binding of the substrate occurs; above that value, even if some more monomer is coordinated, the process levels off and free tBO appears in the spectra;

2) that coordination is so strong that under our experimental conditions no exchange is observed between free and coordinated tBO molecules. Moreover, the addition of n-butanol in a 0.5 ratio to initiator not only does not compete with tBO coordination, but decreases by a factor of 6 the amount of free tBO in the former experiment where tBO/Zn = 2, obviously by dissociating the aggregate and so offering more coordination positions to the monomer;
3) a careful examination of the spectra reveals the existence of 2 coordinated monomer species, different in their environment (CH_3 resonances at 26.8 and 27.4 ppm, with resonances of the other groups overlapping), the second one representing a small minority of the sites, i.e. ca. 5 %.
It is tempting to correlate this observation with "poisoning" experiments using LiCl *(8)*, indicating that high molecular weight polymer (particularly the stereoregular fraction) was produced by a high rate propagation process taking place on at most 4 % of the available Al centers;
4) expectedly, coordination through the oxygen electron-pairs induces extremely important shifts in the resonances of the vicinal carbon atoms (CH_2 and CH); nevertheless, the relative extent of the two shifts is very different and suggests a non-symmetrical coordination, the CH_2 group being more "burried" in, thus more sensitive to, the lipophilic outer layer of the aggregate.

Finally, there are no specific reasons for suspecting that coordination of "faster" monomers such as MO should take place differently than that of tBO.

Conclusions

The picture emerging from these data appears not only coherent, but also fits remarkably well the other available kinetic and structural results. Although the following interpretation still represents a hypothetical mechanism, it is very well supported by the whole set of available experimental data. Thus, on such bigger aggregates (i.e. the n-butyl derivative in heptane, the usual polymerization medium), the oligomeric fraction (always present even if sometimes very small) would be produced by random coordination and growth on the outermost "freer" aluminum atoms, accompanied by significant exchange during growth (always less than 1 oligomer chain per total Al) *(10)*. Conversely, the high molecular weight polymers would be generated on a few (less than 5 %) more hindered sites, located deeper in the aggregate, and on which the propagation would proceed much faster due to less coordinative interference from the growing chain with neighboring metal atoms; the more hindered ones would obviously be more likely to produce stereoregular chains, in response to high steric hindrance and thus orientation during the monomer coordination and insertion process. In that prospect, the exact role of zinc, which is not fully understood as yet, might be contemplated as a "primer" for establishing the rigid aggregate, and also as an electronic relay in the "push-pull" coordinative growth mechanism. These considerations might hopefully also give some clues for interpretating the behavior of other initiators (e.g. the porphyrine-type systems), and will be further developed into a more detailed picture.

Acknowledgments

The authors are very much indebted to M. Osgan (Institut Français du Pétrole) for his pioneering contribution to the field, and to the Services de la Programmation de la Politique Scientifique (Brussels) for general support.

References

1. Pruitt M.E. and Baggett J.M., U.S. Patent 2,706,181 (1955); Colclough R.C. and Gee G., *J. Polym. Sci.*, **34**, 153 (1959)
2. Sakata R. and Tsuruta T., *Makromol. Chem.*, **40**, 64 (1960)
3. Vandenberg E.J., *J. Polym. Sci. A1*, **7**, 525 (1969); Colclough R.O. and Wilkinson K., *J. Polym. Sci*, **4**, 311 (1964)
4. Vandenberg E.J., *Polymer Sci.*, **47**, 489 (1960); see also (3)
5. Osgan M., Silly J.P. and Teyssié Ph., Fr. Pat. 1,478,334 (1965); U.S. Pat. 3,432,445 (1966)

6. Osgan M. and Teyssié Ph., *Polymer Letters,* **5**, 789 (1967); Kollen N., Osgan M. and Teyssié Ph., *J. Polym. Sci., Polymer Letters*, **B6**, 559 (1968)
7. Teyssié Ph., Bioul J.P., Hocks L. and Ouhadi T., *ChemTech*, **7**, 192-194 (1977)
8. Teyssié Ph., Ouhadi T. and Bioul J.P., in *"International review of science", Phys. Chem. Ser.*, 2, **8**, Butterworths, 191-221 (1975)
9. Ouhadi T., Bioul J.P., Stevens Ch., Warin R., Hocks L. and Teyssié Ph., *Inorganica Chimica Acta*, **19**, 203-208 (1976); Hocks L., Bioul J.P., Durbut P., Hamitou A., Marbehant J.L., Ouhadi T., Stevens Ch. and Teyssié Ph., *J. Mol. Cat.,* **3**, 135-150 (1977)
10. Ouhadi T., Ph.D. Thesis, University of Liège (1973)
11. Steinberg H. and Hunter D.L., *Ind. Eng. Chem.,* **49**, 174 (1957); Shearer H.M.M. and Spencer C.B., *Chem. Comm.,* 194 (1966)
12. The authors are indebted to Dr. N. Marchant, BFGoodrich Research and Development Center, for recording and interpretation of these spectra

RECEIVED February 12, 1992

Chapter 12

^{13}C NMR Analysis of Polyether Elastomers

H. N. Cheng

Research Center, Hercules Advanced Materials and Systems Company, Hercules Incorporated, 1313 North Market Street, Wilmington, DE 19894–0001

The ^{13}C NMR analysis of epichlorohydrin-based polyether elastomers are reviewed. Information available from the NMR data includes homopolymer tacticity, copolymer composition, comonomer sequence distribution, and (in low molecular weight polymers) polymer chain ends. Included in this study are the NMR assignments for the homopolymers and copolymers of allyl glycidyl ether. A computerized analytical approach is used to analyze the quantitative NMR data. The result indicates that the epichlorohydrin/ethylene oxide copolymers contain at least three separate components, indicative of at least three catalytic sites.

One of the major contributions of E. J. Vandenberg is the discovery of the polyether elastomers. Most of these polymers are based on epichlorohydrin and are used for automotive applications *(1)*. The catalysts used for this family of polymers contain organoaluminum, water, acetylacetone, and other components. These are unusual but effective catalysts and can produce polymers with stereoregular as well as stereoirregular structures. In the literature they are frequently called "Vandenberg catalysts" in honor of their discoverer.

The microstructure of the polyether elastomers has been studied with thermogravimetric analysis *(2,3)*, solid state NMR *(4)*, and pyrolysis mass spectrometry *(5)*. The most informative technique is high-resolution solution NMR. Steller *(6)* was the first to report the ^{13}C NMR data on polyepichlorohydrin. Using $LiAlH_4$, he reduced the polymer to poly(propylene oxide) and demonstrated through ^{13}C NMR that the commercial polymer is regioregular but stereoirregular. Later, Dworak *(7)* confirmed Steller's findings using the same approach. Cheng and Smith *(8)* used high-field ^{13}C NMR and observed fine spectral features that correspond directly to tacticity. Through curve deconvolution, the different tacticities in polyepichlorohydrin were assigned and quantified. In follow-up work *(9)*, they studied the copolymer of

0097–6156/92/0496–0157$06.00/0

epichlorohydrin and ethylene oxide. Recently, Cheng *(10)* used a computer methodology to analyze the NMR spectral data on polyepichlorohydrin tacticity and ethylene oxide/epichlorohydrin copolymer sequence distribution. Both polymers were found to contain multiple components.

In this work the ^{13}C NMR assignments for polyepichlorohydrin and ethylene oxide/epichlorohydrin copolymer are reviewed. In addition, new data involving allyl glycidyl ether polymers are included.

Experimental Section

The polymers examined in this work were all experimental samples made with Vandenberg catalysts *(1)*. They were dissolved at concentrations of 10-20 % (w/w) in d_6-dimethylsulfoxide. The ^{13}C NMR spectra were obtained on a Nicolet NT360 spectrometer at 120°C, using these conditions: spectrometer frequency, 90.56 MHz; 1-pulse experiment, 70° pulse; repetition rate, 4s; negative linebroadening (EM = -0.2 Hz) to enhance resolution. Curve deconvolution was carried out on the Nicolet 1280 computer by the NMCCAP subprogram.

The computations were done on an IBM PS/2 computer. Quantitative analysis of the NMR data was carried out through the MIXCO.TRIADX program. This program is part of the MIXCO family of programs *(11-12)* designed to analyze multicomponent polymers.

Polyepichlorohydrin

The ^{13}C NMR spectrum of polyepichlorohydrin (PECH) is given in Figure 1. The CH_2Cl substituent resonates at 43 ppm. It is a singlet and shows no effect of tacticity splitting. However, the backbone methine (ca. 78 ppm) and methylene (69 ppm) are indeed sensitive to tacticity. The splittings are small, on the order of 0.2 ppm.

Spectral assignments have been previously made *(8)*. Because the backbone structure is nonsymmetrical, the tacticity splitting is different in the two direction along the polymer chain (i.e., m' ≠ m", r'≠ r" in the structure given below).

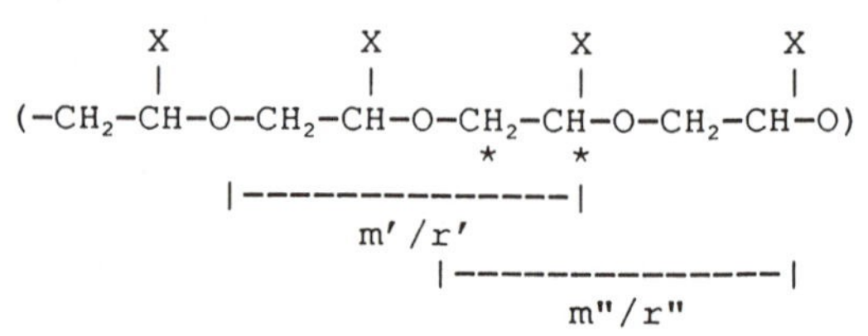

where X = CH_2Cl. The tacticity assignments are shown below. Owing to the small splittings involved, only triads are observed at the magnetic field strength used.

carbon	tacticity	observed shift
CH	m'm"	78.16
	r'm"	78.10
	m'r"	78.10
	r'r"	78.03
CH_2	m'm"	68.76
	r'm"	68.72
	m'r"	68.60
	r'r"	68.54

Ethylene Oxide/Epichlorohydrin Copolymer

The ^{13}C NMR spectrum of the ethylene oxide (EO)/epichlorhydrin (ECH) copolymer is more complex *(9)*. Four distinct regions can be identified with the four types of carbons:

Region 1	ECH	CH (backbone)	78 ppm
Region 2	EO	CH_2	70 ppm
Region 3	ECH	CH_2 (backbone)	68.5 ppm
Region 4	ECH	CH_2Cl	44 ppm

Expanded plots of these regions are given in Figure 2. In each of these regions many overlapping spectral lines are observed. These lines can originate from ECH tacticity as well as EO/ECH sequence placements. Spectral assignments are the easiest for region 4. From the PECH spectrum, we know that this line is not sensitive to tacticity. The splitting we observe must then be due to comonomer sequence placements. Intensity comparisons for samples with different compositions provide the following assignments:

Peak #	^{13}C Shift	Assignment
4a	43.68	EPE
4b	43.60	PPE, EPP
4c	43.52	PPP

For convenience, E = ethylene oxide, P = epichlorohydrin. Through curve deconvolution, we can determine quantitatively the amounts of the P-centered triads (EPE, PPE + EPP, PPP).

It is important to note that because oxygen alternates with ethylene units on the polymer backbone, the PPE and EPP triads are not necessarily equivalent; similarly EEP and PEE may have different ^{13}C shifts.

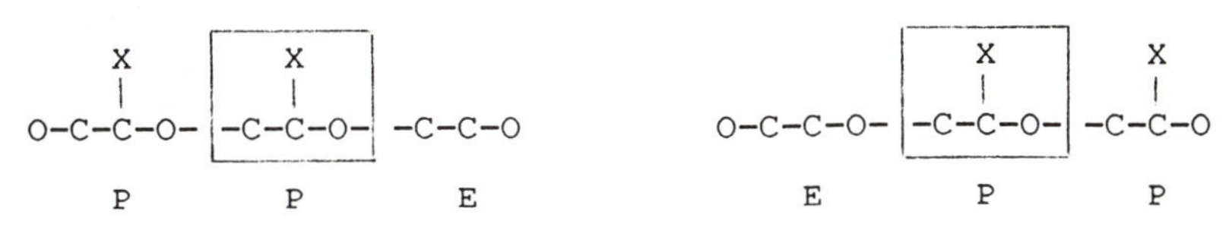

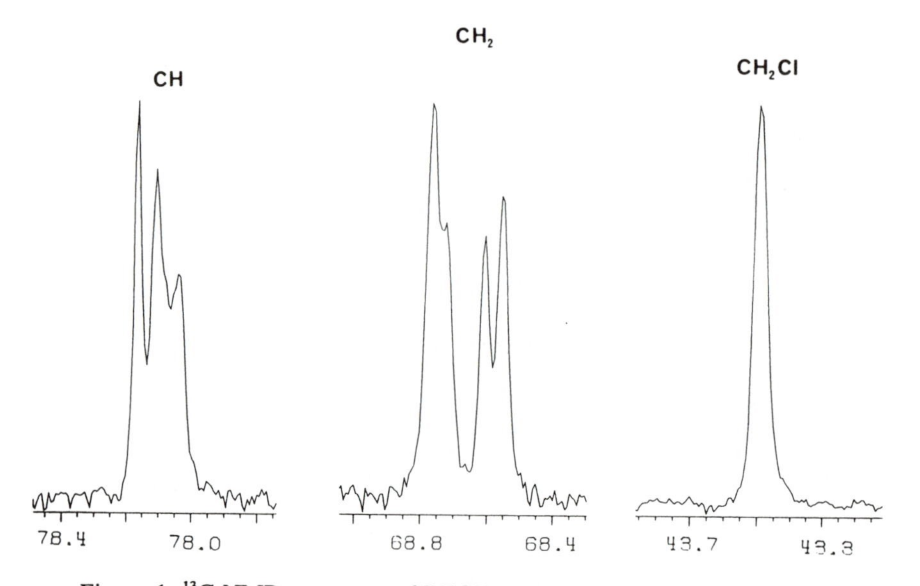

Figure 1. ^{13}C NMR spectrum of PECH, expanded to show the tacticity splittings (adapted from ref. 11).

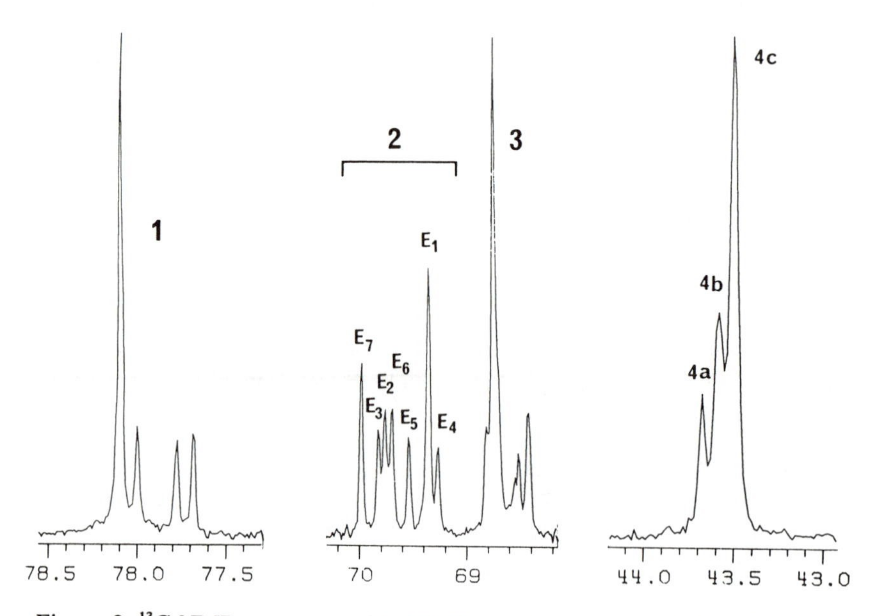

Figure 2. ^{13}C NMR spectrum of EO/ECH copolymer, expanded to show the effects of tacticity and comonomer sequence. The four regions are indicated as in the text.

In this work, we shall indicate the direction of polymer propagation from left to right; the sequence designation likewise follows the same direction (as above).

Determination of the E-centered triads can be obtained from region 2. Seven lines are found in this region, corresponding to the seven different environments whereby an ethylene oxide can be placed.

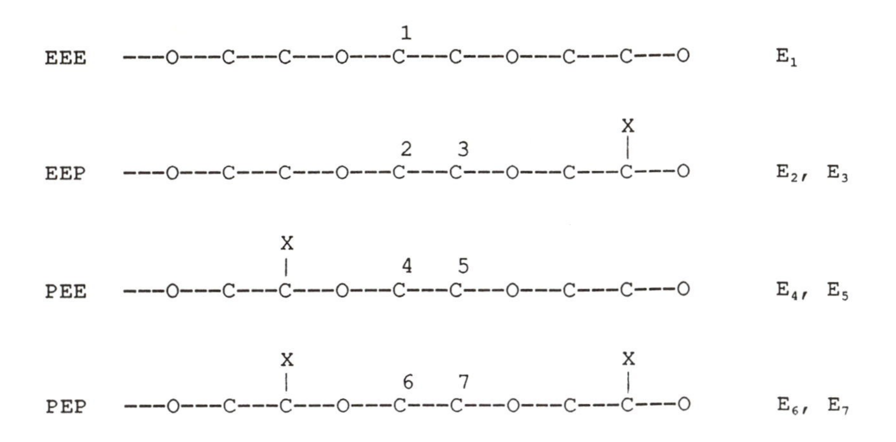

The assignments are given below. Curve deconvolution of the seven lines provides the quantitative results on the E-centered triads.

Peak #	^{13}C Shift	Triad	Assignment
2a	69.98	PEP	E_7
2b	69.83	EEP	E_3
2c	69.78	EEP	E_2
2d	69.71	PEP	E_6
2e	69.54	PEE	E_5
2f	69.36	EEE	E_1
2g	69.28	PEE	E_4

The ^{13}C resonances in region 1 and region 3 have also been assigned *(9)*. They contain information on both E/P comonomer placement and polyepichlorohydrin tacticity. Interested readers may consult the suitable reference *(9)*.

Poly(Allyl Glycidyl Ether)

Allyl glycidyl ether (AGE) is used frequently as a crosslinker in polyether elastomers. The homopolymer of AGE has six distinctive carbons. The ^{13}C NMR spectrum is given in Figure 3.

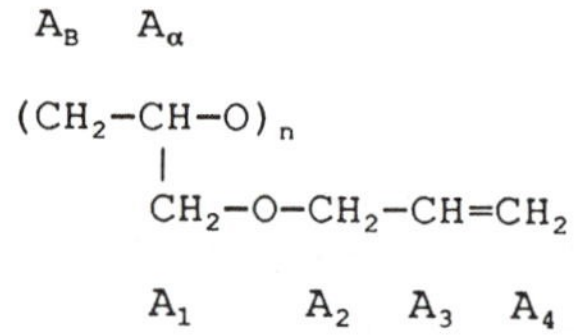

```
A_B  A_α

(CH2-CH-O)n
      |
      CH2-O-CH2-CH=CH2

      A1     A2   A3   A4
```

Through the APT experiment *(13)*, A_α and A_3 carbons can be readily identified. The use of empirical additive shift rules *(14-18)* provides the rest of the assignments.

carbon	shift	carbon	shift
A_α	77.0	A_2	70.3
A_B	68.8	A_3	134.3
A_1	69.1	A_4	115.0

Note that the polymer backbone carbons (A_α and A_B) have complex spectral features due to tacticity. A comparison with the spectrum of PECH suggests that the tacticity assignments in both polymers are similar. The tacticity triads follow the trend, r'r", m'r", r'm", m'm", with increasing field.

Epichlorohydrin/AGE Copolymer

The incorporation of a small amount of AGE in PECH produces a ^{13}C NMR spectrum with additional lines (Figure 4). Three of the AGE lines overlap with the lines due to PECH. The polymer shown in Figure 4 has a low molecular weight, and there are additional lines due to the primary alcohol chain end (designated P^e in the structure below).

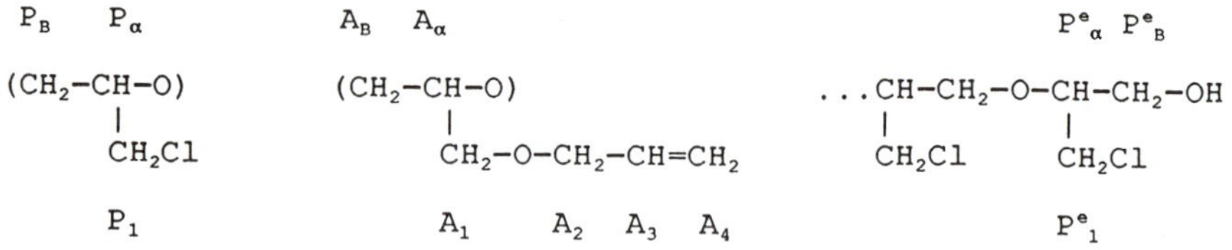

```
P_B    P_α          A_B  A_α                              P^e_α  P^e_B

(CH2-CH-O)          (CH2-CH-O)                   ...CH-CH2-O-CH-CH2-OH
     |                   |                          |          |
     CH2Cl               CH2-O-CH2-CH=CH2           CH2Cl      CH2Cl

     P1                  A1     A2   A3   A4                   P^e_1
```

Number	shift	carbon: ECH	carbon: AGE	ECH end
1	135.3		A_3	
2	116.0		A_4	
3	78.8-80.0			P^e_α
4	77.0-78.8	P_α	A_α	
5	70.0-71.5		A_2	
6	67.8-69.2	P_B	$A_B + A_1$	
7	59.8-60.5			P^e_B
8	43.5	P_1		
9	42.5			P^e_1

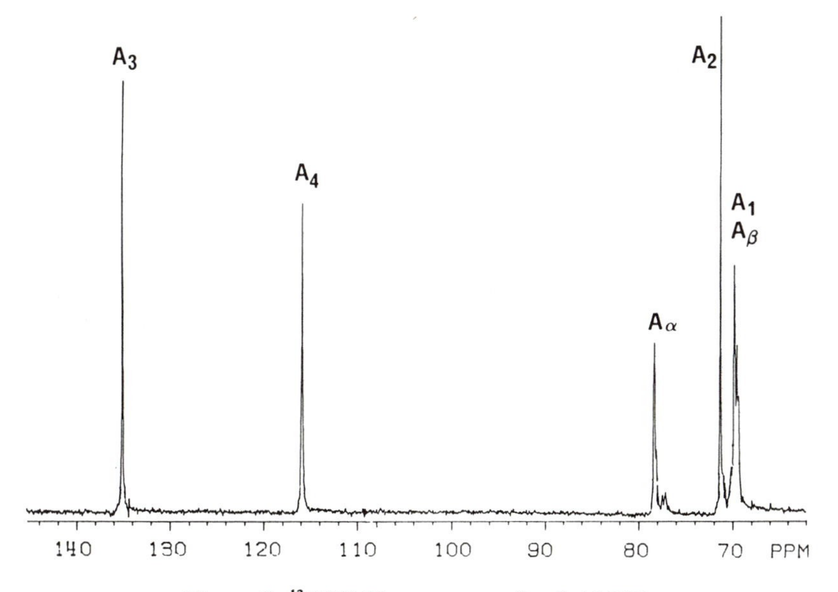

Figure 3. ^{13}C NMR spectrum of poly(AGE).

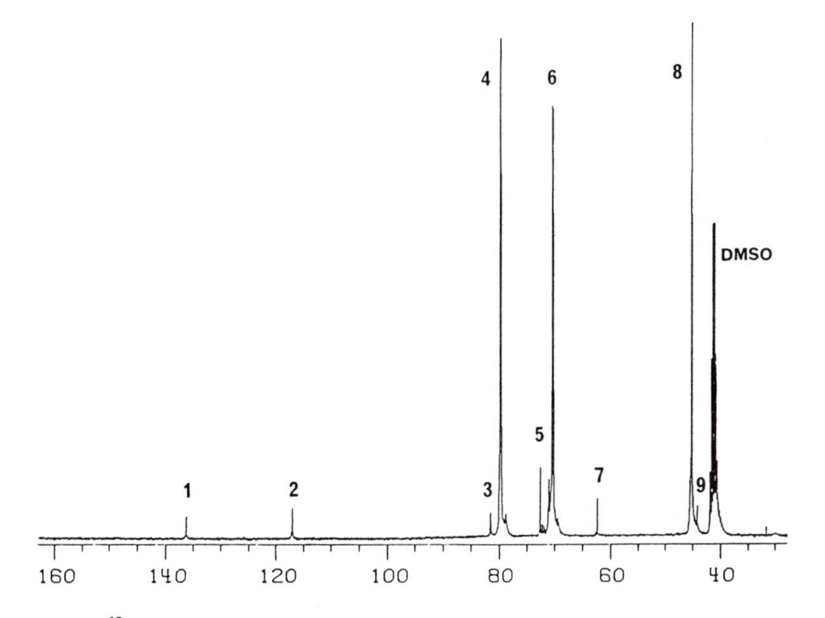

Figure 4. ^{13}C NMR spectrum of ECH/AGE copolymer; line numbering is the same as given in the text.

We can determine the level of AGE incorporation by using the intensity of A_2 (line number 5) and the average line intensities of lines 4, 6, and 7.

$$\text{mole \% AGE} = \frac{I_5}{1/3\,(I_4 + I_6 + I_7)}$$

Ethylene Oxide/Epichlorohydrin/AGE Terpolymer

Owing to the low level of AGE used, this terpolymer has a ^{13}C NMR spectrum (Figure 5) that is very similar to that of the EO/ECH copolymer. As before, three of the AGE lines (A_α, A_B, A_1) overlap the ECH lines. Three lines at 135.3, 116.0, and 70.5 ppm (corresponding to A_3, A_4, and A_2) are indicative of AGE.

```
PB    Pα                  E                    AB   Aα

(CH2-CH-O)            (CH2-CH2-O)            (CH2-CH-O)
      |                                            |
      CH2Cl                                        CH2-O-CH2-CH=CH2

      P1                                           A1     A2   A3   A4
```

Number	shift	carbon		
		ECH	EO	AGE
1	135.3			A_3
2	116.0			A_4
3	77.0-78.8	P_α		A_α
4	71.0-71.5			A_2
5	69.2-70.2		E	
6	67.8-69.2	P_B		$A_B + A_1$
7	43.5	P_1		

The composition of the terpolymer can be obtained by using the line intensities for lines 3-7. Let P', E', and A' be proportional to the concentration of P, E, and A; P', E', and A' can be calculated from the following expressions.

$$\begin{aligned} I_3 &= P' + A' \\ I_4 &= A' \\ I_5 &= 2E' \\ I_6 &= P' + 2A' \\ I_7 &= P' \end{aligned}$$

The molar fractions, (P), (E), and (A), are obtained by normalization:

$$(P) = k'P',\ (E) = k'E',\ (A) = k'A',$$

where $k' = [(P) + (E) + (A)]^{-1}$.

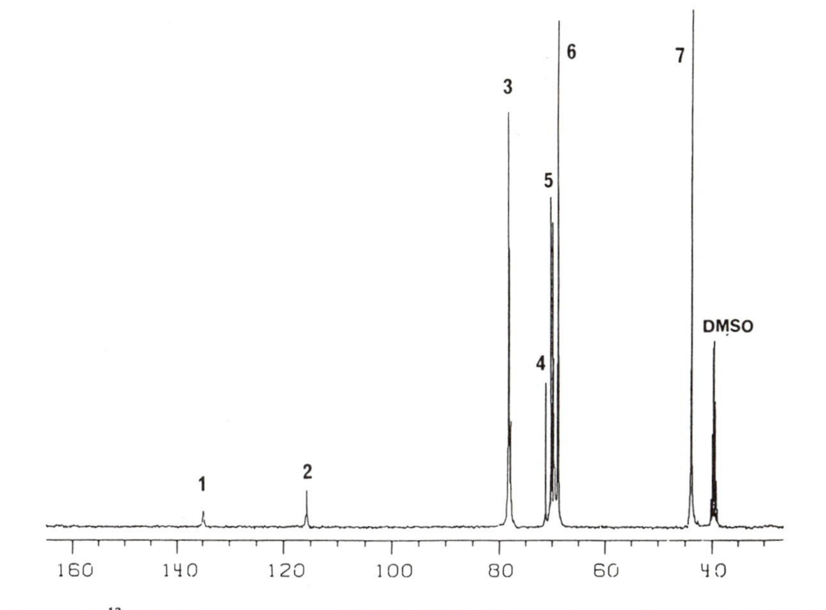

Figure 5. ^{13}C NMR spectrum of EO/ECH/AGE terpolymer; line numbering is the same as given in the text.

Computerized Analytical Approaches

In previous studies *(9,10)* of the tacticity of PECH and the comonomer sequence distribution of EO/ECH copolymers, it appears that these elastomers contain two or three distinctly different components, each of which obeys a different propagation statistics. The results strongly suggest that there are a number of catalytic sites within the catalytic system that can separately polymerize the monomers.

In this work we have sought additional evidence of this multi-site nature of the catalyst system by analyzing additional samples of the EO/ECH copolymer. As before *(19-21)*, the combination of NMR and fractionation gives the most information. Accordingly, a sample of EO/ECH copolymer was separated into toluene-soluble (A) and toluene-insoluble (B) fractions at 25°C. The toluene insoluble fraction was redissolved in toluene at a higher temperature and cooled back down to 25°C. In this way, two more fractions are obtained: the soluble (C) and insoluble (D) fractions. The NMR triad sequence distributions for these fractions are given in Table I.

Table I. Observed and calculated intensities for EO/ECH copolymer fractions and three-component B/B/B model calculation

	toluene extraction				reprecipitation			
	fraction A		fraction B		fraction C		fraction D	
	I_{obsd}	I_{calc}	I_{obsd}	I_{calc}	I_{obsd}	I_{calc}	I_{obsd}	I_{calc}
PPP	16.9	16.5	46.1	45.9	38.9	38.9	64.7	64.7
PPE	22.3	22.9	20.9	20.8	20.5	18.9	8.8	9.8
EPE	11.6	11.6	3.2	5.0	6.8	7.8	1.7	3.3
PEP	12.0	11.4	11.7	10.4	11.6	9.5	6.7	4.9
EEP	23.0	23.3	10.1	10.0	11.9	15.7	7.4	6.6
EEE	14.3	14.3	8.0	7.9	10.3	9.2	10.7	10.7

3-component B/B/B model

		fraction A	toluene extraction	fraction B	fraction C	reprecipitation	fraction D
site 1,	P_{p-1}		0.842			0.942	
	w_1	0.122		0.742	0.373		0.769
site 2,	P_{p-2}		0.483			0.511	
	w_2	0.810		0.131	0.577		0.015
site 3,	P_{p-3}		0.230			0.214	
	w_3	0.067		0.127	0.050		0.216
mean deviation			0.46			1.25	

Analysis was carried out using the MIXCO programs described previously *(11-12,19-21)*. The NMR data of each soluble/insoluble pair are analyzed together. In the multicomponent model, each triad intensity (I_{total}) is taken to be the sum of the intensities (I_i) of several components:

$$I_{total} = \Sigma w_i I_i$$

where w_i is the weight factor for the *i*th component. The theoretical intensity of each component can be calculated from the reaction probabilities that are appropriate for the component. In the actual computation, guess values of the reaction probabilities and weight factors are initially fed into the program, and theoretical values for the sequences calculated. The observed intensities are compared with the theoretical intensities for the twelve triad sequences (six for each fraction) through the simplex algorithm. The iteration continues until the best-fit values of the parameters are obtained.

With this methodology, the data for the initial EO/ECH copolymer fractions (fractions A and B) were found to fit very well to a three-component Bernoullian (B/B/B) model. Each of the components can be characterized by a Bernoullian probability, P_{p-i}, where the subscript p refers to the epichlorohydrin addition and i refers to the *i*th component. The results of the model-fitting procedure are shown in Table I. The mean deviation between the observed and calculated values is very small (0.5).

The same model fitting procedure was used to analyze the reprecipitated fractions (C and D). Again, the data can be fitted to a three-component B/B/B model (Table I). The mean deviation is somewhat larger (1.3). It is of interest that the reaction probabilities for the three components in this case are very similar to the (A + B) case. The compatibility of the two analyses further confirms the multicomponent nature of this copolymer.

Note that the Bernoullian probabilities [P_{p-i}] are very different for the three components, suggesting that the components originate from different catalytic sites. Site 1 (with $P_{p-1} \approx 0.9$) favors the propagation of ECH, and will produce polymers high in ECH content. In contrast, site 3 (with $P_{p-3} \approx 0.2$) favors the propagation of EO. Site 2 shows approximately equal preference for ECH and EO ($P_{p-2} \approx 0.5$). If one assumes that in a continuous polymerization the molar ratio (ECH:EO) in the monomer feed is 7:1, then the reactivity ratios can be estimated:

site 1: $r_{EO} = 0.87$, $r_{ECH} = 1.16$;
site 2: $r_{EO} = 7.00$, $r_{ECH} = 0.14$;
site 3: $r_{EO} = 24.8$, $r_{ECH} = 0.04$;

where $r_{EO} \cdot r_{ECH} = 1$ because the Bernoullian model has been assumed.

The analytical methodology used also permits us to estimate the relative importance of each of these catalytic sites in overall polymer production. This is done by multiplying the weight % of each fraction (obtained from fractionation) with the weight factor of each site (obtained from NMR analysis) and summing up the products for each site. The result for fractions A and B is shown in Table II. It appears that the catalyst site 2 makes the most polymer (69 %), followed by site 1 (23 %) and then site 3 (8 %).

Table II. Contribution of catalytic sites to overall polymer yield

Fr	wt. of fraction	weight factor			wt. frac. * wt. factor		
		site 1	site 2	site 3	site 1	site 2	site 3
A	81.9%	0.122	0.810	0.067	10.0	66.3	5.5
B	18.1%	0.741	0.131	0.127	13.4	2.4	2.3
sum					23.4	68.7	7.8

It may be noted that in general if the NMR/fractionation data can be analyzed by a n-site model, the data can also be fitted to a (n+1)-site model. The fact that the EO/ECH copolymer data can be analyzed via a three-site model implies that there are *at least* three catalytic sites in the catalyst. To determine if indeed there are four active sites, one can repeat the analysis using a more detailed fractionation scheme, e.g., the combination of TREF and NMR *(21)*.

Acknowledgment

The author wishes to thank Dr. Wendell P. Long for providing the polymer samples used in the NMR analysis and for stimulating the initial phase of this work, Dr. E. J. Vandenberg for useful comments on the manuscript, and David A. Smith for technical assistance.

Literature Cited

1. Vandenberg, E. J. In *Kirk-Othmer Encyclopedia of Polymer Science and Technology*, 3rd Ed.; Wiley: New York, NY, 1978, Vol. *8*; pp. 568-582.
2. Jaroszynska, D.; Kleps, T.; Gdowska-Tutak, D. *J. Therm. Anal.* 1980, *19*, 69.
3. Day, J.; Wright, W. W. *Br. Polym. J.* 1977, 66.
4. Sullivan, M. J. *Trends Anal. Chem.* 1987, *6*, 31.
5. McGuire, J. M.; Bryden, C. C. *J. Appl. Polym. Sci.* 1988, *35*, 537.
6. Steller, K. E. *Am. Chem. Soc. Symp. Ser.* 1975, *6*, 136.

7. Dworak, A. *Makromol. Chem., Rapid Commun.* 1985, *6*, 665.
8. Cheng, H. N.; Smith, D. A. *J. Appl. Polym. Sci.* 1987, *34*, 909.
9. Cheng, H. N.; Smith, D. A. *Makromol. Chem.* 1991, *192*, 267.
10. Cheng, H. N. *ACS Polym. Preprint* 1991, *32*(1), 551.
11. Cheng, H. N. *J. Appl. Polym. Sci.* 1988, *35*, 1639.
12. Cheng, H. N. *Am. Chem. Soc. Symp. Ser.* 1989, *404*, 174.
13. Patt, S.; Shoolery, J. *J. Magn. Resonance* 1982, *46*, 535.
14. Grant, D. M.; Paul, E. G. *J. Am. Chem. Soc.* 1964, *86*, 2984.
15. Carman, C. J.; Tarpley, Jr., A. R.; Goldstein, J. H. *Macromolecules* 1973, *6*, 719.
16. Cheng, H. N.; Ellingsen, S. J. *J. Chem. Inf. Computer Sci.* 1983, *23*, 197.
17. Cheng, H. N.; Bennett, M. A. *Makromol. Chem.* 1987, *188*, 135.
18. Cheng, H. N.; Bennett, M. A. *Anal. Chim. Acta* 1991, *242*, 43.
19. Cheng, H. N. *J. Appl. Polym. Sci.: Appl. Polym. Symp.* 1989, *43*, 129.
20. Cheng, H. N. *Polym. Bull.* 1990, *23*, 589.
21. Cheng, H. N.; Kakugo, M. *Macromolecules* 1991, *24*, 1724.

RECEIVED November 15, 1991

Chapter 13

Calorimetric Study of Cure Behavior of an Amine–Epoxy Resin under Ambient Conditions

Suk-fai Lau

Research Center, Hercules Advanced Materials and Systems Company, Hercules Incorporated, 1313 North Market Street, Wilmington, DE 19894–0001

Epoxy resin prepregs cure slowly under ambient conditions. Both temperature and humidity accelerate this process. Such resin advancement limits the time that a prepreg can be held in manufacturing. Otherwise its tack or flow is reduced too much to allow a void-free part to be made.

New calorimetric techniques were used to follow the cure of an epoxy-amine system under ambient conditions. The reaction was found to be autocatalytic with the activation energy being a function of resin conversion. The kinetic approach of Van Geel using a modified activation energy term, allowed the calorimetric data to be accurately reduced. A precise estimation of the remaining time before a prepreg had to be cured could then be made.

Fabrications of composite parts from prepreg are generally performed at ambient temperatures and variable humidity. A long out-time is usually required for fabricating large parts. Amine-epoxy resins in particular, advance readily under these conditions. As a result, prepregs tend to lose tackiness and increase in viscosity during the fabrication process. Consequently, delaminations and voids are found between ply interfaces and in the resin matrix which reduce its impact strength.

To test for the suitability of a prepreg lot, one of the current practices is to construct a part by going through the lay-up and cure cycles. The cured composite is then machine-cut, polished, and a cross-sectional area is examined for flaws (delamination and void formation). This practice is time consuming and cost ineffective. The test results may not reflect the behavior of the remaining prepreg lot since it advances during the test time.

Both tack and flow properties of the prepreg are directly related to how far the resin has advanced. Its time-temperature-humidity-conversion profile (t-T-H-α) is thus important for its out-time prediction. The t-T-H-α relationship can be established for a resin if its cure kinetic parameters (viz., activation energy and pre-exponential factor) and the concentration function

0097–6156/92/0496–0170$06.00/0

of reactants are known. Or, one can use a time-temperature-superposition technique (t-T-S) to set up the t-T-α relationship [1-3]. This technique is based on the Arrhenius rate equation describing the same reaction kinetics at two isothermal temperatures.

$$\exp\{[E(T_1 - T_2)] / (RT_1T_2)\} = t_2 / t_1 \quad (1)$$

where E = activation energy
T_i = isothermal temperature at temperature i
t_i = exposure time at temperature T_i
R = Gas constant

A conversion-time curve at one temperature is shifted along the time scale to a reference temperature by utilizing a measured activation energy describing the reaction. Barton and Prime had successfully applied the t-T-S shift on epoxy systems using a constant activation energy term [2, 3]. Due to the difference in reactivity of the functional groups (in both the reactants and the intermediates) as well as the humidity of environment, the apparent activation energy may not necessarily be a constant.

In-situ methods such as Infrared (IR) and Differential Scanning Calorimetry (DSC) have been widely used to follow the cure kinetics at elevated temperatures [4-8]. In these cases, extrapolations are made from the high temperature results to ambient or use conditions, and humidity (moisture levels) may not be controlled. Because the reaction mechanism at high temperatures may be different from that at ambient conditions, these extrapolations may not be valid.

In the present paper, the cure reaction of an epoxy-amine resin at low temperatures (70°C or less) and at constant humidity (50% RH) is investigated. The profile of its t-T-α relationship at ambient conditions is established. The effect of humidity on the resin cure will be discussed in the next paper.

Experimental Section

A ThermoMetric 2277 Thermal Activity Monitor (TAM) was used to follow the resin advancement isothermally. The monitor consists of four power-compensation-calorimeters which are about 1,000 times more sensitive than the conventional DSC if the same sample weight is used. A further gain in sensitivity can also be achieved by using larger sample size since the TAM sample container can hold 2.5-3.0 g resin sample. A fixed sample weight of 500 mg, which provides a constant exposure area, is used in the study. This translates to an additional 10 times gain in sensitivity when compared to the DSC analysis since the DSC sample container can hold less than 50 mg sample. Each calorimeter was calibrated separately with a known input thermal power and was equilibrated to a test temperature before the sample analysis.

An epoxy resin with a sulfone-type amine hardener was used in this study. The frozen resin sample was allowed to warm up to room temperature in a sealed container before weighing in order to reduce the moisture absorption by the resin. A fixed sample weight of 0.5 gram was weighed directly into 3 ml glass ampoules, the sample container for the TAM cell. A glass tube (its outside diameter of approximately 3 mm) filled with a salt solution was inserted into the glass ampoule. Salt solutions of lithium chloride, magnesium chloride, magnesium acetate, magnesium nitrate and potassium acetate were used to provide controlled humidity environments (from 18% to 75% RH) within the glass ampoule. The relative humidity provided by the salt solution at constant temperature and under the static condition was measured separately using a digital humidity meter. The glass ampoule containing the resin and the salt solution was tightly sealed with a lid. The sample was then equilibrated in a preheated oven at the test temperature (30, 50, 60, or 70°C) for 30 minutes before it was introduced into the TAM calorimetric cell. The heat flow generated from the resin during curing at the test temperature was recorded with respect to time. The accumulated heat "H_i" generated up to a time "t_i" was extracted from the heat flow curve by integrating the area under the curve. The ratio of "H_i" to the total reaction heat "H_{tot}" from an uncured resin defines the resin conversion, α_i, at time "t_i".

$$\alpha_i = H_i / H_{tot}, \qquad \text{at time } t_i \tag{2}$$

At least three calorimetric scans were performed on the resin at each test temperature and humidity. The data (i.e., α_i at each t_i) generated at 30°C also serves as a reference curve for the construction of the t-T-α diagram. The approach will be presented in detail on the later part of this paper.

The predicted resin advancement at ambient conditions was checked by chemical methods [9]. A set of resin samples was aged in a controlled environmental chamber at 23°C/50%RH for different exposure times. The % advancement of the aged samples was studied by titration (residual epoxide group) and by DSC (residual heat of reaction). The titration method required the resin sample to be dissolved in methylene chloride. The solution was then reacted with tetraethylammonium bromide and potentiometrically titrated with perchloric acid. This titration yielded the total amounts of epoxide and amine that were present in the solution. The amine content was determined separately by another titration, using the same titrant but without the presence of tetraethylammonium bromide. The epoxide content was determined from the difference between the two titrations.

Results And Discussions

I. Resin Cure Behavior. The epoxy-amine resin shows autocatalytic cure behavior. The reaction rate gradually increases with time until a maximum rate is attained, as shown in Figure 1. The time between the start of measurement and the time at maximum rate is the "lag time". The increase in the exposure temperature shortens the "lag time" significantly. The autocatalytic behavior is a result of the hydroxyl group generated during the amine-glycidyl ether reaction [10]. The hydroxyl group accelerates the amine-glycidyl ether reaction markedly through hydrogen bonding with the oxygen in the epoxide ring. This aids the opening of the epoxide ring in a transition state. The autocatalytic behavior participates throughout the entire advancement range but the "overall reaction rate constant" varies. Moisture in the resin plays a similar role to the hydroxyl group.

A basic rate equation, Equation (3), has been widely used to describe this behavior [11]. This equation contains two rate constants (a catalytic constant, k_c, and an autocatalytic constant, k_a) and a concentration function of reactants. The autocatalytic constant may be considered to be dependent upon an apparent frequency factor, Z_a, and an activation energy, E_a. Both parameters change simultaneously if the reaction mechanism changes. They are also considered to be a function of α and H, the degree of advancement and the relative humidity, respectively. Thus, it is important to this approach that k_a may vary as a function of conversion and relative humidity.

$$\text{Reaction rate} = d\alpha/dt = (k_c + k_a\alpha)(1 - \alpha)^2 \tag{3}$$

$$\text{Reduced reaction rate} = (d\alpha/dt)(1-\alpha)^{-2} = k_c + k_a\alpha \tag{4}$$

where α = fraction converted, or degree of conversion = 0 to 1
t = reaction time, in hours
k_a, k_c = in hours

The reduced reaction rate, Equation (4), is derived from Equation (3). This form allows separation of rate data into the desired k_c and k_a. A plot of Equation (4), a reduced rate plot of the resin at 60°C and 50%RH, is shown in Figure 2 as an example. Three different regions can be identified: an initial linear portion, a second linear section leading to a maximum, and finally, a decreasing rate at conversion around 50%. From the first region ($\alpha<12\%$), we take the intercept, at $\alpha=0$, as k_c and the slope as k_a. The second region ($12\%<\alpha<\sim50\%$) is dominated by the autocatalytic reaction. The rate expression can be simplified by setting $k_c=0$ in Equation (4). At about 50% conversion, the reaction rate reduces gradually which marks the onset of vitrification. In this region ($\alpha>50\%$) the reaction shifts from kinetic control to diffusion control and the reaction rate drops markedly. The cure reaction, however, does not stop at this point, but continues at a slower rate. The shift from kinetic control to diffusion control, at different isothermal temperatures, is not iso-conversional. This change occurs when the glass transition temperature of the resin (Tg) equals to the temperature of cure (T_{cure}). At lower cure temperature, the onset of diffusion control shifts to a lower conversion level. This phenomenon have been shown by Hale and Wisanrakkit through the studies of Tg-conversion relationship on two epoxy systems [12, 13]. As a result, an equation describing the diffusion characteristic of the resin after vitrification is required for this region.

To relate advancement to time and temperature, one has to integrate the appropriate rate equation for each advancement region. Equations discussed above, however, are not manageable since the frequency factor (Z_a) as well as the activation energy (E_a), included in the autocatalytic rate constant (Equation 5), are advancement and humidity dependent. We assume that an apparent activation energy, E_a, and an apparent frequency factor, Z_a, can be used to represent some average of the reaction. This does not mean that only one reaction is occurring. The equation is difficult to use because the nature of the reaction(s) is constantly changing with conversion and humidity. Setting up the apparent Z_a and the apparent E_a functions are thus difficult or may be impossible.

$$k_a = Z_a \exp(E_a/RT) \tag{5}$$

where E_a = in kcal/mole

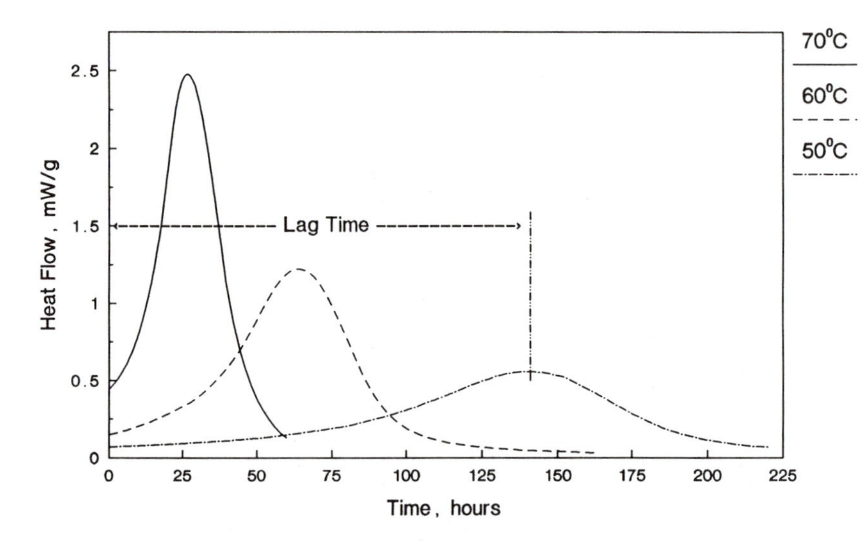

Figure 1. Heat generation versus time during curing of an epoxy–amine resin at isothermal temperatures of 50 °C, 60 °C, and 70 °C, and at 50%RH.

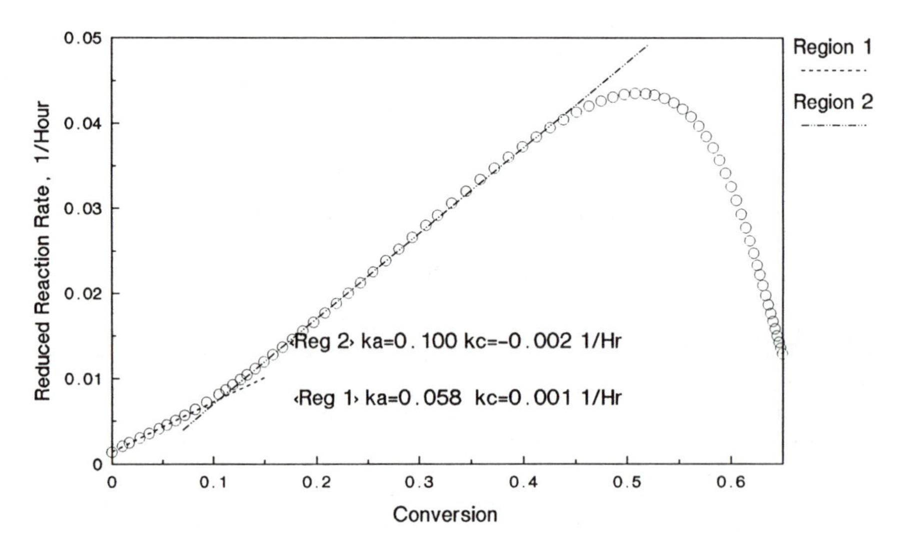

Figure 2. Reduced reaction rate plot of an epoxy-amine resin at 60°C and 50%RH.

Z_a = in 1/min
R = Gas constant = 1.987 cal/mole/K
T = Exposure temperature, in K

Van Geel provided an approach to relate conversion to exposure time and temperature, without knowing the concentration function of reactants and the apparent Z_a function [14]. Based on his studies on the decomposition of nitrate ester propellants, he concluded that similar information (α and t_2) can be derived from Equation (6) at any temperature (T_2) as long as the conversion (α) and exposure time (t_1) at temperature T_1 are known. At iso-conversion:

$$q_1 / q_2 = [\exp(-E_a/RT_1)] / [\exp(-E_a/RT_2)] = t_2 / t_1 \qquad (6)$$

where q_i = heat generation rate at temperature i and at same conversion level, in mcal/sec/g

Through simplification of the second term, Equation (6) can be reduced to Equation (1), the time-temperature-superposition technique described by Prime [1].

$$\exp\{[E(T_1 - T_2)] / (RT_1T_2)\} = t_2 / t_1 \qquad (1)$$

Equation (6) was derived based on an assumption that the apparent activation energy of decomposition of nitrate ester was constant throughout the reaction. For the epoxy-amine reaction, the above equation has to be modified such that it also takes the conversion effect, the autocatalytic effect as well as the humidity effect into consideration. The approach is to establish the empirical relationship between the activation energy and the overall conversion, a combination of autocatalytic effect and the neat amine-epoxide reaction. This relationship can then be used, together with the last two terms in Equation (6) for the prediction of t-T-α relationship. Or, one can use Equation (6) to establish the time-temperature-(heat-generation-rate)-α diagram. For the rest of the paper, we shall concentrate on the t-T-α relationship.

II. Construction of Time-Temperature-Conversion Diagram. The apparent activation energy, E_a, at each conversion level and constant humidity is obtained from the slope of ln(time) versus 1/Temperature plot (Figure 3). The same reaction mechanism is evidenced in the test temperature range 50°C to 70°C since all data points fall onto a straight line at each advancement level. The apparent activation energies, for different conversion levels, are also derived from the slope of heat-generation-rate versus 1/Temperature plot from the same isothermal studies. The results obtained from both methods agree with each other. This indicates that no induction period is involved (e.g., consumption of inhibitor) and the catalytic reaction starts right away. E_a of the cure process, however, is conversion dependent, as indicated in Figure 4. When there is no change in humidity, the linear decrease in E_a along the reaction course is the result of hydroxyl group formation. The change in E_a due to the accumulation of hydroxyl groups, before the diffusion control region, is found to be in the order of 1 kcal/mole. The sudden increase in activation

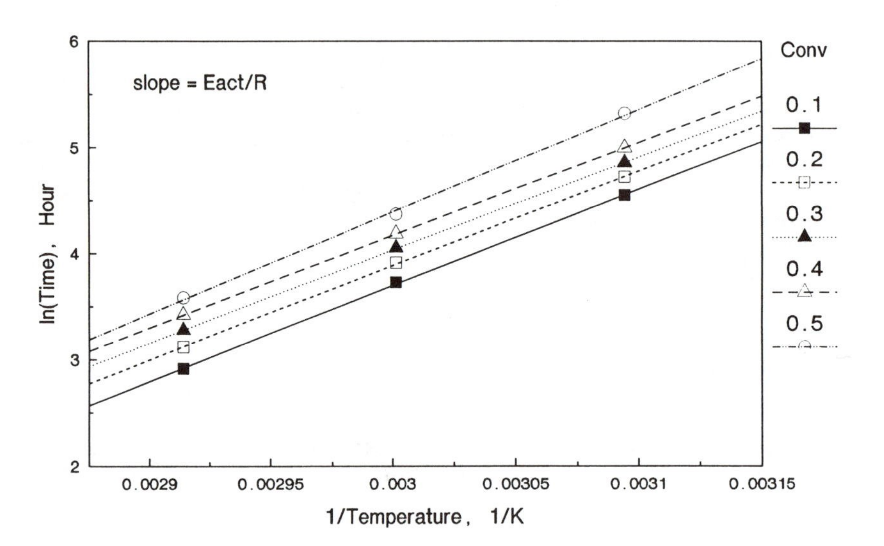

Figure 3. Reaction time versus 1/temperature plot of an epoxy-amine resin at constant conversion levels and at 50%RH.

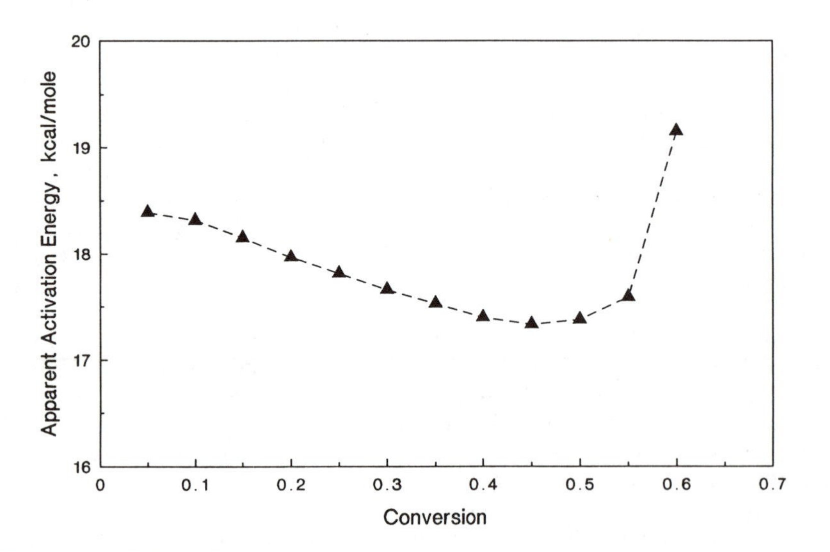

Figure 4. Effect of % advancement on apparent activation energy of an epoxy-amine resin at 50%RH.

energy at $\alpha > 0.50$ marks the onset of the diffusion control. A similar trend holds for resin cured at different iso-humidities.

With the derived activation energies for different conversion levels, and a reference heat flow curve (experimental data at 30°C), a series of isothermal curves, which relate advancement to exposure time, can be derived. The reference heat flow curve provides information on how the conversion changes with time. This reference curve, together with the derived activation energy function is then shifted to time at high or low temperatures through the relationship described in Equation (6) for each conversion level. The predicted time for each α at different temperatures (30, 50 and 70°C) are plotted in Figure 5. These predicted values (solid curves) agree well with the experimental heat flow results (squares or circles) at different temperatures.

III. Validity of Time-Temperature-Conversion Prediction. The time-conversion prediction is extrapolated outside the test temperatures to simulate the actual exposure conditions (for example, at 23°C and 50%RH) at a manufacturing site (Figure 6). To test the validity of these predictions, two parallel analyses were performed on a set of samples which had been aged under the same conditions for different periods of time. The % advancement of the aged samples was studied by titration (residual epoxide group) and by DSC (residual heat of reaction). Within the experimental error, % advancement measured by both methods (squares in Figure 6) agreed with the time-conversion prediction.

VI. Application of Time-Temperature-Conversion Relationship. <a> Resin advancement after any isothermal history can be directly obtained from the time-temperature-conversion curves. For any partially advanced resin or prepreg system, its initial advancement needs to be known. This can be measured by either chemical titration and/or resin glass transition temperature (Tg). The use of Tg is particularly recommended [12, 13]. We constructed an empirical relationship between resin advancement and Tg at 23°C (Figure 7). A one-to-one relationship was found between Tg and resin conversion. Figure 7 allowed any resin sample, regardless of form, to be quickly measured for advancement by DSC (Tg). The initial advancement is then used as the starting point on the time-temperature-conversion curves for the isothermal treatment segment.

<b> The residual out-time of the prepreg can be estimated from the relationship, if its initial advancement and its maximum allowable resin advancement limit (MARAL) are known (Figure 8). The limit defines an extreme for resin advancement at which the prepreg will pass the porosity test. The time difference between these two points designates the out-time of the prepreg. If the estimated out-time is shorter than the requirement, the prepreg should either be scrapped or used for laying up smaller parts.

<c> To ensure prepreg lots meet the out-time requirements, the total allowable resin advancement limit (TARAL) of all process segments should be defined. The process segments include resin blending, filming and prepreg processing. The TARAL can be obtained by subtracting the maximum allowable time at which the prepreg fails the porosity test from the out-time requirement (Figure 9). The corresponding advancement will be allocated for

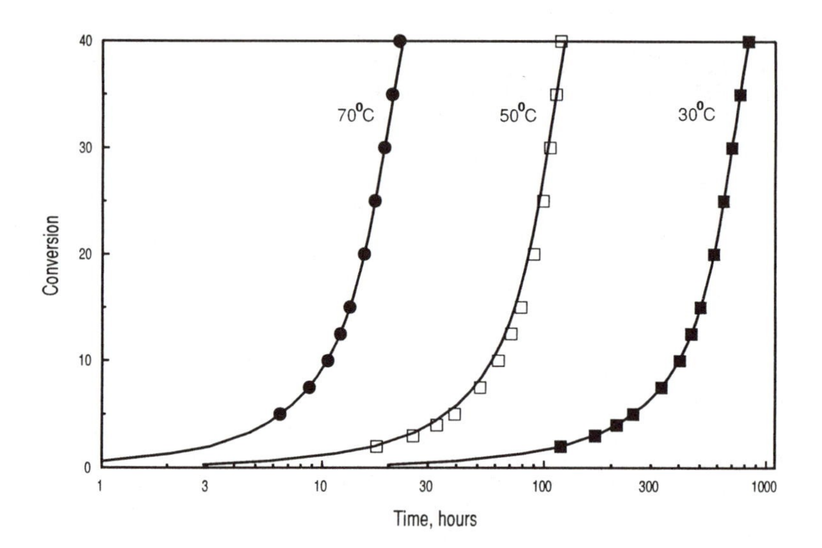

Figure 5. Comparison of experimental data to predicted results. Predicted results, ———; Experimental data: ■, 30°C; □, 50°C, and ●, 70°C.

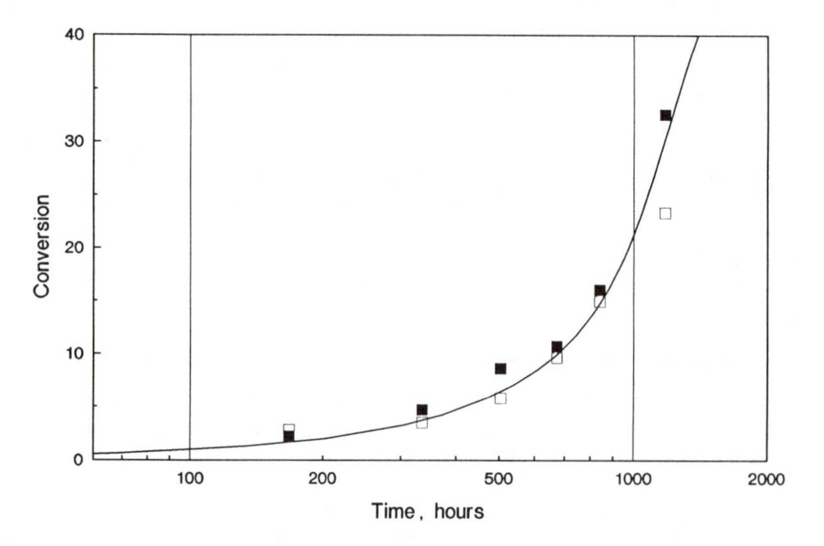

Figure 6. Comparison of results from independent analysis to predicted values at 23°C and 50%RH. Predicted results, ——— ; Independent analysis: ■, by DSC; □, by titration.

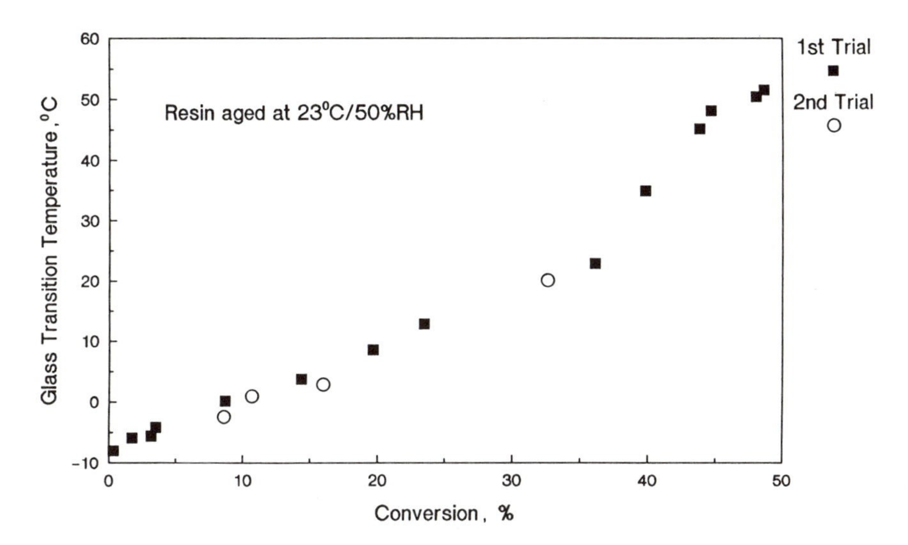

Figure 7. Relationship of glass transition temperature and conversion of an epoxy-amine resin.

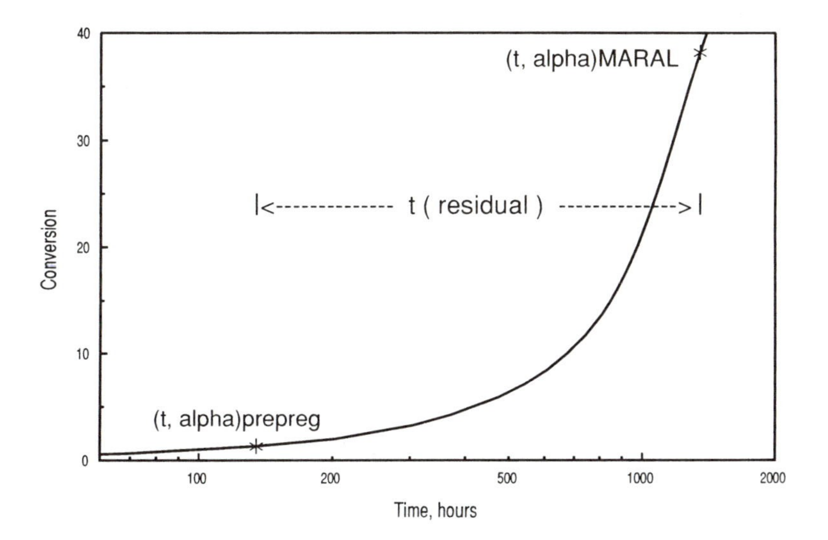

Figure 8. Application of time-temperature-conversion diagram for the prediction of residual out-time of the prepreg.

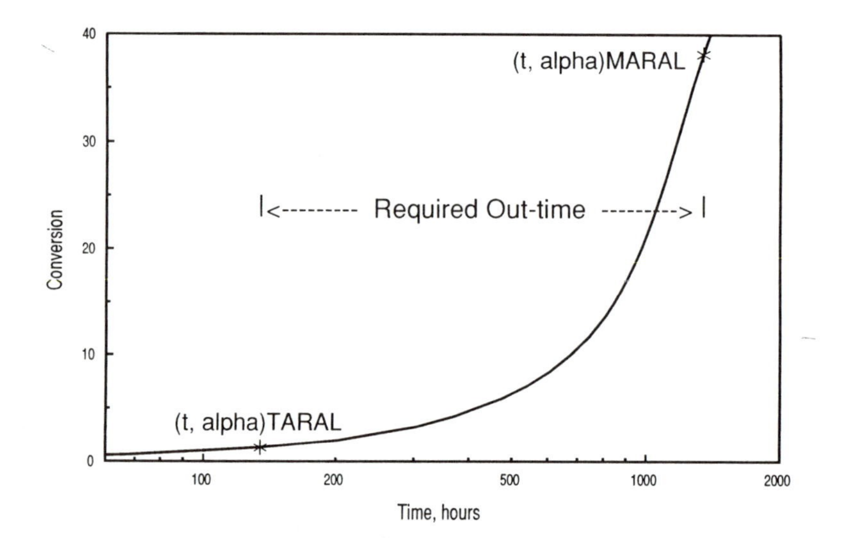

Figure 9. Application of time-temperature-conversion diagram for the prediction of maximum allowable resin advancement limit for process segments.

the process segments. If the allocated advancement is not tolerable for the process segment, the process conditions should be revised and modified.

Conclusions

The cure kinetics of an epoxy-amine resin has been followed using isothermal calorimetry. A high sensitivity calorimeter has proved to be valuable in following the low temperature reaction so that extensive extrapolation is not required.

Three conversion regions were identified, each dominated by a different mechanism. At low conversion, reaction was catalytic, due to impurities present, and autocatalytic, due to moisture and hydroxyl group reaction intermediates. In the medium region, the reaction was autocatalytic. At high conversion, the reaction was diffusion controlled.

The resin advancement depends on time and temperature. Its time-temperature-conversion relationship can be easily established via the modified Van Geel approach without any knowledge of the apparent frequency factor, which is conversion dependent.

The time-temperature-conversion relationship can be used to predict resin advancement, residual out-time of prepreg and to set the maximum advancement limit for the process segments (resin blending, filming and prepregging).

Acknowledgments

The author gratefully acknowledges the helpful suggestion and comments of Dr. R. B. Prime. The author also thank Dr. H. Jabloner for the research funding and Dr. A. A. Duswalt for technical discussion on the Van Geel approach. Thanks are also extended to J. F. Carre and A. J. Sosnowik for performing the calorimetric experiments, and Dr. P. K. Kuo and J. M. Hoffmann for the titration results.

Literature Cited

1. Prime, R. B. Chapter 5 in *Thermal characterization of Polymeric Materials,* Editor: E. A. Turi, Academic Press, **1981,** pp. 478-480.
2. Barton, J. M. *Brit. Polym. J,* **1979,** 11, 115.
3. Prime, R. B. *Proceedings 14th NATAS Conference,* **1985,** 137.
4. Garton, A. *J. Macromol. Sci.-Chem.,* **1989,** *A26,* 17.
5. Mertzel E.; Koenig, J. L. *Adv. Polym. Sci.,* **1985,** *72,* 75.
6. Antoon, M. K.; Koenig, J. E. *J. Polym. Sci., Chem. Ed.,* **1981,** *19,* 549.
7. Horie, K.; Hiura, H.; Sawada, M.; Mita, I.; Kambe, H. *J. Polym. Sci., Part A-1,* **1970,** *3, 1357.*
8. *Chang, S.-S. J. Thermal Anal.,* **1988,** *34, 135.*
9. *Jay, R. R. Anal. Chem.,* **1964,** *36,* 667.
10. Shechter, L.; Wynstra, J.; Kurkjy, R. P. *Ind. Eng. Chem.,* **1956,** *48,* 94.
11. Sourour, S.; Kamal, M. R. *Thermochim. Acta,* **1976,** *14,* 41.
12. Wisanrakkit, G.; Gillham, J. K. *J. Coating. Tech.,* **1990,** *62,* 35.
13. Hale, A.; Macosko, C. W.; Bair, H. E. Macormol., **1991,** 24, 2610.
14. Van Geel, J. L. C. *Thesis disertation,* **1969.**

RECEIVED January 8, 1992

Chapter 14

X-ray Fiber Diffraction Study of a Synthetic Analog of Cellulose

Poly[3,3-bis(hydroxymethyl)oxetane]

J. M. Parris and R. H. Marchessault

Pulp and Paper Center, Department of Chemistry, McGill University, Montreal, Quebec H3A 2A7, Canada

An investigation based on x-ray fiber diffraction and conformational analysis methods has provided a proposed chain conformation and crystalline structure for poly [3,3-bis(hydroxymethyl) oxetane]. The analysis of the fiber diagram led to a monoclinic unit cell of dimensions a = 11.85Å, b = 10.39Å, c (fiber axis) = 4.79Å, and γ = 90.0°. The probable space group is $P2_1$, where two monomers of P(BHMO) comprise the asymmetric unit and there are four monomers per unit cell. A planar zigzag backbone configuration was derived from a comparison of experimental and calculated fiber repeats. A systematic conformational analysis of the hydroxymethyl side groups identifies conformations with intra- and intermolecular hydrogen bonding. The initial refinement of the geometrical and packing parameters produced results that favor the extended sidechain conformation which allow only intermolecular hydrogen bonds.

Poly [3,3-bis(hydroxymethyl) oxetane], P(BHMO), is a member of the series of polymers with the general form $[\text{-O-CH}_2\text{-CR}_1\text{R}_2\text{-CH}_2\text{-}]_n$ (Figure 1). The first and chemically simplest of this series is polyoxetane, with R_1 and R_2 both being hydrogen atoms. The substitution of R_1 and R_2 by OH or CH_2OH groups introduces important changes in the properties of this family of polymers. The effect of symmetrical pendant side groups on the conformation and their influence on chain-chain interactions is of major interest in the study of this group of polymers.

The polyoxetanes are semicrystalline polymers and the crystalline structure of the first three members, polyoxetane (PTO) (*1*), poly (3,3-dimethyl oxetane) P(DMO) (*2-4*) and poly (diethyl oxetane) P(DEO) (*5*) have been described previously. However, no crystallographic information about poly [3,3-bis(hydroxymethyl) oxetane] P(BHMO) has been reported.

P(BHMO) is an interesting synthetic analog of cellulose because of its substituent hydroxyl groups and oxygen in the backbone. It also has a very

0097–6156/92/0496–0182$06.00/0

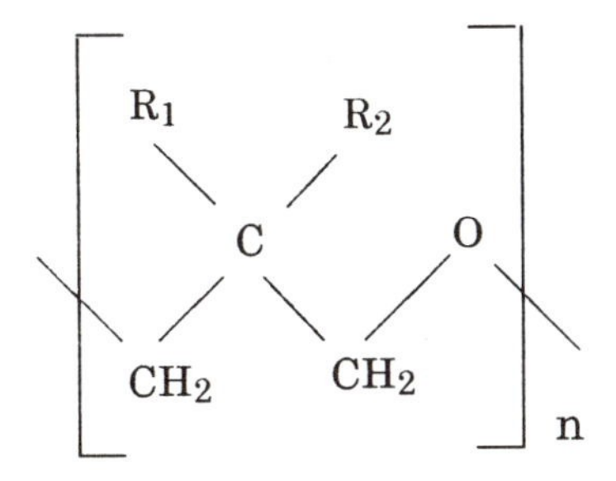

Figure 1. General composition of the 3,3 substituted polyoxetane family

high melting point (314°C) and is insoluble in most organic solvents. There are important differences compared to cellulose, namely: the hydroxyls are all primary, the main chain oxygens are ethers and not acetals and there are no asymmetric carbons.

In previous work (*6*), the thermal transitions and the influence of the crystallization temperature on the melting behavior were studied for P(BHMO) using differential scanning calorimetry (d.s.c.) and X-ray powder diffraction. The occurrence of one melting peak in the d.s.c. curves indicates that only one crystal form exists and this is confirmed by X-ray analysis (*6*).

The present work is concerned with the structural study of poly [3,3-bis(hydroxymethyl) oxetane] and with the comparison of the chain conformation of this polymer with other members of the series of polyoxetanes. In addition, this polyether with hydroxymethyl substituents presents hydrogen bonding opportunities, intra- and inter-molecular, reminiscent of cellulose, but with the simplification that only primary hydroxyls are present. Similarly, like cellulose, P(BHMO) can be derivatized.

Experimental

Material and Specimen Preparation. P(BHMO) was prepared by E.J. Vandenberg (*6*). The polymer was synthesized by polymerizing the trimethylsilylether of 3,3-bis(hydroxymethyl) oxetane with i-Bu_3Al-0.7 H_2O cationic catalyst at low temperatures, followed by acid promoted hydrolysis. Reduced viscosity measurements indicated that P(BHMO) of high molecular weight was obtained (η_{inh} up to 5.2).

Films and fibers suitable for X-ray measurements were prepared by solution casting of P(BHMO)-diacetate in $CHCl_3$, air drying, then cold drawing. Hydrolysis in 1.0M NaOH/EtOH then yielded P(BHMO). Density was measured at 25°C by the air pycnometer method and found to be 1.28g/mL(*6*).

X-ray Fiber Diffraction. X-ray diffraction photographs were recorded under vacuum using Ni filtered CuK_α radiation. Pinhole collimation and a flat-plate, "Statton" camera (3.0, 5.0 and 17 cm film to sample distances) were used to record the X-ray patterns. The specimens were dusted with finely powdered crystalline NaF to calibrate the diffraction patterns. From these X-ray patterns, the *d*-spacings of each different spot and l ayer line assignments could be obtained accurately and a unit cell was derived. Specimens were tilted towards the incident beam at appropriate angles to register each *00l* reflection at maximum intensity on the meridian and any systematic absences for *00l* reflections determined.

Intensities were determined by visual comparison. Correction factors for Lorentz, polarization, multiplicity were applied.

Results & Discussion

The x-ray fiber diffraction pattern of P(BHMO) is well oriented (Figure 2a) but shows poor order (only ten reflections) similar to cellulose II (rayon). Like cellulose II there is a strong second layer line reflection. P(BHMO) diacetate showed great improvements in the order and twenty six reflections were observed (Figure 2c). From the FTIR spectra the two hydroxyl groups in P(BHMO) displayed overlapping stretching bands between 3650 - 3100 cm^{-1} with a sharp maximum at 3359 cm^{-1} . The absence of absorption near 3650 cm^{-1} indicated that there were no free hydroxyls in solid P(BHMO) and all assignments must be in terms of hydrogen-bonded groups. Band assignments proposed by Marchessault and Liang for cellulose II were used (*8*). They found that OH intermolecular hydrogen bonding occurred between 3175 - 3350 cm^{-1} as broad bands while intramolecular hydrogen bonding involving sharp bands occurred between 3450 - 3490 cm^{-1}. It would seem that samples of P(BHMO) may consist of intermolecular hydrogen bonded hydroxyls and perhaps some intramolecular hydrogen bonds. This is evidenced from the broad band centered at 3350 cm^{-1}. The relative intensities of OH and CH stretching bands are very similar to the cellulose IR spectra.

Unit Cell Determination. Ten reflections were observed and indexed by a monoclinic cell with parameters a = 11.85Å, b = 10.39Å, c (fiber axis) = 4.79Å, and γ= 90.0°. An agreement factor between observed and calculated spacings of 2.4% was found. By assuming that there are four monomers per unit cell, the density was calculated as 1.33 g/ml. This value is acceptable in comparison with the observed one (1.28 g/ml). Systematic absences for odd valued *00l* reflections were determined from tilted fiber diagrams (Figure 2b) and indicated the possible space group $P2_1$ which contains two general positions. From modeling (below) it was determined that P(BHMO) adopts a planar zigzag backbone configuration. For the $P2_1$ space group this means the molecule cannot sit on a symmetry element and the asymmetric unit contains two monomers.

Molecular Conformations. Calculations were performed using the commercial molecular modeling system SYBYL 5.21 (Tripos Associated, St. Louis, MO) using a VAX 11/750. The overall conformation of P(BHMO) was separated into backbone and sidechain conformations.

A stepwise procedure was employed for the determination of the allowed low-energy conformations of P(BHMO). The first step involved constructing the monomer of P(BHMO) including associated hydrogen atoms (Figure 3). Only hydroxyl hydrogens are shown for clarity.

The monomer's energy was then minimized using the versatile energy minimization program called MAXIMINN. This uses a conjugate gradient approach to determine the energy minimum of a given structure. The potential energy is calculated from:

a

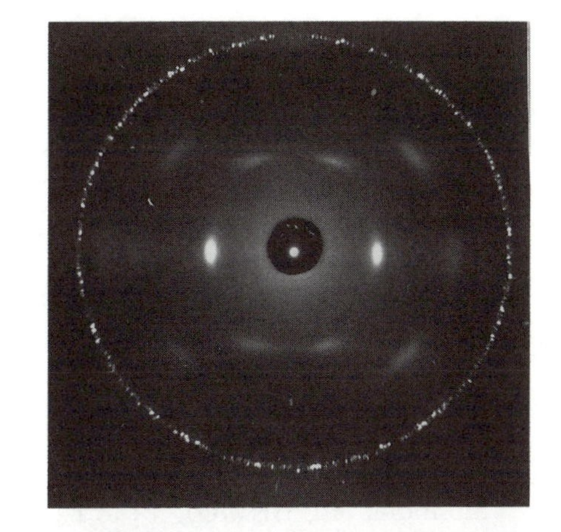

b

c

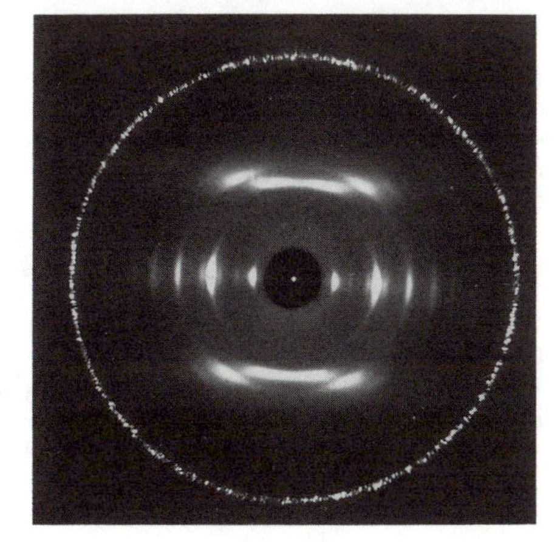

Figure 2. Fiber photographs of (a) P(BHMO) (b) tilted P(BHMO) and (c) P(BHMO) diacetate.

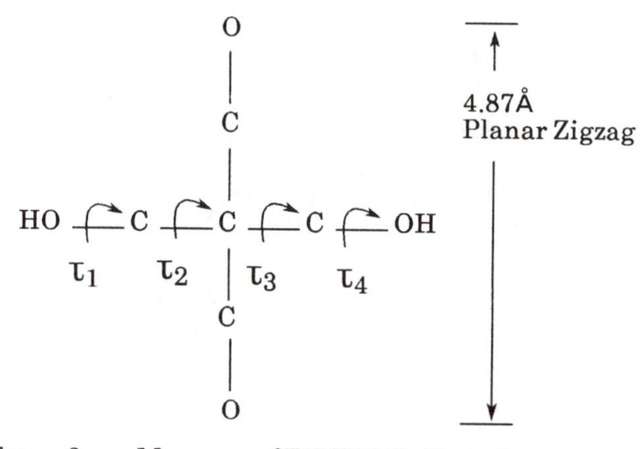

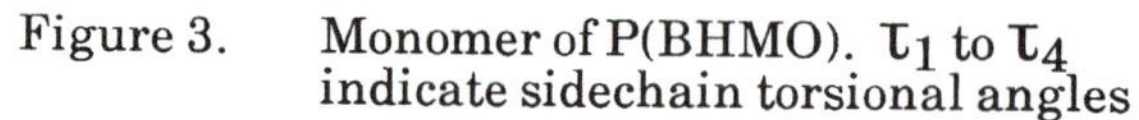
Figure 3. Monomer of P(BHMO). τ_1 to τ_4 indicate sidechain torsional angles

$$\varepsilon = \omega_{str}\,\varepsilon_{str} + \omega_{ang}\,\varepsilon_{ang} + \omega_{tor}\,\varepsilon_{tor} + \omega_{ele}\,\varepsilon_{ele} + \omega_{VDW}\,\varepsilon_{VDW}.$$

where the ω 's represent constants and ε the energy terms for bond-stretching energy (str), angle-bending energy (ang), torsional contact energy (tor), electrostatic energy (ele) and van der Waals energy (VDW) (including hydrogen bonding). Bond lengths and angles (Table I) generated from the program were comparable with values obtained from the *Tables of Interatomic Distances and Configurations* (7).

Table I. Structural Parameters P(BHMO)

Bond Length (Å)		Bond Angle (deg.)	
C - C	1.54	C - C - C	109.5
C - O	1.44	C - C - O	111.0
O - H	0.95	C - O - C	113.0

Backbone Conformation. The minimized monomer backbone torsional angles were found to be 180° (planar zigzag) with a corresponding calculated fiber period of 4.87Å. This value agrees well with the observed fiber period of 4.79Å. Therefore, P(BHMO)'s backbone conformation was determined to be planar zigzag.

Hydroxymethyl Conformations. The identification of permissible hydroxymethyl conformations involved the use of a systematic conformational analysis program called SEARCH . This program checks for VDW contacts among the nonbonded atoms by scanning all possible torsional angles around the rotatable bonds. Conformations were eliminated based solely on unfavorable VDW interactions. Three forms of VDW interactions were monitored: nonbonded interactions, which allow for thermal vibrations of atoms, 1,4 interactions, and hydrogen bond interactions. Four bonds (τ_1 to τ_4) were surveyed in 30° increments over 360°.

SYBYL scanned 10^4 possible combinations and required 3 hrs. of CPU time. Four solutions were found (Figure 4). These were then minimized using the MAXIMINN program, allowing all the atoms to relax. The results for the four models are given in Table II.

Table II. Torsional Angles of Side Chains P(BHMO)

No.	Energy Kcal/mole	τ_1	τ_2	τ_3	τ_4	Fiber Repeat
I	4.9	175.6	59.1	54.0	-180.0	4.87
II	4.9	176.7	-63.1	53.6	53.6	4.87
III	6.3	-60.5	60.5	-179.9	179.9	4.87
IV	6.6	179.5	-179.5	53.0	-53.0	4.87

Structure Determination. The intensities obtained from the diffraction diagrams could be resolved into 10 individual intensity groupings. A total of 36 crystallographic planes were predicted to contribute to these intensities as calculated from the unit cell. An additional 38 reflections were predicted but not observed, and their intensities were estimated at one half of the minimum observable intensity at the corresponding diffraction angle. Therefore, a total of 74 individual reflection intensities could be used in the structure refinement.

Initial Refinement Each model was packed into the $P2_1$ space group and its energy minimized, using as variables only chain rotations about the c-axis and translation along the *a* and *b* axis. No water molecules were included in the unit cell and an isotropic temperature factor was arbitrarily set at 5. The primary refinement criteria was the usual residual:

$$R = \{ \Sigma \mid |F_i|_o - |F_i|_c \mid \div \Sigma |F_i|_o \}$$

where $|F_i|_o$ and $|F_i|_c$ are the observed and calculated structure factors. Observed intensities, I_o, were estimated visually and given a corresponding numerical value. They were then corrected for Lorentz and polarization factors. Reflections with multiple indices were decomposed according to the ratios of I_c (calculated intensity) values of each unique *hkl* component. The most intense reflection has an interplanar spacing of 5.2Å (figure 2b) and yields multiple indices of (200) (020) (210) and (120). Model II was immediately discarded as a possible solution because of the high residual (R = 0.52). Similarly, Models I and IV are unlikely as they yield residuals of 0.35 and 0.39, respectively. Model III seems most probable

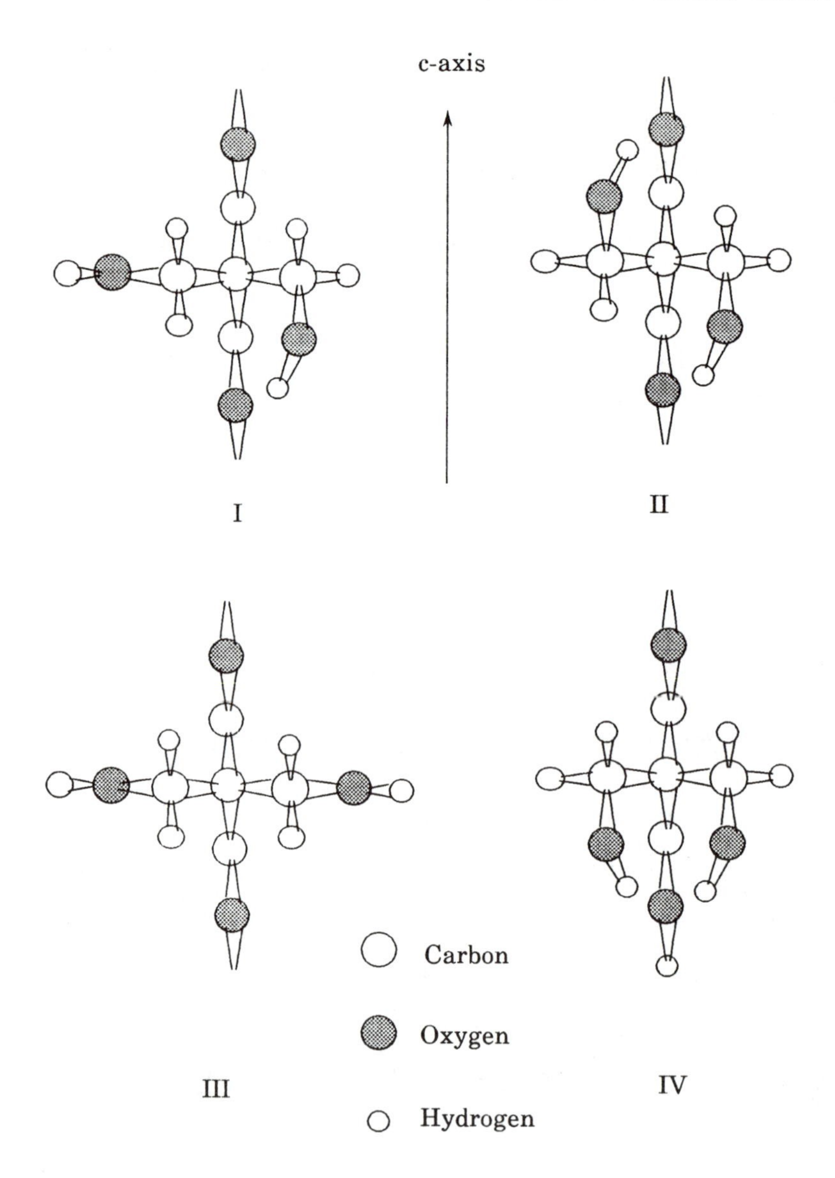

Figure 4. Hydroxymethyl conformations of P(BHMO). Planar zigzag backbone in all cases.

with R = 0.25, and, in addition, structurally it is very similar to P(DEO) which also has extended side chains. Further refinement of Model II yielded a residual of 12.7%. In Figure 5 we see a schematic baseplane projection of four monomers of P(BHMO) in the $P2_1$ space group. The

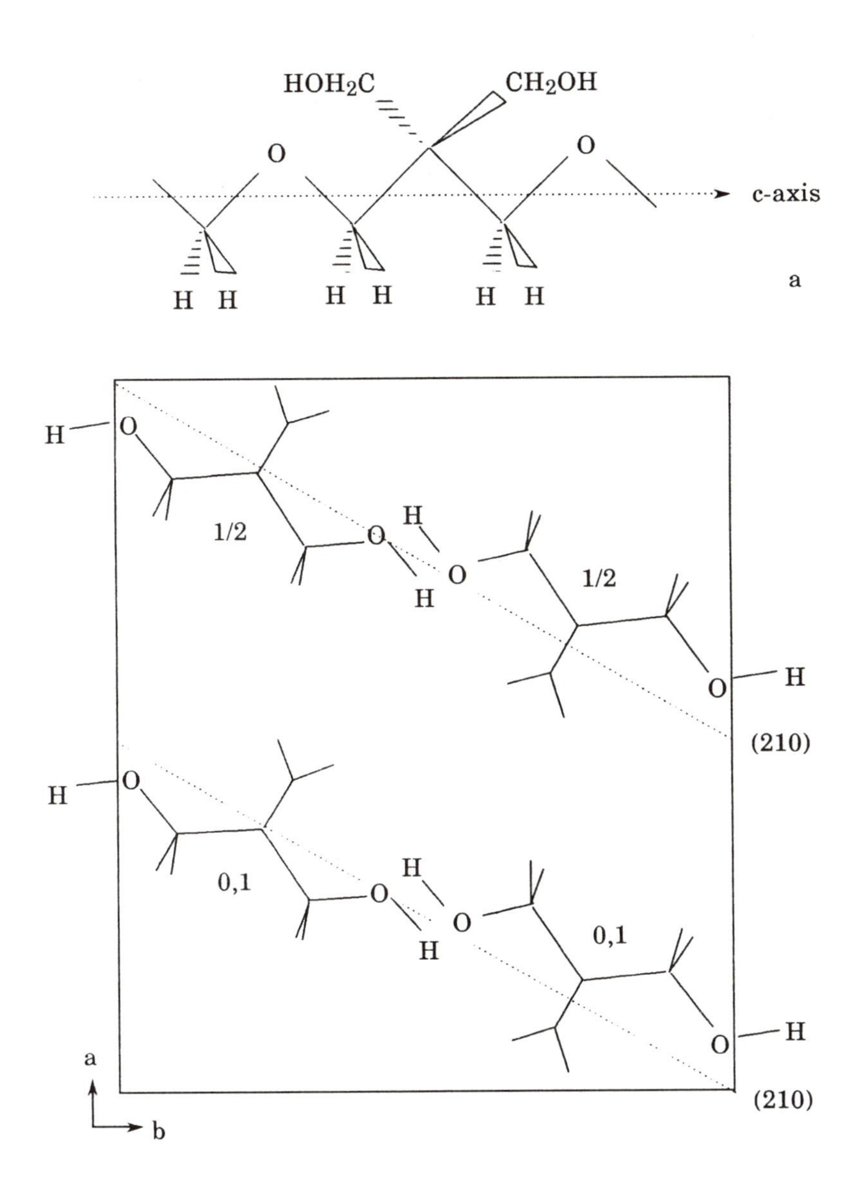

Figure 5. Schematic baseplane projection of Model III of P(BHMO) unit cell. The four monomers are arranged so that a pair (i.e. 1/2-1/2 or 0,1-0,1) comprise the assymetric unit.

notations (0,1) and (1/2) indicate the translations of the sidegroups along the *c*-axis. The asymmetric unit contains two monomer's . The oxygen/oxygen distance between closest hydroxyl groups was 2.9 Å at an angle of approximately 90° while the O---H hydrogen bond distance was 2.69Å. The hydroxyl groups within an asymmetric unit take part in intermolecular hydrogen bonding and form a four member cyclic ring. The fact that Model III has both hydroxyl groups involved in intermolecular hydrogen bonding accounts for the high melting point (314°C). Also, one can see that the (210) plane contains most of the electron density and therefore is expected to give the most intense reflections. The final model was optimized by using a least squares refinement program against $|F_o|^2$ while minimizing the hydrogen bonding energy.

Conclusions

The x-ray results and preliminary structure analysis of P(BHMO) allows for comparison with other members of the oxetane family. In the cases of PTO, P(DMO) and P(DEO), all were found to have a planar zigzag backbone conformation. P(DEO) is roughly the same size as P(BHMO) and has comparable unit cell dimensions. The ethyl sidegroups are extended and take part in increased interactions with the backbone as well as improved intermolecular forces. When ethyl is replaced by a hydroxymethyl group, where the intermolecular interactions are the much stronger hydrogen bonding, the close packing achieved by the chains allows the van der Waals forces to act cooperatively to provide additional stability to the crystallite. The melting temperatures of PTO and P(DEO), 20°C and 57°C, respectively, are considerably lower than for P(BHMO). The replacement of the ethyl group by polar groups such as ethoxymethyl in P(BEMO) or azidomethyl groups in P(BAMO) result in only modest increases in the melting temperature of 92°C and 102°C . The hydroxymethyl groups in P(BHMO) give improved interactions in the form of hydrogen bonding between chains in the crystal lattice which provides strong cohesion and accounts for the high melting point compared to other oxetanes.

Acknowledgments

Thanks are due to Prof. E.J. Vandenberg for providing the motivation and samples. Crystallographic support and advice from Prof. F. Brisse of Université de Montréal was most helpful. Finally, the Clinical Research Institute of Montreal provided molecular modeling support under MRCC Grant (MR-10131). Xerox and NSERC provided financial support.

Literature cited

1. Tadokoro, H., Takahashi, Y., Chatani, Y. and Kakida, H. *Makromol. Chem.* **1967**, *109*, 96
2. Kakida, H., Makino, D., Chatani, Y., Kobayashi, M. and Tadokoro, H. *Macromolecules* **1970**, *3*, 569
3. Takahashi, Y., Osaki, Y., and Tadokoro, H. *J. Polym. Sci., Polym. Phys. Edn.* **1980**, *18*, 1863

4. Takahashi, Y., Osaki, Y., and Tadokoro, H. *J. Polym. Sci., Polym. Phys. Edn.* **1981**, *19*, 1153
5. Gomez, M.A., Atkins, E.D.T., Upstill, C. Bello, A., and Fatou, J.G. *Polymer* **1988**, *29*, 224
6. Vandenberg, J., Mullis, J.C., and Juvet Jr., R.S. *J. Poly. Sci. Part A: Polymer Chemistry*, **1989**, *27*, 3083
7. *Tables of Interatomic Distances and Configurations in Molecules and Ions*; Bowen, H.J.M., Sutton, L.E., Special Publication No. 11, The Chemical Society: London, **1958**.
8. Liang, C.Y., Marchessault, R.H., *J. Polymer Sci.*, **1959**,*37*, 385
9. Murphy, C.J., Henderson, G.V.S., Murphy, E.A., Sperling, LH., *Polymer Engineering and Science*, **1987**,*27*, 781

RECEIVED August 21, 1991

Anionic Polymerization

ORDINARY ANIONIC POLYMERIZATION of many vinyl monomers and epoxides and other ring-opening polymerizations has been widely practiced for many years. Anionic catalysts do vary widely in their effectiveness and performance depending on monomer type and other factors. Living anionic polymerization has been a very important research area since the mid-1950s as a result of the pioneering work of Szwarc and collaborators. The exact nature of the anionic catalysts plays an important role in determining the structure of the final products as well as determining important mechanism details (such as whether living or not), transfer reactions, and stereoregularity where possible. In this section, Inoue describes his unique metalloporphyrin (Al and Zn) catalysts that work on a variety of vinyl monomers and ring-opening monomers, usually characterized by a living or the novel concept of an "immortal" polymerization. The mechanism of this unusual system is discussed by Inoue, but it is unclear if the exact mechanism is known. Thus, we have arbitrarily placed it in this section because it behaves so differently from most coordination catalysts. Spassky et al. present results on some aluminum complexes of Schiff bases as catalysts that appear somewhat analogous to the Inoue systems and are useful for the oligomerization of oxiranes. Matyjaszewski describes his studies on catalysts to control the structure of inorganic backbone polymers. Specifically, new anionic routes to polyphosphazenes are presented together with the anionic polymerization of strained cyclosilanes as "the first possibility of tacticity control in polysilanes."

Chapter 15

Metalloporphyrin Catalysts for Control of Polymerization

Shohei Inoue

Department of Synthetic Chemistry, Faculty of Engineering, University of Tokyo, Hongo, Bunkyo-ku, Tokyo 113, Japan

Aluminum and zinc porphyrins are excellent initiators for the living ring-opening and addition polymerizations of a wide variety of monomers such as epoxide, episulfide, lactone, and (meth)acrylic esters. In some cases, the polymerization in the presence of a protic compound proceeds with a chain transfer reaction, but the molecular weight distribution of the polymer remains narrow. Hence, a novel concept of 'immortal polymerization' is presented. Taking advantage of the wide applicability, various block copolymers and end-reactive polymers can be synthesized. The copolymerizability of epoxide and episulfide can be controlled by visible light in the copolymerization with a zinc porphyrin. Stereochemistry of the ring-opening of epoxide in the polymerization provides a novel insight into the polymerization mechanism.

The Vandenberg catalyst, which is composed of trialkylaluminum, water, and acetylacetone, is a well-known, excellent system for the polymerization of epoxide to give a polymer of high molecular weight(*1*). A related catalyst is the dialkylzinc-water system which also gives high polymer from epoxide. The formation of the product with very high molecular weight indicates that only a fraction of the metallic species in the catalyst is effective for polymerization reaction. This is due to the highly associated nature of these systems, involving metallic species of various coordination states. In such systems, a ligand bound to the metal (e. g. , initiating and propagating species of polymerization) can have various reactivities. The broad molecular weight distribution of polymers obtained by these systems corresponds to the existence of reactive species or catalytic sites with various reactivites.

Recently, we have found that metalloporphyrins of aluminum

0097–6156/92/0496–0194$06.00/0

and zinc are excellent catalyst for the living polymerization of epoxide. The molecular weight distribution of the polymer is narrow, and the molecular weight can be controlled by changing the monomer-to-catalyst ratio. Metalloporphyrin is characterized by a metal surrounded by a rigid macrocycle, where a high extent of association is not likely. Thus, all metallic species exhibit very similar reactivity to each other, resulting in a homogeneity in the initiation and propagation reactions of all polymer molecules. Living polymerization with metalloporphyrin catalyst has been extended to other monomers than epoxide, such as episulfide, lactone, and acrylic monomers.

First a brief account will be made of the polymerization with metalloporphyrin catalyst. Description of some of the recent advances on the polymerization of epoxides will follow.

Wide Applicability to Ring-Opening and Addition Polymerizations

Some aluminum porphyrins such as (TPP)AlX (**1**, X=Cl, OR, OAr, O_2CR, etc.; TPP: tetraphenylporphinato) are excellent catalysts for the living ring-opening polymerization of epoxide(*2*).

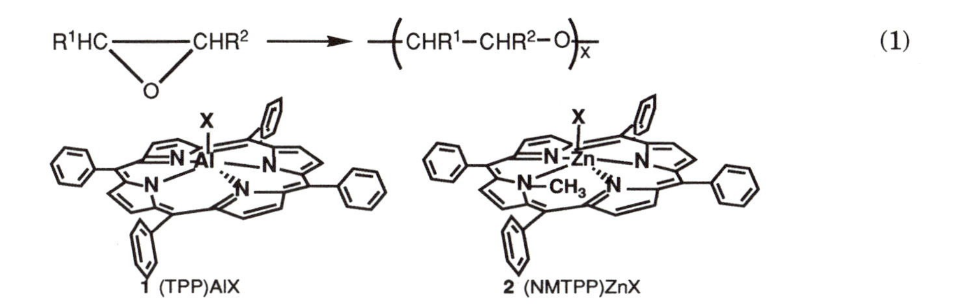

Similar aluminum porphyrins, (TPP)AlX, X= Cl, O_2CR, are also effective for the living polymerization of β - lactone (*3*). On the other hand, as for δ – and ε- lactones, (TPP)AlOR is an active initiator for the living polymerization, but not (TPP)AlCl (*4*).

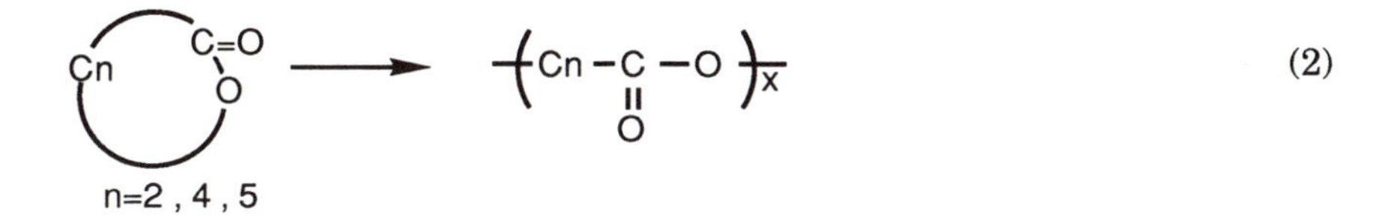

(TPP)AlOR is also effective for the living polymerization of lactide (*5*).

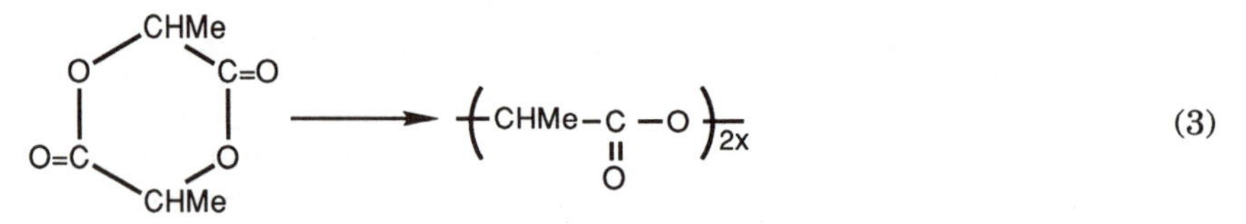

As to episulfide, aluminum porphyrins including (TPP)AlSR are not effective, while a zinc porphyrin (NMTPP)ZnSR (**2**, X= SR) is effective for the living polymerization(*6*).

$$CH_2\text{—}CHMe\text{ (—S— bridged)} \longrightarrow \left(CH_2\text{-}CHMe\text{-}S\right)_x \qquad (4)$$

Aluminum porphyrins when coupled with an appropriate quaternary salt (R_4PX, R_4NX) are effective for the living, alternating copolymerization of epoxide and carbon dioxide (*7*) or cyclic acid anhydride (*8*).

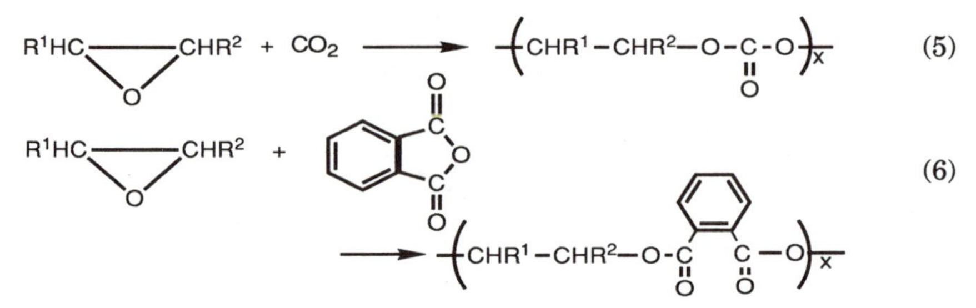

Aluminum porphyrins are also effective for the living addition polymerizations of methacrylic esters (*9*), acrylic esters (*10*), and methacrylonitrile (*11*), where (TPP)AlR, (TPP)AlSR, and (TPP)Al(O-C=C) (enolate) are active catalysts.

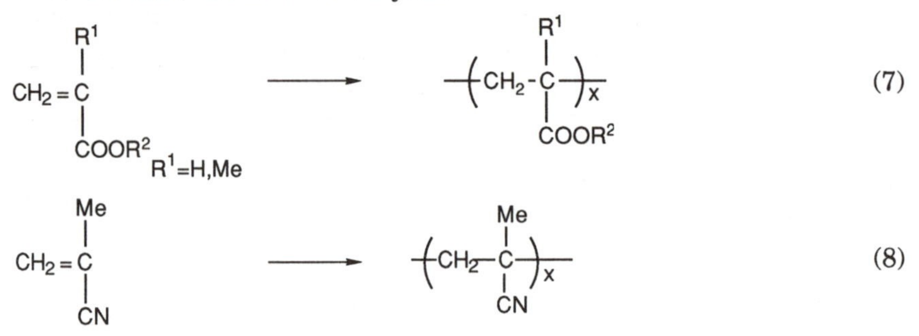

All these polymerization reactions proceed by the repeated insertions of the monomer to the (porphinato)metal-axial ligand (X) bond, and demonstrate the interesting reactivity of metalloporphyrins. First, high nucleophilic reactivity of the metal-X bond is of primary importance. Second, in some cases, Lewis acidity of the metal plays a decisive role. And third, the effect of visible light is most characteristic of the porphyrin compound.

Immortal Polymerization

Immortal polymerization is a novel concept we have proposed (*12*). In immortal polymerization, polymer with narrow molecular weight distribution is formed even in the presence of chain transfer reaction, being essentially different from 'living' polymerization.

For example, the polymerization of epoxide catalyzed by aluminum porphyrin proceeds readily even in the presence of alcohol. The molecular weight of the polymer decreases, or the number of the polymer molecules increases, with the increasing amount of alcohol, indicating the participation of alcohol as a chain transfer agent. Of particular interest is the fact that the molecular weight distribution is retained narrow even under these conditions.

In the polymerization system containing alcohol, the rapid and reversible exchange takes place between the growing species, a (porphinato)aluminum alkoxide, and alcohol.

$$\text{(TPP)AlOR} + \text{R'OH} \rightleftharpoons \underset{\mathbf{3}}{\text{ROH}} + \underset{\mathbf{4}}{\text{(TPP)AlOR'}} \qquad (9)$$

The forward reaction is the usual chain transfer reaction, to give the 'dead' polymer **3** and the new initiating species **4**. In contrast to the usual chain transfer reaction, the polymer molecules which appear once 'dead' can 'revive' to the growing species (immortal), since the above reaction is reversible. When this reaction is much faster than the propagation reaction, the molecular weight distribution of the polymer becomes narrow, as actually observed.

In living polymerization, the number of the polymer molecules formed is, at most, equal to that of the initiator. In contrast, immortal polymerization can give the polymer of narrow molecular weight distribution, with the number of polymer molecules more than that of the initiator. Immortal polymerization by metalloporphyrin has been extended to epoxide, episulfide, and lactones, in the presence of an appropriate protic compound such as alcohol, thiol, and carboxylic acid. Immortal polymerization provides new, very useful methods for the syntheses of block copolymers and end reactive polymers with controlled molecular weight.

Block Copolymers

By the virtue of the wide applicability of aluminum and zinc porphyrins as initiators for the living and immortal polymerizations of various monomers, a wide variety of block copolymers with controlled chain lengths can be synthesized (Table I).

TABLE I. Examples of Block Copolymers Obtained with Aluminum and Zinc Porphyrins as Initiators

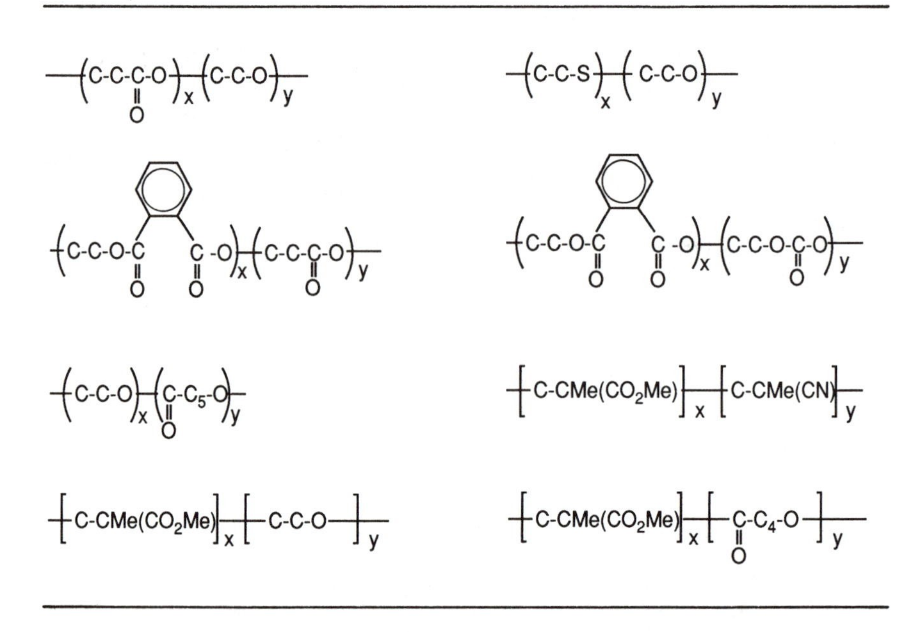

Of particular interest is the synthesis of block copolymers from methyl methacrylate and propylene oxide or δ-valerolactone by the direct, successive addition and ring-opening polymerizations (*13*)

End-Reactive Polymers

A variety of end-reactive polymers and oligomers can be obtained taking advantage of the wide applicability of aluminum porphyrin to living (*14*) and immortal polymerizations (Table II).

Table II. Examples of End-Reactive Polymers Obtained by the Immortal Polymerization of Epoxide or Lactone with Aluminum Porphyrin as Initiator

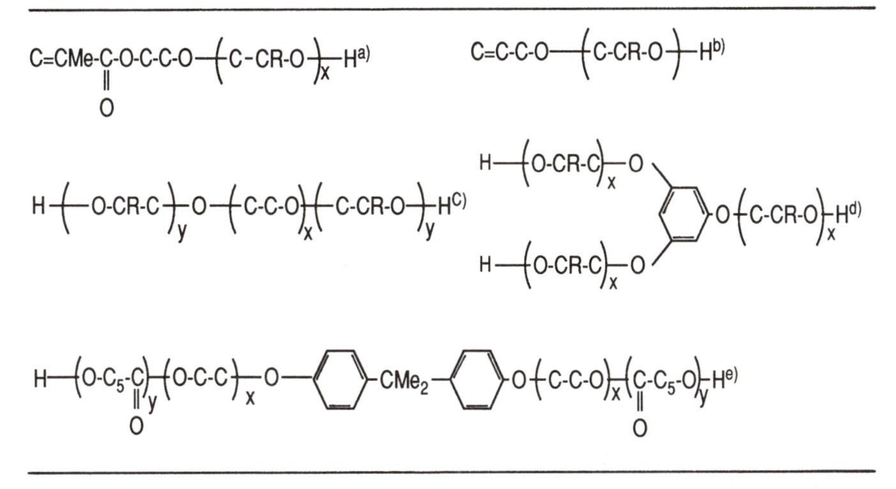

Protic Compound : a) 2- hydroxyethyl methacrylate, b) allyl alcohol, c) polyethylene glycol, d) phloroglucinol, e) Bisphenol-A.

Control of Copolymerizability by Light. Epoxide-Episulfide Copolymerization with Zinc Porphyrin

Although there are a number of examples of photopolymerization, in most cases light acts only to generate the initiating species and the propagation proceeds regardless of irradiation. In this respect, the living polymerization of methacrylic esters with aluminum porphyrin (*9*) and that of epoxide with zinc porphyrin (*15*) are of particular interest, since not only the initiation step but also the propagation steps are accelerated by visible light. On the other hand, the effect of visible light is not observed in the polymerization of episulfide with zinc porphyrin (*6*). Thus, it is of much interest to examine the effect of light on the copolymerization of epoxide and episulfide with zinc porphyrin, which is expected to involve cross propagation reactions.

Copolymerization of propylene sulfide (PS) and ethylene oxide (EO) with a zinc porphyrin (NMTPP)ZnSPr (**2**, $X=SC_3H_7$) was investigated in benzene under irradiation by visible light (>420nm) or in the dark (*16*). As shown in Figure 1, the reactivity of episulfide is much higher than epoxide at room temperature in the dark, as usually observed for conventional initiators. Of interest to note is the fact that the reactions of both epoxide and episulfide are accelerated by light, in contrast to the homopolymerization of episulfide (see above). An elevated temperature also accelerates the reaction of both monomers as expected.

However, it should be emphasized that the effects of light and temperature are essentially different. As illustrated in Figure 2, the relative reactivity of the monomers does not change by elevating the temperature, but visible light enhances the reactivity of epoxide resulting in a decrease in the difference in reactivities between these monomers. Increase in the contents of cross sequences (PS ⟶ EO and EO ⟶ PS) in the copolymer by light was confirmed by ^{13}C-NMR analysis. Thus, cross propagation reactions are considered more accelerated by light than homo propagation reactions. It is possible by the present system to synthesize epoxide-episulfide random copolymer which is otherwise difficult to obtain. To our knowledge, the present result is the first example of the control of copolymerizability by light.

Stereochemical Aspect of Ring-Opening of Epoxide

When compared with other organometallic catalyst systems for the polymerization of epoxide such as $AlEt_3$-H_2O and $ZnEt_2$-H_2O systems with highly associated structures probably containing Al-O-Al or Zn-O-Zn linkages, Al and Zn porphyrins are characterized by their isolated metal atoms surrounded by a large, planar, rigid ring. In the polymerization of epoxide with Al-alkyl and Zn-alkyl based systems, the configuration of carbon atom at the C-O bond to be cleaved has been reported to be inverted. Since the reaction taking place on a single metal atom appears more likely to lead to retention of configuration in the ring scission of epoxide, a mechanism involving two metal atoms has been postulated to explain inversion (*1*). In this respect, it is a subject of much interest to investigate the stereochemistry of ring opening of epoxide on metalloporphyrin(*17*).

Previously, we observed the inversion of configuration in the ring opening of 2,3-butene oxide in the polymerization with (TPP)AlCl (**1**, X=Cl), by the cleavage of the polymer to the glycol units (*1*). In order to explain the observed stereochemistry, a mechanism involving two Al-porphyrin molecules is considered possible, one of which acts as a Lewis acid to accommodate the monomer (Scheme 1). In fact, a similar mechanism involving two Al-porphyrin molecules has been proposed for the polymerization of δ-lactone (*4*).

On the other hand, in the polymerization of epoxide with zinc N-methylated porphyrin, the back side of the axial ligand (initiating and propagating species) is protected by the methyl group, and the above mechanism is excluded. In this respect, it is of particular interest to examine the stereochemistry of ring opening of epoxide with zinc N-methylated porphyrin. In the present study, the reaction of 2,3-butene oxide (BO) with (NMTPP)ZnX (X = SPr, OMe) was examined in the presence of an excess of propanethiol or methanol, that is, under the conditions of immortal polymerization, in order to obtain unimeric addition product **6** corresponding to the structural unit of the polymer in initiation and propagation steps.

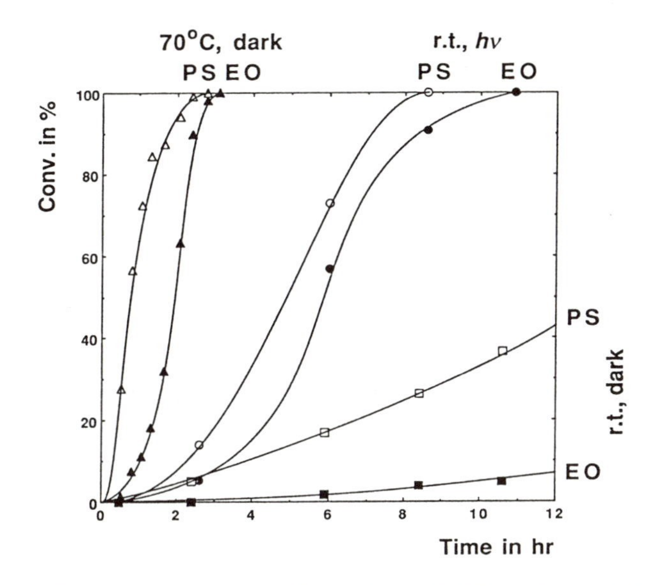

Figure 1. Copolymerization of Propylene Sulfide (PS) and Ethylene Oxide (EO) with (NMTPP)ZnSPr. $[PS]_0 / [EO]_0 / [Zn]_0 = 50/50/1$ in C_6D_6. Time-Conversion Profile.

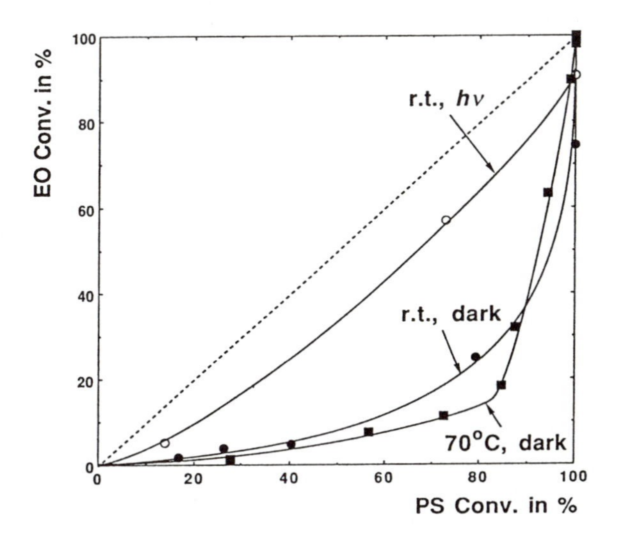

Figure 2. Copolymerization of Propylene Sulfide (PS) and Ethylene Oxide (EO) with (NMTPP)ZnSPr. $[PS]_0 / [EO]_0 / [Zn]_0 = 50/50/1$ in C_6D_6. Relative Reactivity. Adapted from ref. 16.

$$\text{MeCH}\underset{\text{O}}{\text{—}}\text{CHMe} + (\text{NMTPP})\text{Zn}-\text{X} \longrightarrow (\text{NMTPP})\text{Zn}-\text{O}-\text{CHMe}-\text{CHMe}-\text{X} \quad (10)$$

5

$$5 + \text{H}-\text{X} \longrightarrow (\text{NMTPP})\text{Zn}-\text{X} + \text{HO}-\text{CHMe}-\text{CHMe}-\text{X} \quad (11)$$

X = SPr
OMe

6

As summarized in Table III, cis and trans isomers of 2,3-BO exclusively gave threo and erythro products respectively in all cases examined, indicating the inversion of configuration in the ring opening.

Thus, the mode of ring opening of epoxide on N-methylated zinc porphyrin is the same as that in the reaction on Al-porphyrin, in spite of the presence of a methyl group shielding the other side of axial ligand. In order to explain inversion, another type of participation of two Zn-porphyrin molecules may be considered (Scheme 2, A). Otherwise, a mechanism involving a single Zn-porphyrin can be a possibility in which the Zn-X bond becomes so loosened or nearly dissociated in transition state (Scheme 2, B).

Experimental

Aluminum complexes of tetraphenylporphyrin (TPP)AlX (**1**) can be readily prepared by the reaction of an organoaluminum compound with tetraphenylporphyrin $(TPP)H_2$ as shown in equation (12).

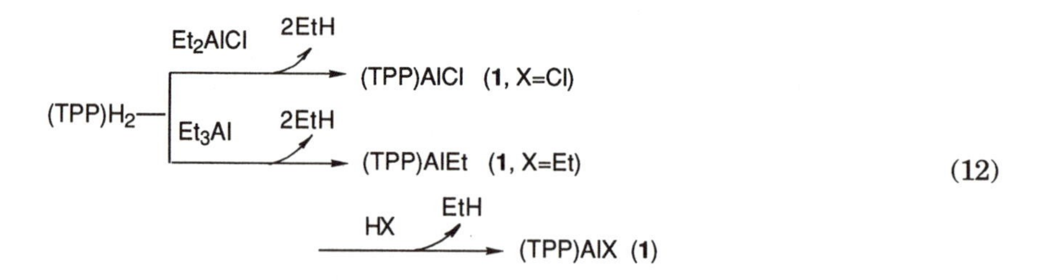

(12)

The reaction proceeds rapidly and quantitatively at room temperature, to give (TPP)AlCl from Et_2AlCl(*2*) and (TPP)AlEt from $AlEt_3$, respectively. The alkyl complex such as (TPP)AlEt can be converted to alkoxide (*18*), phenoxide (*19*), carboxylate (*18*), etc. by the reaction with the corresponding protic compound, respectively. The zinc porphyrin (NMTPP)ZnX (**2**) can be prepared similarly by the reaction of N-methyltetraphenylporphyrin with dialkylzinc to give (NMTPP)ZnR, followed by the reaction with a protic compound such as thiol(*6,15*).

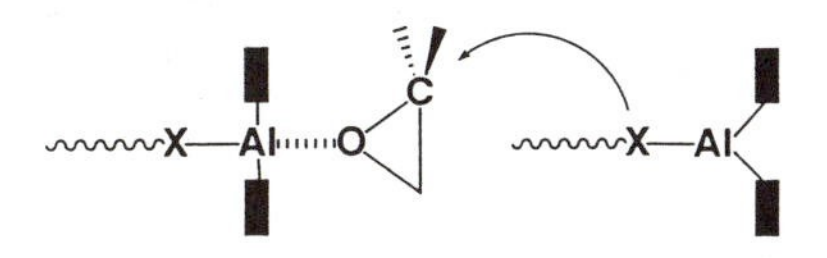

Scheme 1

Table III Ring-Opening Reaction of 2,3-BO with (NMTPP)ZnX (2) / HX System

X	2,3-BO	$[Zn]_0/[HX]_0/[BO]_0$		temp., °C	time, day	conv., %	product	config.
SPr	cis	1 / 0 / 0.5	*hv*	35	9	64	threo	inversion
	trans				19	52	erythro	inversion
	cis	1 / 50 / 25	*hv*	35	11	100	threo	inversion
	trans	1 / 100 / 50			11	17	erythro	inversion
OMe	cis	1 / 50 / 25	*hv*	35	33	3.2	threo	inversion
	trans				58*	17	erythro	inversion
OMe	trans	1 / 50 / 25	dark	80	1	43	erythro	inversion

* *hv* 11 days plus dark 47 days

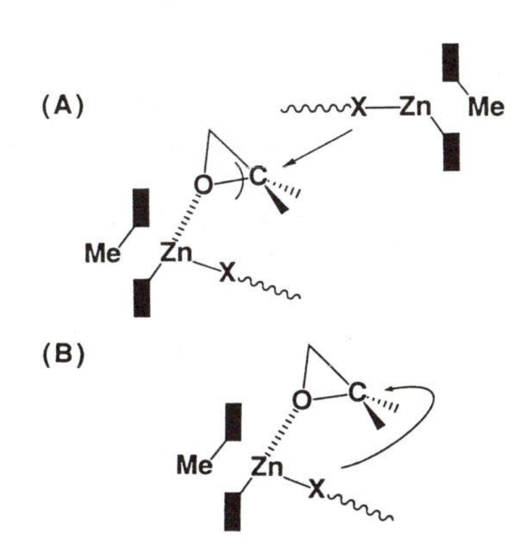

Scheme 2

The polymerizations are usually carried out by a simple procedure such as the reaction in a tear-drop type flask with a three-way cock with magnetic stirring, at a moderate temperature in nitrogen atmosphere. The monomer and the catalyst solution can be transferred to and from the flask by a hypodermic syringe.

Acknowledgment

The author is grateful to Dr. T. Aida and Mr. Y. Watanabe for their collaboration.

Literature Cited

1. Vandenberg,E.J. *J.Polym.Sci.*, **1969**, , *A-1,7*, 525.
2. Aida,T.; Inoue,S. *Macromolecules*, **1981**, *14*, 1162.
3. Sugimoto,H.; Aida,T.; Inoue,S. *Macromolecules*, **1990**, *23*, 2870.
4. Shimasaki,K.; Aida,T.; Inoue,S. *Macromolecules*, **1987**, *20*, 3076.
5. Trofimoff,L.; Aida,T.; Inoue,S. *Chem. Lett.*, **1987**,*991*.
6. Aida,T.; Kawaguchi,K.; Inoue,S. *Macromolecues*, **1990**, *23*, 3887.
7. Aida,T.; Ishikawa,M.; Inoue,S. *Macrmolecues*, **1986**, *19*, 8.
8. Aida,T.; Inoue,S. *J. Am. Chem. Soc.*, **1985**, *107*, 1358.
9. Kuroki,M.; Aida,T.; Inoue,S. *J. Am. Chem. Soc.*, **1987**, *109*, 4737.
10. Hosokawa,Y.; Kuroki,M.; Aida,T.; Inoue,S. *Macromolecules*, **1991**,*24*, ,824.
11. Inoue,S.; Aida,T.; Kuroki,M.; Hosokawa,Y. *Makromol. Chem., Macromol. Symp.*, **1990**, *32*, 255.
12. Aida,T.; Maekawa,Y.; Asano,S.; Inoue,S. *Macromolecues*, **1988**, *21*, 1195.
13. Kuroki,M.; Nashimoto,S.; Aida,T.; Inoue,S. *Macromolecules*, **1988**,*21*, 3114.
14. Yasuda,T.; Aida,T.; Inoue,S. *J. Macromol. Sci.-Chem.*, **1984**, *A21*, 1035.
15. Watanabe,Y.; Aida,T.; Inoue,S. *Macromolecules*, **1990**, *23*, 2612.
16. Watanabe,Y.; Aida,T.; Inoue,S. *Macromolecules*, **1991**, *24*, 3970.
17. Watanabe,Y.; Yasuda,T.; Aida,T.; Inoue,S. *Macromolecules*, **1992**, *25*, 1396.
18. Asano, S.; Aida,T.; Inoue,S. *Macromolecules*, **1985**,*18*, 2507.
19. Yasuda, T; Aida,T; Inoue,S. *Bull. Chem. Soc. Jpn.*, **1986**, *59*, 3931.

RECEIVED February 20, 1992

Chapter 16

Oligomerization of Oxiranes with Aluminum Complexes as Initiators

V. Vincens, A. Le Borgne, and N. Spassky

Laboratoire de Chimie Macromoléculaire, URA 24, Université Pierre et Marie Curie, 4, Place Jussieu, 75252 Paris, Cédex 05, France

Bifunctional oligoethers with defined structure have been prepared by catalytic oligomerization of substituted oxiranes using aluminum complexes as initiators. Almost pure dimer, trimer and tetramer from methoxymethyloxirane were obtained using (porphinato)aluminum chloride complex. A new series of initiators derived from Schiff's bases were developed and successfully tried with differently substituted oxiranes leading to defined α-chloro-ω-hydroxyoligoethers. Chiral Schiff's base complexes allow the preparation of enantiomerically enriched oligomers starting from racemic monomeric mixtures.

Bifunctional oligoethers having a controlled structure, i.e. stereochemistry (regular enchainments, chirality), functionality (reactive end groups) and given degree of polymerization are potential precursors for the synthesis of a wide range of materials (liquid crystalline polymers, biomaterials, crown ethers, ...).

$$\mathrm{H[\text{-}O\underset{\displaystyle R}{C}HCH_2\text{-}]_n\,X}$$

[1]

X = Cl, OH
R = alkyl, aryl
n < 10

They can be prepared by oligomerization of oxiranes using suitable initiators allowing the control of molecular weight, functionality and direction of ring-opening.

Porphinatoaluminum derivative initiators, developed by Inoue et al. *(1)*, fulfill these requirements. Living polymerization of several classes of monomers (heterocycles : oxiranes, lactones, ..., acrylic monomers) has been achieved. In addition, a novel concept of "immortal" polymerization *(2)* was recently introduced in which protic compounds (water, alcohols, phenols, ...) act as transfer agents

0097–6156/92/0496–0205$06.00/0

and in that case polymers with narrow molecular weight distribution and controlled molecular weights may be obtained.

We have found that the synthesis of chiral bifunctional oligoethers from methyloxirane (MO) is possible when using (tetraphenylporphinato)aluminum chloride (TPPAlCl) as initiator system *(3)*. Recently, we have reported that a chiral aluminum complex of a Schiff's base is an efficient initiator for the oligomerization of MO leading to bifunctional oligoethers *(4)*.

In this paper, we present results on oligomerization of oxiranes with aluminum complexes : oligomerization of methoxymethyloxirane (MMO) with TPPAlCl initiator system will be discussed and the synthesis of bifunctional oligoethers (derived from MO and MMO) using new initiators obtained by reaction between Schiff's bases and $AlEt_2Cl$ is reported. The latter initiators are presumed to have the structure **[2]**.

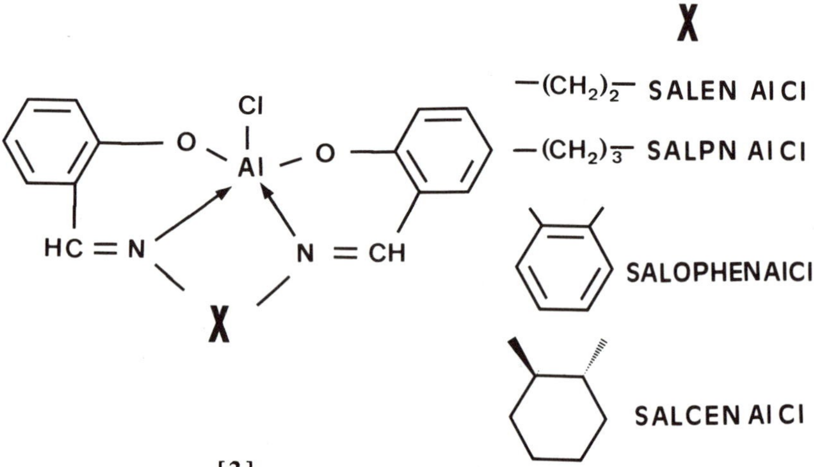

[2]

The chiral initiator SALCEN-AlCl has a stereoelective character and leads to a preferential oligomerization of one enantiomer from a racemic monomer mixture.

Experimental

All the oligomerizations were carried out in apparatus sealed under high vacuum. All monomers used were distilled, dried under vacuum on CaH_2 and distributed in graduated tubes with breakseals.

The preparation of initiators, the oligomerization reaction and the isolation of oligomers have been described elsewhere *(3,4)*. Schiff's base SALCEN has an enantiomeric purity of 94 %.

Elemental analysis of the samples were performed by "Service Central d'Analyse du C.N.R.S." IR spectra were recorded using a Bruker IRFT IFS45 spectrometer. NMR spectra were recorded with a Bruker apparatus (^{1}H 250 MHz, ^{13}C 62.89 MHz). Molecular weights were determined by GPC in THF (Waters apparatus equipped with μ-styragel columns : 10^5, 10^4, 10^3, 500, 100 Å ; detection : refractometric and with UV (254 nm) ; flow rate : 1 ml/mn ; polystyrene standards and by vapor phase osmometry in $CHCl_3$ (Knauer apparatus). The positive FAB mass spectrum was performed on a VG ZAB 2-SEQ mass spectrometer using 3-nitrobenzyl alcohol as a matrix.

Oligomerization of methoxymethyloxirane with (tetraphenylporphinato)-aluminum chloride

MMO was chosen in order to obtain oligomers with some hydrophilic character. The oligomerization with TPPAlCl was carried out in bulk at 60°C during four days using a molar ratio (monomer) / (initiator) of 10. Yield : 61 %, M_n (VPO) = 500 ($M_{n\ th}$=570).

The structure and the functionality of the oligomer were established by IR spectroscopy, elemental analysis, ^1H and ^{13}C NMR. IR spectrum shows several characteristic bands of absorption. The large band at 3470 cm^{-1} corresponds to a hydroxyl group. A series of bands located at 2820, 2880, 2910 and 2980 cm^{-1} is related to CH vibrations (CH_3, CH_2, CH). The strong band at 1110 cm^{-1} is characteristic of the ether group. A small doublet at 700 - 747 cm^{-1} is assigned to CH_2Cl group in the α-position on a substituted carbon.

The nature of terminal hydroxyl group was determined by ^1H NMR (250 MHz) after derivatization with trichloroacetylisocyanate (TAI) *(5)*. The resonance peaks of methylene and methine protons adjacent to the terminal hydroxyl group shift to 4.0-4.3 and 5.0-5.2 ppm corresponding respectively to the primary and secondary alcohol groups. In the present case, more than 90 % of terminal hydroxyl groups are secondary, indicating a predominant ring-opening in the β-position :

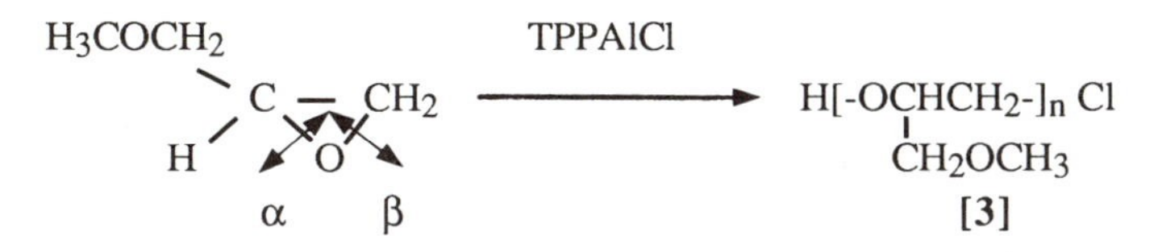

Based on NMR, M_n was found to be 560.

Elemental analysis shows the presence of chlorine (5.89 %) in the oligomer. Considering one Cl atom per molecule, one finds a value of 600 for M_n. The satisfactory agreement between the different determinations of M_n confirms the bifunctionality of oligo MMO which appears to have the structure **[3]** (average n ≅ 6.2). These results are substantiated by ^{13}C NMR (62.89 MHz) spectrum presented in Figure 1. Four main patterns of peaks corresponding to $\underline{C}H_3$, $\underline{C}H_2$ (lateral), $\underline{C}H_2$ (chain), $\underline{C}H$ are observed around δ = 59, 70, 73 and 79 ppm (respectively). The additional peak at 43.5 ppm corresponds to the terminal carbon atom CH_2Cl. Terminal carbon $\underline{C}HOH$ appears at 69.6 ppm nearly lateral $\underline{C}H_2$ carbons and can be differentiated using JMSET technique. No unsaturation is detected between δ = 100 and 140 ppm.

A more detailed study was performed on oligo MMO prepared in dichloromethane solution at 60°C during 7 days using the same initiator system. After elimination of initiator residues, the mixture of oligomers was fractionated by distillation under high vacuum (≅ 10^{-5} mmHg) and three fractions enriched in different n-mers described in Table I have been obtained.

A fairly good agreement between the different determinations of molecular weights is observed.

The composition of each fraction is determined by GPC and by FAB mass spectrometry. The mass spectrum of fraction 2 is shown in Figure 2. Each fraction is highly enriched in one of the n-mers (n = 3, 4, 5) with some proportion of n - 1 and n + 1 mers.

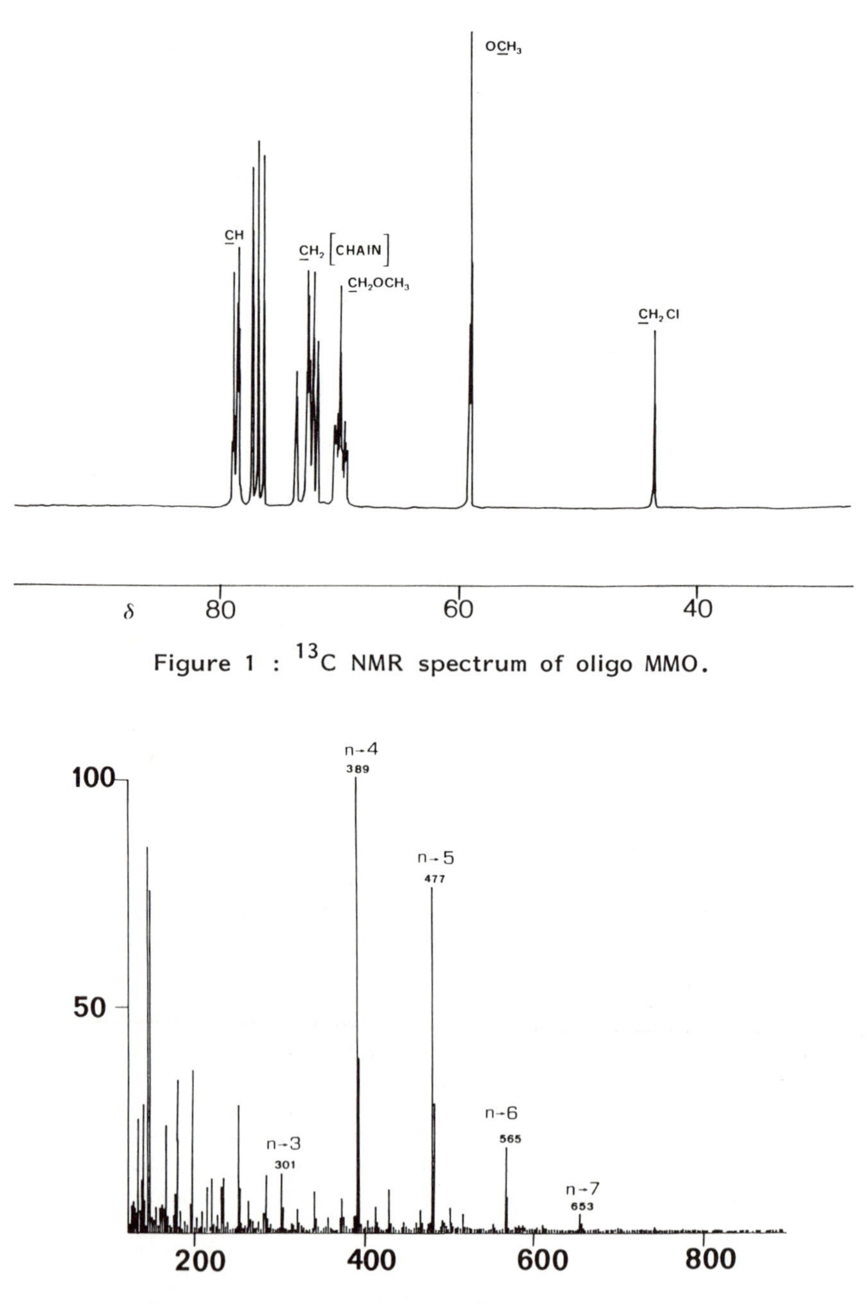

Figure 1 : ^{13}C NMR spectrum of oligo MMO.

Figure 2 : FAB mass spectrum of oligo MMO.

Table I. Molecular weights of different fractions of oligo MMO

	Yield (%)	*Cl* (%)	M_n (EA)	M_n (NMR)	M_n (VPO)	M_n (GPC)	DP_n
Fraction 1	18.5	12.0	300	300	290	300	2.95
Fraction 2	24	8.1	440	410	380	400	4.25
Fraction 3	9	6.8	520	440	380	420	4.6

MW by : EA = elemental analysis ; NMR = ^{1}H NMR after derivatization with TAI ; VPO in $CHCl_3$; GPC in THF

The first oligomers (dimer, trimer and tetramer) were isolated by column chromatography (silica gel, ethyl acetate as eluent for the dimer, ethyl acetate/methanol (98/2) for the trimer and the tetramer). These oligomers have the expected structure **[3]** with n = 2, 3, 4. The ^{13}C NMR (62.89 MHz) spectra of the dimer and of the trimer are presented in Figure 3.

From above results, one can conclude that the initiation of oligomerization of MMO occurs with a ring-opening of oxirane in β-position followed by insertion of monomeric unit in AlCl bond leading to α-chloro-ω-hydroxy oligomers. Individual oligomers may be isolated and fully characterized.

Oligomerization of various oxiranes with aluminum complexes of Schiff's bases as initiator systems

In the synthesis of oligomers described in the previous section, considerable amounts of tetraphenylporphyrin are required. Moreover, the obtained products are often colored and the complete elimination of initiator residues is difficult. In this context, we started the investigation of a new family of initiators which appears to be easier to handle.

Recently, Goedken et al *(6)* had shown that the reaction between Schiff's bases (SB) and $AlEt_3$ leads to the formation of aluminum complexes in which aluminum atom is pentacoordinated like in porphyrin aluminum complexes. Following the same procedure, we have prepared initiators by metallation of Schiff's bases with $AlEt_2Cl$ (SB-AlCl structure **[2]**) and we have tested their ability for oligomerization of a series of oxiranes : methyloxirane (MO), methoxymethyloxirane (MMO) and epichlorohydrin (ECH). Results are presented in Table II.

ECH appears to be the most reactive monomer. This can be probably explained by the activation of CH_2-O bond, resulting from electronegativity of chlorine atom.

Among the aluminum complexes used for oligomerization of MO, the order of reactivity is SALOPHEN AlCl < SALPN AlCl ≅ SALCEN AlCl, indicating that ligands with distorted geometries are the most efficient.

Characterization of prepared oligomers was again carried out by IR, NMR (^{1}H, ^{13}C), elemental analysis in order to determine their structure and their functionality.

In IR spectra of all oligomers the following characteristic bonds were observed : a broad band at 3470 cm^{-1} for the hydroxyl stretching and a doublet (747 and 705 cm^{-1}) corresponding to CH_2Cl group.

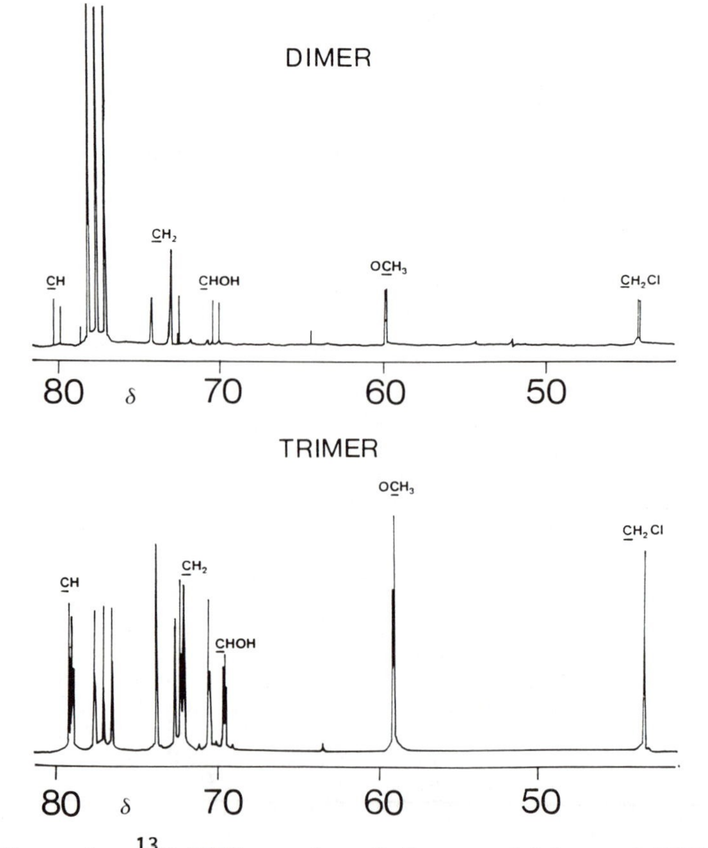

Figure 3 : ^{13}C NMR spectra of dimer and trimer of MMO.

Table II. Bulk oligomerization of substituted oxiranes with aluminum complexes of Schiff's bases (SB-AlCl) at 80°C

Monomer	SB in initiator	C/M (mole %)	Time (h)	Conv. (%)	M_n (calc.)	M_n (VPO)	(M_w/M_n) (GPC)
MO	SALEN	1	480	74	4300	2450	1.15
MO	SALPN	10	24	97	600	510	---
MO	SALPN	1	240	61	3500	1370	1.14
MO	SALOPHEN	8.3	130	82	610	540 a)	---
MO b)	SALCEN	5	29	51	610	460 a)	---
MMO	SALPN	10	144	62	580	500 [500a)]	---
ECH b)	SALCEN	20	1	93	1750	930 c)	---

a) by elemental analysis, based on Cl amount
b) temperature of oligomerization : 60°C
c) by ^{1}H NMR, after derivatization with TAI

Elemental analysis indicates the presence of chlorine atom in oligomers derived from MO and MMO. Molecular weights determined, considering one chlorine atom per oligomer molecule, are in rather good agreement with the calculated ones (Cf. Table II).

^{1}H NMR spectra were recorded at 200 MHz after derivatization of oligomers with trichloroacetylisocyanate. Ratio of integrations of $C\underline{H}_2OH$ and $C\underline{H}OH$ protons (respectively shifted downfield at 4.4 ppm and 5.05 ppm) shows that more than 95 % of hydroxyl groups are secondary indicating that ring-opening of oxirane by aluminum complexes of Schiff's bases occurs predominantly in β-position.

^{13}C NMR spectra were recorded at 62.89 MHz. The spectrum of oligo MO (M_n = 680) is presented in Figure 4. Three main patterns of peaks corresponding to $\underline{C}H_3$, $\underline{C}H_2$ and $\underline{C}H$ carbons are located around 17, 73 and 75 ppm, respectively. These patterns are complex : in addition to major peaks corresponding to central units, minor peaks arising from end groups are observed. The peak located at δ = 47.5 ppm can be assigned to the terminal carbon $\underline{C}H_2Cl$ while the set of two peaks at δ = 65.8 and 67.2 ppm corresponds to the terminal carbon $\underline{C}HOH$ with influence of the penultimate unit (dyad effect). No unsaturation is detected between δ = 100 and 140 ppm.

The spectrum of oligo ECH (M_n = 930) is shown in Figure 5. The major peaks at 43.5, 69.6 and 79.0 ppm correspond, respectively, to $\underline{C}H_2Cl$, $\underline{C}H_2O$ and $\underline{C}H$ carbons of the central units. Minor signals at 71.2 and 71.5 ppm correspond to carbon CHOH (dyad effect due to the influence of the penultimate unit) while the peak at 45.4 ppm is that of the methylene carbon atom :

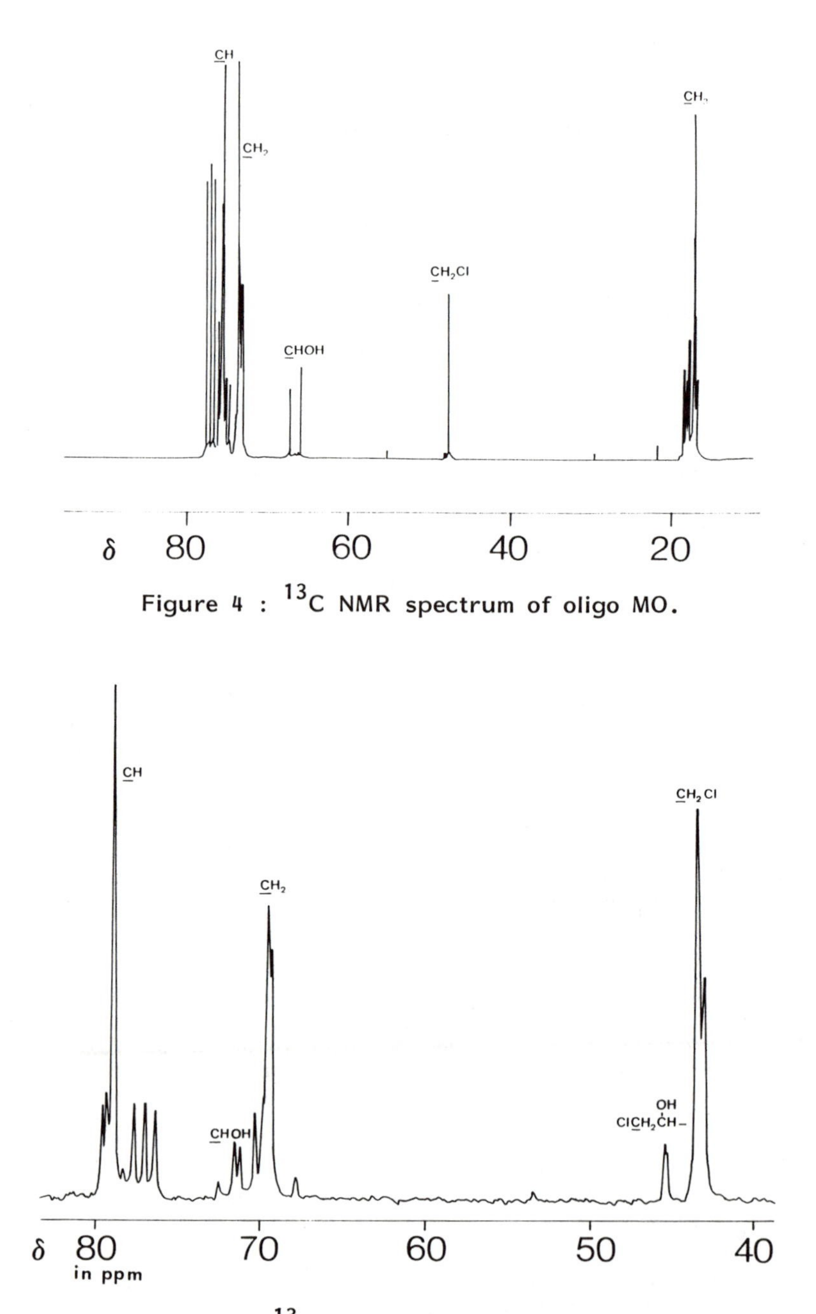

Figure 4 : ^{13}C NMR spectrum of oligo MO.

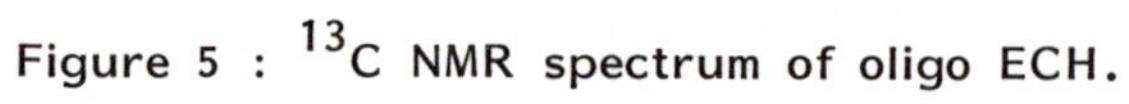

Figure 5 : ^{13}C NMR spectrum of oligo ECH.

$ClCH_2CH(OH)\sim\sim\sim$

No unsaturation is detected between 100 and 140 ppm.

Thus, the oligomer derived from ECH appears to have the following structure :

$$HO\underset{CH_2Cl}{\underset{|}{C}}HCH_2[-O\underset{CH_2Cl}{\underset{|}{C}}HCH_2-]_n\ Cl$$

indicating that the ring-opening reaction occurs again in β-position.

From the above results it appears that in the case of SB-AlCl initiators the initiation reaction proceeds by insertion of the oxirane monomer into the Al-Cl bond leading to an aluminum alcoholate. During the propagation, the monomer inserts into the Al-O bond, with ring-opening in β-position. The behavior of these aluminum complexes of Schiff's bases towards oxiranes is therefore reminiscent of the initiator systems, based on porphyrins developed by the group of Inoue *(1)*.

Stereoelective oligomerization of methyl oxirane with SALCEN-AlCl

In a further step, we have examined the possibility to obtain optically active oligomers starting from racemic monomer and using a chiral initiator. Such a process, named "stereoelective or enantioasymmetric" has been already observed in the polymerization of oxiranes using chiral zinc derivatives *(7)*. In the latter case significant enantiomeric enrichments were obtained, but the polymerization proceeds in heterogeneous conditions, the efficiency of initiators (fraction of metal atom used for polymerization) being rather low and the control of molecular weight not possible. It appears attractive to find initiator systems allowing homogeneous oligomerizations and having high efficiency.

We have carried out oligomerization of racemic MO using an aluminum complex of a chiral Schiff's base as initiator *(4)*. Initiator was obtained by reaction between N,N'-bis(2-hydroxybenzylidene) (1R, 2R)-1,2 cyclohexanediamine (SALCEN) and $AlEt_2Cl$. The oligomerization reaction was performed in bulk at 60°C using a ratio [initiator]/[monomer] : [C]/[M] = 5 %. Shortly after the beginning of reaction (~ 30 mn) the oligomerization mixture becomes homogeneous. Experimental results are reported in Table III.

Table III. Stereoelective oligomerization of racemic MO with (R, R)-SALCEN-AlCl initiator system

Polymerization time in hours	Polymer yield in %	Optical rotation of unreacted monomer α_D^{25}, neat [a]	E.e. in % [b]	stereoelectivity ratio (r(S))	M_n [c]
22	40	+ 0.76	6.3	1.28	380
62	70	+ 1.85	15.4	1.29	680
107	80	+ 2.22	18.5	1.26	590

[a] α_D^{25} of (S)-MO : - 12° (neat) ; [b] E.e. : enantiomeric excess ;
[c] by ^1H NMR after derivatization with TAI.

The sign of unreacted monomer indicates that (S)-(-)-MO is preferentially incorporated in the polymer chain. The magnitude of the preferential consumption of the (S)-enantiomer by the initiator is measured by the stereoelectivity ratio $r_{(S)}$, which corresponds to the ratio of consumption rate constants of the two enantiomers $r_{(S)} = k_{(S)}/k_{(R)}$. Assuming a first-order law for enantiomer consumption, the following relationship correlating α/α_o (optical purity of unreacted monomer) with conversion x may be used for the determination of the stereoelectivity ratio *(7)* :

$$(1 - x)^{r(S)-1} = \frac{(1 + \alpha/\alpha_o)}{(1 - \alpha/\alpha_o)^{r(S)}}$$

The average value found for $r_{(S)}$ is close to 1.28, which corresponds to an optical purity of unreacted monomer at half-conversion of 9 %. The enantiomeric enrichments observed are not as high as those obtained with chiral initiator zinc systems (r = 1.8) *(7)*, but the synthesis is performed in homogeneous conditions and it is possible to control both functionality and degree of polymerization of oligomers.

The stereoelective oligomerization proceeds presumably through preferential coordination, resulting from steric effects, of one of the enantiomers on the aluminum atom.

Further studies in order to improve the efficiency of initiator and to establish the mechanism of oligomerization (in particular by the technique of ^{27}Al NMR) are presently in progress.

Conclusion

Bifunctional oligoethers having a controlled structure and defined functionality can be prepared using aluminum complexes as initiators. It was possible to isolate in the case of methoxymethyloxirane the first terms of the series i.e. dimer, trimer, tetramer under an almost pure form. Similarly, defined α-chloro-ω-hydroxy oligoethers may be prepared from other substituted oxiranes, e.g. epichlorohydrin.

Porphyrin type aluminum complexes may be used as initiators. A new series of initiators derived from Schiff's bases were successfully tried for oligomerization. These new initiators seem to have a potential interest because of their easy handling, the large variety of available structures and particularly the possibility of preparation of chiral complexes leading to stereoelective effects.

Literature Cited

1. Aida, T. ; Mizuta, R. ; Yoshida, Y. ; Inoue, S. Makromol. Chem. **1981**, 182, 1073
2. Asano, S. ; Aida, T. ; Inoue, S. J. Chem. Soc., Chem. Comm., **1985**, 1148
3. Jun, C.L. ; Le Borgne, A. ; Spassky, N. J. Polym. Sci., Polym. Symp., **1986**, 74, 31
4. Vincens, V. ; Le Borgne, A. ; Spassky, N. Makromol. Chem., Rapid Commun., **1989**, 10, 623
5. Goodlett, V.W. Anal. Chem. **1965**, 37, 431
6. Dzugan, S.G. ; Goedken, V.L. Inorg. Chem., **1986**, 25, 2858
7. Coulon, C. ; Spassky, N. ; Sigwalt, P. Polymer, **1976**, 17, 821

RECEIVED February 12, 1992

Chapter 17

Catalysts and Initiators for Controlling the Structure of Polymers with Inorganic Backbones

Krzysztof Matyjaszewski

Department of Chemistry, Carnegie Mellon University, 4400 Fifth Avenue, Pittsburgh, PA 15213

Three general methods for the synthesis of inorganic polymers: condensation polymerization, ring opening polymerization, and modifications thereof are described. New catalyzed polymerization of phosphoranimines provides quantitatively low polydispersity, high molecular weight, linear polyphosphazenes at temperature below 100 °C. Polysilanes with low polydispersities are prepared by sonochemical reductive coupling. Ring-opening polymerization of strained cyclotetrasilanes provides the first possibility of the tacticity control in polysilanes. Modification of polysilanes with triflic acid and various nucleophiles allows incorporation of functional groups to polysilanes and also synthesis of graft copolymers.

Inorganic and organometallic polymers belong to a new class of advanced materials located at the interphase of organic polymers, ceramics and metals (*1*):

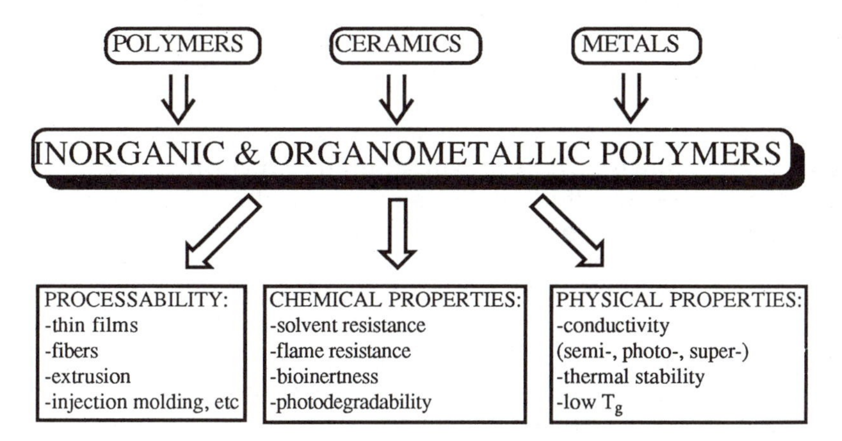

0097–6156/92/0496–0215$06.00/0

Their unique chemical and physical properties and processability stimulate new applications. Some inorganic polymers behave as low temperature elastomers, some are fire and solvent resistant, some are bioinert, some have special surface properties, others show unusual electronic and optical effects:

However, at present, most inorganic and organometallic polymers are available as ill-defined materials with extremely high polydispersities (often with polymodal molecular weight distribution), non-controlled degrees of polymerization and unknown structure of end groups. Some applications, particularly those in biomedical and optoelectronic fields, require well-defined polymers. Such polymers are also necessary for structure-property correlations and as special model compounds. Some phenomena, such as formation of one dimensional superlattices (*2*) based on conducting and semiconducting polymers, can be observed only for very regular systems. Also, a microphase separation occurs in regular block and graft copolymers with controlled dimensions of segments. In this article possibilities of the improvement of the structural control of the inorganic and organometallic polymers will be discussed. Therefore, the correct choice of catalysts and initiators, which can improve control of macromolecular structure, is very important.

Most polymers with inorganic backbones are based on silicon: polysiloxanes, polysilanes, polysilazanes, and polycarbosilanes can serve as examples. Polyphosphazenes form another, commercially important, class of inorganic polymers. These linear inorganic polymers are true macromolecules with molecular weights above $M_n>100{,}000$ and sometimes even above $M_n>1{,}000{,}000$. Usually they are soluble in common organic solvents and can be processed in a way typical for organic polymers (depending on structure of substituents). High molecular weight catenates of other elements such as germanium and sulfur are also known. Dialkyl substituted polygermanes resemble polysilanes. Polysulfur is thermodynamically stable only above its floor temperature ($T>160\ ^{o}C$). In most other cases, cyclics dominate over linear chains. This is true not only for pure catenates (e.g. polyphosphines) but also for polymers with alternating structure of elements BN, BP, SiS, SiP, etc. Linear polymers are known for silazanes with small substituents (sometimes as small as hydrogen), cyclics are preferred even for permethylated systems. Polymerization of potentially strained four membered rings is not possible (silathianes) because larger elements can absorb angular deformations much easier than carbon. In some systems a polymer can be formed as a kinetic product which can be degraded to strainless cyclics under appropriate conditions, but, which can be sometimes sufficiently stable for some useful applications.

There are three general methods leading to inorganic polymers. They resemble classic methods known for organic polymers: condensation polymerization (usually step growth), ring opening polymerization (chain growth), and modifications of polymers formed in the previous two ways. Double bonds between two elements larger than carbon are not common (a few compounds with double bonds and with bulky substituents, e.g. $R_2Si{=}SiR_2$, are known (*3*)only for non-polymerizable systems) and, therefore, olefin-type polymerization is not applicable to inorganic polymers.

Some condensation reactions occur as pure step growth process (hydrolytic condensation of disubstituted dichlorosilanes to form polysiloxanes). However, in some other cases, reactivity of the end groups in growing polymer chains are very different from reactivities of a monomer and chain growth model operates (reductive coupling of dichlorosilanes with sodium to form polysilanes, thermal and catalyzed condensation of phosphoranimines). The main obstacle for the condensation process is formation of strainless cyclic oligomers.

Under special conditions strained rings can be formed and subsequently opened in the chain growth process. Here again, back-biting and formation of strainless cycles is the main challenge in the formation of well defined polymers. Correct choice of initiators and catalysts can lead, however, to polymers with regular structures. Anionic polymerization of hexamethylcyclotrisiloxane is probably the best example of living inorganic polymerization. Ring-opening polymerization of hexamethylcyclodisilazane yielded first known linear permethylated polysilazane (*4*). Ring-opening of hexachloro-cyclotriphosphazene is a commercially important process leading to polymers with molecular weights $M_n \approx 1{,}000{,}000$.

Modification of inorganic polymers is often used to exchange substituents, introduce special functional groups such as hydrophilic, lipophobic, electron reach/poor, mesogenic, bioactive, etc. Most modifications have been performed on polyphosphazenes (nucleophilic displacement of chlorines) and on partially hydrogenated siloxanes (hydrosilylation). Modifications of polysilanes is also known.

In this article the main emphasis will be put on the synthesis of new synthetic routes towards polyphosphazenes and polysilanes with improved structural control and regularity which are being developped in our laboratories.

Polysiloxanes

Polysiloxanes are the first type of polymers with an inorganic backbone in which a high degree of structural control has been achieved. The synthetic method best suited for the preparation of well-defined polysiloxanes is the anionic polymerization of hexamethylcyclotrisiloxane (D_3) with a lithium counterion (*5*). High molecular weight polymers ($M_n \cong$ 100, 000) with low polydispersity ($M_w/M_n<1.1$) can be prepared in this way. Use of less strained monomer (octamethylcyclotetrasiloxane) and more bulky counterions (Na^+, K^+) is less chemoselective and a linear polymer is accompanied by various cyclics. Cationic polymerization is also accompanied by the formation of macrocyclics, mostly due to the end-biting reaction of the electrophilic active site with the silanol terminal groups (*6,7*).

Poly(dimethylsiloxane) is the only symmetrically substituted siloxane polymer which does not form a mesophase and has only one first order thermal transition. Longer and larger organic substituents (diethyl, dipropyl, diphenyl) provide various crystalline phases and, in addition, a columnar mesophase typical for polymers with inorganic backbone(*8,9*). The range of the mesophase increases strongly with the size of the organic substituent.

The living nature of D_3 anionic polymerization allows formation of various block copolymers which phase separate very easily and produce materials with new morphologies (*10*).

Polyphosphazenes

Polyphosphazenes are interesting inorganic polymers which find applications as advanced materials with flexibility at a very low temperature, superb solvent and oil resistance, flame resistance, toughness, and vibration damping properties and also as potential biomaterials (*11,12*). Polyphosphazenes are used as seals, gaskets, o-rings, fuel hoses, and as vibration shock mounts, serviceable over a wide range of temperatures and in harsh environments.

There are two main synthetic routes towards polyphosphazenes. The first is based on the ring-opening polymerization of hexachlorocyclotriphosphazene at approximately 250 °C and the subsequent displacement of chlorine atoms by alkoxy and aryloxy substituents (*13*).

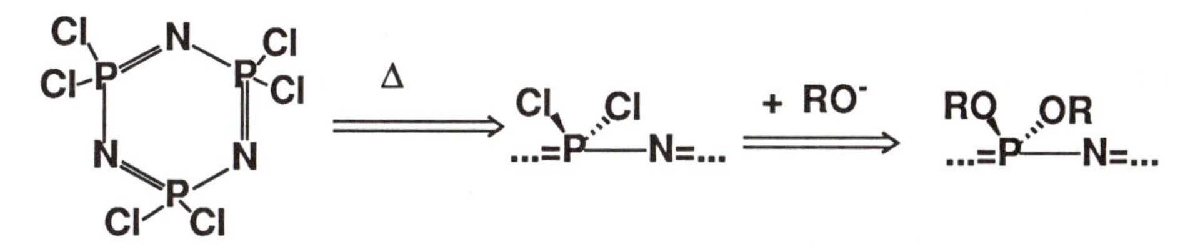

At present, this is the only process utilized commercially. Unfortunately, this method leads to polymers with high polydispersities (*14*). Application of Lewis and protonic acids as catalysts reduces polydispersities but also reduces molecular weights (*15*).It is possible that milder reaction conditions reduce branching and crosslinking.

The alternative preparation of polyphosphazenes employs the high temperature (200 °C) condensation of N-silylphosphoranimines and provides polymers with alkyl and aryl substituents directly bound to phosphorus (*16*).

$$Me_3Si{-}N{=}P{-}OCH_2CF_3 \xrightarrow{-CF_3CH_2OSiMe_3} ...{=}N{-}P{=}N{-}P{=}...$$

R= Me, Me
Me, Ph
Et, Me ...

Monomer synthesis is quite cumbersome and requires four steps with limited yields. The synthesis of phosphoranimines can be considerably improved in the case of trialkoxyderivatives which can be formed in one step from phosphites and trimethylsilyl azide (*17,18*). This Staudinger-type reaction provides in high yields phosphoranimines which can be thermally (200 °C) polymerized to polyphosphazenes, identical as formed in the classic Allcock's ring-opening/modification process:

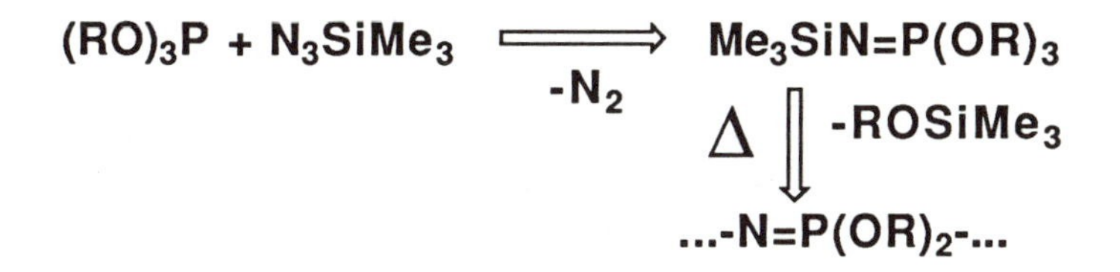

Polyphosphazenes obtained by thermal polymerization of phosphorainimines give linear products with molecular weights in the range of 10,000 to 100,000. No control of molecular weights and no control of end-groups is possible in the thermal polymerization of phosphoranimines.

Catalyzed Polymerization of Phosphoranimines. Thermal polymerization of phosphoranimines has a character of the chain reaction with high polymer formed at low conversion and with noticeable induction periods. We decided to explore possibility of the catalysis in this process. The induction periods indicate that in the initial stage some type of active species is slowly formed which later becomes responsible for the chain growth nature of the process. It is known that silyl group in phosphoranimines has high mobility and that cations of a type $R_2P\text{-}N^+\text{-}PR_2$ are quite stable due to strong delocalization of the positive charge. Thus, we first assumed that polymerization may have an anionic character and that we should provide initiators which will carry bulky cations and anions with a high affinity towards trimethylsilyl group. The obvious choice was tetra(n-butyl)ammonium fluoride which tremendously accelerated reaction and provided a high polymer within one hour at 100 °C (compare with a few days at 200 °C for a noncatalyzed process). Polymers with molecular weights ranging from M_n=20,000 to 100,000 were prepared using tetrabutylammonium fluoride and other anionic initiators which include various salts with crowned and cryptated cations. The simplified plausible reaction mechanism is presented in the scheme below (*19*):

$$(RO)_3P{=}N\text{-}Si(CH_3)_3 + {}^+NBu_4F^- \longrightarrow (RO)_3P{=}N^-\,{}^+NBu_4 + (CH_3)_3SiF\uparrow$$

$$(RO)_3P{=}N^-\,{}^+NBu_4 + (RO)_3P{=}N\text{-}Si(CH_3)_3 \longrightarrow$$

$$(RO)_3P{=}N\text{-}P(OR)_2{=}N^-\,{}^+NBu_4 + (CH_3)_3SiOR \xrightarrow{+\,n(RO)_3P{=}N\text{-}Si(CH_3)_3}$$

$$(RO)_3P{=}N\text{-}[P(OR)_2{=}N]_n\text{-}P(OR)_2{=}N^-\,{}^+NBu_4 + n(CH_3)_3SiOR$$

$$R = OCH_2CF_3$$

Trimethylsilyl fluoride was detected, indeed, at the early stages of the polymerization as a volatile product. However, it is possible that the actual propagation step may involve not purely ionic species but active centers with pentacoordinated silicon atom. Preliminary kinetic data indicate low external orders in initiators and catalysts. This suggets that catalysts may be not involved in the rate determining step. In such a case, the rate determining step may involve the formation of a complex between monomer and non-charged chain ends. The complex is then attacked by the catalyst in the fast step leading to the chain extension. More detailed analysis of changes of the molecular weight with conversion indicate that some other reactions also play an important role. Reaction between end groups, not important at the beginning of the reaction (large excess of monomer), leads to increase of molecular weight by interchain condensation:

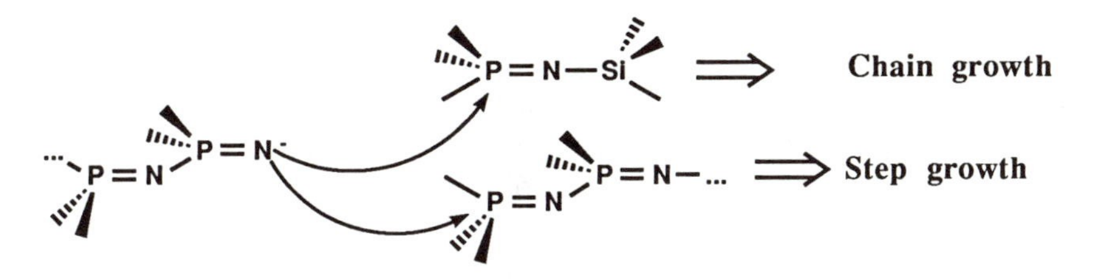

Another class of effective catalysts is based on salt-free systems. N-Methylimidazole, known for its high affinity towards silyl group has been successfully used in polymerization of phosphoranimines. The role of a cation is then played by the silylated imidazol. This species might be in a dynamic equilibrium with the non-silylated species and intermediate species with a pentacoordinated silicon atom:

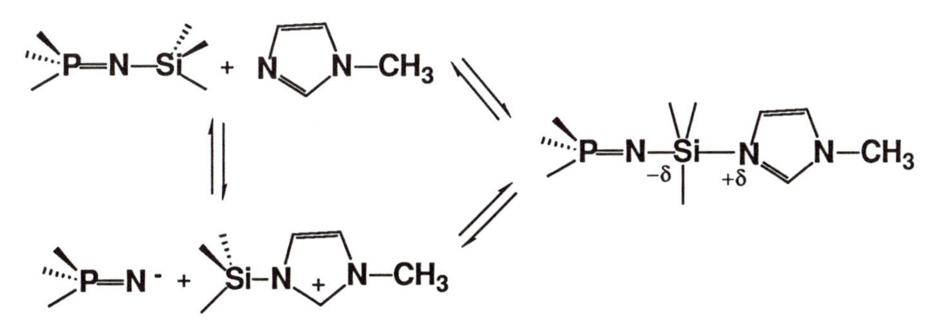

Therefore, imidazole will act more as a catalyst than as an initiator, since it will regenerate in each propagation act. Qualitatively, polymerizations initiated by salts and by imidazole behave in a similar way. Yields are quantitative and NMR indicates a linear polymer structure (>95%). Molecular weights from 20,000 to 200,000 were obtained. Molecular weight distribution are in the range of $M_w/M_n \cong 1.2$ to 1.5.

Termination of polymerization of the phosphoranimines with electrophilic reagents such as benzyl bromide leads to the quantitative incorporation of benzyl moiety as the end group into polymer chains. This enables formation of end-functionalized polymers and oligomers based on polyphosphazenes, polymers with unique properties.

Thus, the catalyzed polymerization of phosphoranimines is a potential route to the formation of well defined polyphosphazenes with controlled molecular weight and controlled structure of end groups.

Polyphosphazenes by Direct Reaction of Silyl Azides with Phosphonites and Phosphinites. Reactions of phosphonites and phosphinites with silyl azide have to be carried out with extreme care. In addition to the phosphoranimine, the formation of phosphine azides is observed. These compounds may explosively decompose to polyphosphazenes and nitrogen in a chain process. The relative proportion of path A (phosphine azide) and B (phosphoranimine) depends on substituents at phosphorous. Thus, phosphites do not form phosphine azides, phosphonites yield a mixture of products, and phosphinites give nearly exclusively phosphine azides and no phosphoranimines.

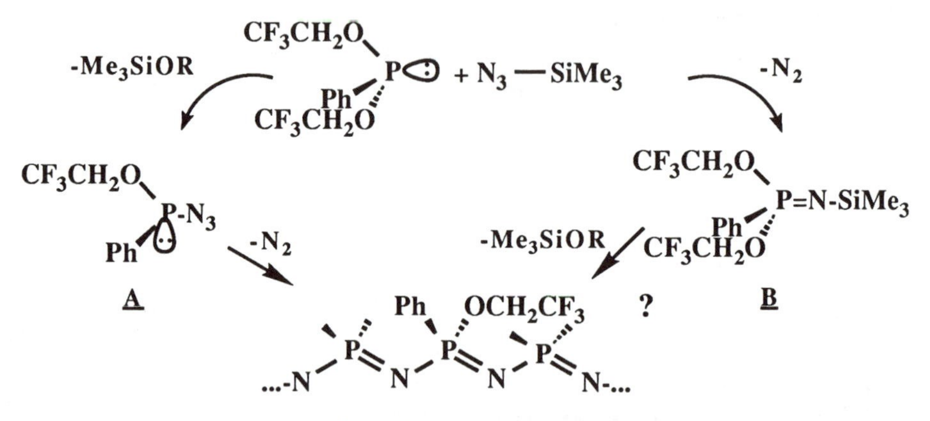

Random polymer

Reaction of trifluoroethyl diphenylphosphinite occurs explosively in bulk but can be mediated in solution. Poly(diphenylphosphazene) precipitates as a white powder from etheral solutions. Polymerization is quite sensitive to solvent effects and in alcohols and chlorinated solvents provides mostly oligomers. Similar reaction with diphenylchlorophosphine leads preferentially to cyclics (*20*).

The examples presented of the polymerization of phosporanimines and phosphine azides which were formed *in situ* demonstrate the possibility for the preparation of polyphosphazenes in previously inaccessible range of molecular weights, polymers with increased control of macromolecular structure (modalities, polydispersities) and new polymers with e.g. diphenyl and phenyl/trifluoroethoxy substituents. It is believed that more thorough studies on new synthetic routes towards polyphosphazenes will enable additional control of microstructure, macromolecular structure and supramolecular structure in these important materials.

Polysilanes

Polysilanes are macromolecules with a linear Si-Si catenation in the main chain and with two organic substituents at each silicon atom. They have interesting physical and chemical properties and are of a potential commercial importance (*21,22,23*). Poor compatibility of inorganic and organic segments provides morphologies with clear mesophases for symmetrically substituted polymers (*24*). Photosensitivity of polysilanes leads to applications in microlithography (*22,25*). Strong delocalization of electrons in the backbone provides materials with extremely interesting electronic properties: semiconductors, photoconductors, and nonlinear optical materials (*21,22,26,27,28*). However, the synthetic aspect of polysilanes is not yet developed to the level corresponding to their physicochemical characterization. First, and still the most common preparative technique, is based on the reductive coupling of disubstituted dichlorosilanes with alkali metals (*29,30*). The polymodality of the obtained polymers limits some characterization techniques and may also disable some applications, since low molecular weight polymer have poor film forming properties (*21,22*). The dehydrogenative coupling in the presence of transition metals usually provides low molecular weight materials (*31,32*). There are two other routes to polysilanes based on the anionic polymerization of "masked disilenes" (*33*) and on the ring-opening polymerization (*34*). These techniques may provide additional control of the microstructure of the nonsymmetrically substituted systems. In this paper some of our recent activities in the synthesis and characterization of polysilanes will be discussed. These include mechanistic studies of the sonochemical reductive coupling process, ring opening polymerization, and reactions on polysilanes, i.e. modification and grafting, as well as characterization of some copolysilanes.

Sonochemical Reductive Coupling Process. Polymerization of disubstituted dichlorosilanes with alkali metals via reductive coupling has a strong character of a chain (not a step) process. Molecular weights are very high at low conversions and independent of the $[Mt]_0/[Si\text{-}Cl]_0$ ratio. Several intermediates such as silylene, radicals, and anions have been proposed as potential chain carriers. Yield of high molecular weight polymer and polydispersities depend very strongly on reaction conditions (stirring rate, solvents, temperature, reducing agent, substituents at silicon atom). The correct choice of initiator and reaction conditions has tremendous influence on the polymerization process.

Under sonochemical conditions (ambient temperature) polymerization has mostly ionic character (*34,35,36*). Chloroterminated chain ends (...-SiR_2-Cl) participate in two one-electron transfer steps to form silyl anions (...-SiR_2^-, Mt^+). Anions react with a monomer in the nucleophilic substitution process which is rate determining. The exact nature of the Si-Mt bond is still obscure and, under some conditions, it may have a covalent character. Silyl radicals are formed as short living intermediates on the pathway from chloroterminated chain ends to silyl anions. They can be trapped but do not react directly with a monomer and are not responsible for the chain growth.

In toluene, using sodium as a reducing agent, monomers with aryl groups react much faster than dialkylsubstituted dichlorosilanes (*37*). The latters require elevated temperatures (above 80 °C), although they react readily with Na/K and K.

On contrary, methylphenyldichlorosilane does not react with potassium within 2 hours under similar conditions. This apparent discrepancy has been solved by the analysis of the product of the reaction of phenylmethyldichlorosilane with K at longer reaction times. The resulting polymer ($M_n \approx 2{,}000$) is not a polysilane. It does not absorb above 300 nm and it contains a large amount of toluene moieties, in contrast to any other polysilanes. GC/MS analysis of the first products formed in this reaction indicates the presence of Cl-SiMePh-PhMe species, formed via reaction of a monomeric radical with toluene. This result confirms the chain nature of the polymerization in which an electron transfer to a polymer chain occurs much faster than to the monomer. Of course, an electron transfer to methylphenyldichlorosilane from potassium is much faster than from sodium, but initiation is still a few orders of magnitude slower than propagation. Moreover, an electron transfer from potassium may occur from a much longer distance than from sodium (in a way analoguous to Grignard reagent formation) (*38*). Therefore, the monomeric radicals may be separated by two or three solvent molecules from the metal surface and, instead of the coupling process or the second electron transfer, they react with toluene or diffuse to the bulk solvent. Reaction of triethylsilyl radicals with toluene is very fast ($k=1.2 \times 10^6$ $mol^{-1} \cdot L \cdot s^{-1}$) (*39*), a polymeric radical may react with a similar or slightly lower rate.

Reductive coupling at ambient temperatures in the presence of ultrasound leads to monomodal polymers with relatively narrow molecular weight distributions (M_w/M_n from 1.2 to 1.5) and relatively high molecular weights (M_n from 50,000 to 100,000) (*37*). There are two phenomena responsible for the more selective polymerization. First, lower polymerization temperature and continuous removal of the sodium chloride from the sodium surface suppresses formation of the low molecular weight polymer (M_n from 2,000 to 10,000). This polymer might be formed via some side reactions (transfer or termination). Second, ultrasound mechanically degrades polysilanes with molecular weights above 50,000. This limit might be set by the chain entanglement and Si-Si bond energetics. Polysilanes prepared in separate experiments could also be selectively degraded. It seems that degradation in toluene in the presence of alkali metals is slightly accelerated, but no low molecular weight cyclooligosilanes are formed. On the other hand, in THF and diglyme (or in toluene in the presence of cryptands and potassium) polymer is completely degraded to cyclohexasilanes and cyclopentasilanes. The anionic intermediates have been observed spectroscopically in this degradation.

Copolymerization of various dialkylsubstituted dichlorosilanes by a reductive coupling process usually leads to statistical copolymers (*40*). The copolymer composition usually corresponds to the monomer feed and the distribution of various triads, pentads, and heptads roughly corresponds to Bernoullian statistics.

Polysilanes are usually prepared by the reductive coupling of disubstituted dichlorosilanes with alkali metals in non polar solvents such as toluene or hexane. No substantial degradation of the polymer is observed in the presence of an excess of alkali metal in these solvents. Polymers with higher molecular weights are prepared in solvents of lower solvating ability. Polysilanes can be degraded by an excess of alkali metal in THF or in diglyme. Cyclic oligomers are the only degradation products. The rate of degradation depends on the substituents at the silicon atom, solvent, alkali metal, and temperature. Degradation is much faster in more polar THF than in toluene. Addition of cryptand [2.2.2] or THF to toluene increases the degradation rate. The rate of degradation strongly increases with the reactivity of the metal. Electron transfer from potassium is much easier than from sodium or from lithium. The effect of the counterion seems to be less important, since degradation initiated by sodium or lithium naphthalides proceeds nearly as fast as with potassium. On the other hand, the initial electron transfer process is strongly metal dependent. The application of the ultrasound helps to clean the metal surface by cavitational erosion but does not change the rate of degradation. Degradation is much faster for poly(methylphenylsilylene) than for poly(di-n-hexylsilylene). This is in agreement with the general trend in the stability of silyl anions discussed in the previous section. Aryl groups on the silicon atom stabilize anionic intermediates. A counterion effect on the degradation is observed. Degradation with K^+ is faster than with Na^+ or Li^+. Moreover, lithium can not start the degradation process. Even in THF solution the first electron transfer is not possible. This may be due to surface phenomena, since lithium naphthalide degrades polysilylenes efficiently. During chemical degradation cycles are formed but the molecular weight of polysilylenes does not change. This indicates a chain process with slow initiation and fast propagation.

The first step in which a polymeric radical anion is formed should be rate limiting. It is facilitated in more polar solvents due to the energy gain by the solvation of alkali metal cations. The intermediate radical anion cleaves to a radical and an anion. The fate of the radicals is not known. They may react with a solvent, take a second electron to form a silyl anion, or recombine, although radical concentration is usually very low. The silyl anions may now start cleaving Si-Si bonds:

$$...\text{-}(SiR_2)_n\text{-}(SiR_2)_m\text{-}... + Na \rightarrow ...\text{-}(SiR_2)_n\text{-}(SiR_2)^{\cdot -}{}_m\text{-}...,{}^{+}Na \rightarrow$$
$$\rightarrow \quad ...\text{-}(SiR_2)_n^{-} \quad + \quad \cdot(SiR_2)_m\text{-}...$$

$$...\text{-}(SiR_2)_n^{-} \quad \rightarrow \quad ...\text{-}(SiR_2)_{n-5}^{-} \quad + \quad \text{-}(SiR_2)_5\text{-}$$

The rate of the intramolecular reaction must be much faster than the rate of the intermolecular process due to entropic effects. The anchimeric assistance (or neighboring group participation) is the highest for the least strained five- and six-membered rings . Therefore, the degradation process leads to cyclopentasilanes and cyclohexasilanes. The back-biting process for larger or smaller rings is retarded for enthalpic reasons. The back-biting process has to be distinguished from the end-biting process which may occur during the synthesis of polysilylenes. Additional evidence for different mechanisms of end-biting and back-biting processes are provided by the structure of cyclic oligomers which are formed during the synthesis of poly(di-n-hexylsilylene). Above 80% of octa-n-hexylcyclotetrasilane (δ= -20.3 ppm in ^{29}Si NMR) is formed under the reductive coupling process in a mixture of toluene and isooctane. On the other hand, the degradation leads to the formation of deca-n-hexylcyclopentasilanes (δ= -34.6 ppm).

The present results clearly indicate that back-biting (or degradation) is not important in toluene in the absence of any additives. It is also not very important for poly(di-n-hexylsilylene) in THF. However, polymerization of methylphenyldichlorosilane with any alkali metal in THF will result in cyclics rather than in a linear polymer.

Thus, the correct choice of the conditions of reductive coupling and the structure of the reducing agent is extremely important in determining yields and molecular weight distribution of final products.

Ring-Opening Polymerization. Reductive coupling of disubstituted dichlorosilanes with alkali metals is usually accompanied by the formation of cyclooligosilanes. Under some conditions, a linear polysilane can be completely degraded to cyclics. Thus, a polysilane can be considered as a kinetic product, whereas cyclic oligomers are true thermodynamic products. The majority of known cyclooligosilanes are thermodynamically stable and cannot be converted to linear polymers. Some potentially strained rings such as octaphenylcyclotetrasilanes can be prepared in high yield since the repulsive interactions between phenyl groups present at each silicon atom have a greater influence than the angular strain in the four membered ring. Additionally, low solubility and high melting point (mp. =323 ^{o}C) precludes polymerization of octaphenylcyclotetrasilanes at higher concentrations. We have discovered rapid and clean Si-Ar bond cleavage with trifluoromethanesulfonic acid (*41*). Reaction of octaphenylcyclotetrasilanes with four equivalents of the acid leads to 1,2,3,4-tetra(trifluoromethanesulfonyloxy)-1,2,3,4-tetraphenylcyclotetrasilane. Subsequent reaction with either methylmagnesium iodide or methyl lithium yields four stereo isomers of 1,2,3,4-tetraphenyl-1,2,3,4-tetramethylcyclotetrasilane.

A Si-Si bond is quite labile in the presence of strong electrophiles and nucleophiles. Reaction with silyl anions leads to ring-opening and to the regeneration of silyl anions[42]. This is the propagation step. Silyl anions may also attack the Si-Si bonds in the same polysilane chain and form macrocycles and strainless cyclooligosilanes. Rates of polymerization and degradation depend on solvent, temperature, and alkali metals. With 1 mol% of silyl potassium or butyl lithium initiator, only cyclooligosilanes have been found after less than 2 minutes at room temperature in pure THF. In benzene, with less than 3% THF, polymerization is completed after more than 1 hour. In mixtures of 60 % THF with benzene polymerization is completed within less than 2 minutes, but degradation starts after 1 hour. Polymers with molecular weights from 10,000 to 100,000 have been prepared via the anionic ring-opening polymerization of cyclotetrasilanes. This technique provides a pathway to various functional polymers and block copolymers.

In this system the proper choice of solvent, temperature, and counterion determines yields and distribution of products. Non polar solvents provides higher yields and higher molecular weights of polymers. More polar solvent lead to much faster polymerization but also to much faster degradation to strainless cyclic oligomers.

Modification of Poly(phenylmethylsilylene). The severe conditions of the reductive coupling process and the anionic process allow only alkyl and aryl substituents at silicon. There are only a few polysilanes known with substituents other than alkyl and aryl. However, the Si-Ph bond can be easily cleaved by strong protonic acids such as triflic acid. The rate of the dearylation is strongly influenced by the presence of an electron withdrawing group at the neighboring Si atoms. Model studies on dearylation of α,ω-diphenylpermethyloligosilanes with triflic acid indicate that the displacement of the first phenyl group is always faster than that of the second, even for pentasilanes:

$$Ph\text{-}(SiMe_2)_n\text{-}Ph \xrightarrow[k]{+HOTf} Ph\text{-}(SiMe_2)_n\text{-}OTf \xrightarrow[k']{+HOTf} TfO\text{-}(SiMe_2)_n\text{-}OTf$$

k/k'=23, 13, 10, 7 for n=2, 3, 4, 5

Apparently, the reactivity of the oligosilanes increases with the chain length in contrast to the electron density on the *ipso*-C atom which is attacked in the rate determining step. This indicates that the transition states rather than the ground states control reactivities of polysilanes.

The dearylation process applied to polysilanes containing phenyl substituents provides polymers with strong electrophilic silyl triflate moieties. Silyl triflates belong to the strongest known silylating reagents. They react with ketones 10^8 times faster than silyl chlorides do. They can react with any nucleophiles such as alcohols, amines, carbanions, organometallics, etc. This opens a new synthetic avenue towards various functional polysilanes (*35,43*). The reactivity of silyl triflates is so high that they can initiate cationic polymerization of some alkenes and heterocyclics to form graft copolymers.

Silylated triflate backbone is very reactive and triflate groups can be easily displaced by a variety of nucleophiles such as alcohols, amines, organometallic reagents, etc. This chemistry resembles the modification of chlorinated polyphosphazenes. A variety of functional groups lead to new materials which are lipophobic, hydrophilic, and some of them show additional phase ordering. For example, the attachment of a p-methoxybiphenyl mesogen via long flexible spacer with six methylene units to partially triflated poly(methylphenylsilylene) leads to modified polysilanes with liquid crystalline behavior. Shorter spacers and weaker mesogens form only isotropic materials. Probably inherent incompatibility of the main chain and side groups as well as the rod-like behavior of polysilanes prevent ordering of side chain mesogen groups in most systems.

Copolysilanes with Chiral Substituents

It is well established that polysilanes with two identical alkyl groups form at ambient temperatures regular crystalline structures (*24*). Various analytical techniques identified the extended zig-zag conformation for di(n-hexyl) derivative, but 7/3 helical structure for di(n-pentyl) and di(n-butyl) derivatives. The ordered crystalline phase is still accompanied by a typical chain folding, since single crystals with a typical screw-dislocation are formed from poly(di-n-pentylsilylene) (*44*). A micrograph of a larger amount of these crystals indicates (as expected) that an equal amount of crystals with left-handed and right-handed dislocation are present.

It was of interest to control supramolecular structure of these crystals and to prepare all crystals with one-sense dislocation. This was attempted by incorporation of 5 to 20% of chiral 2-methylbutyl (isopentyl) groups into poly(di-n-

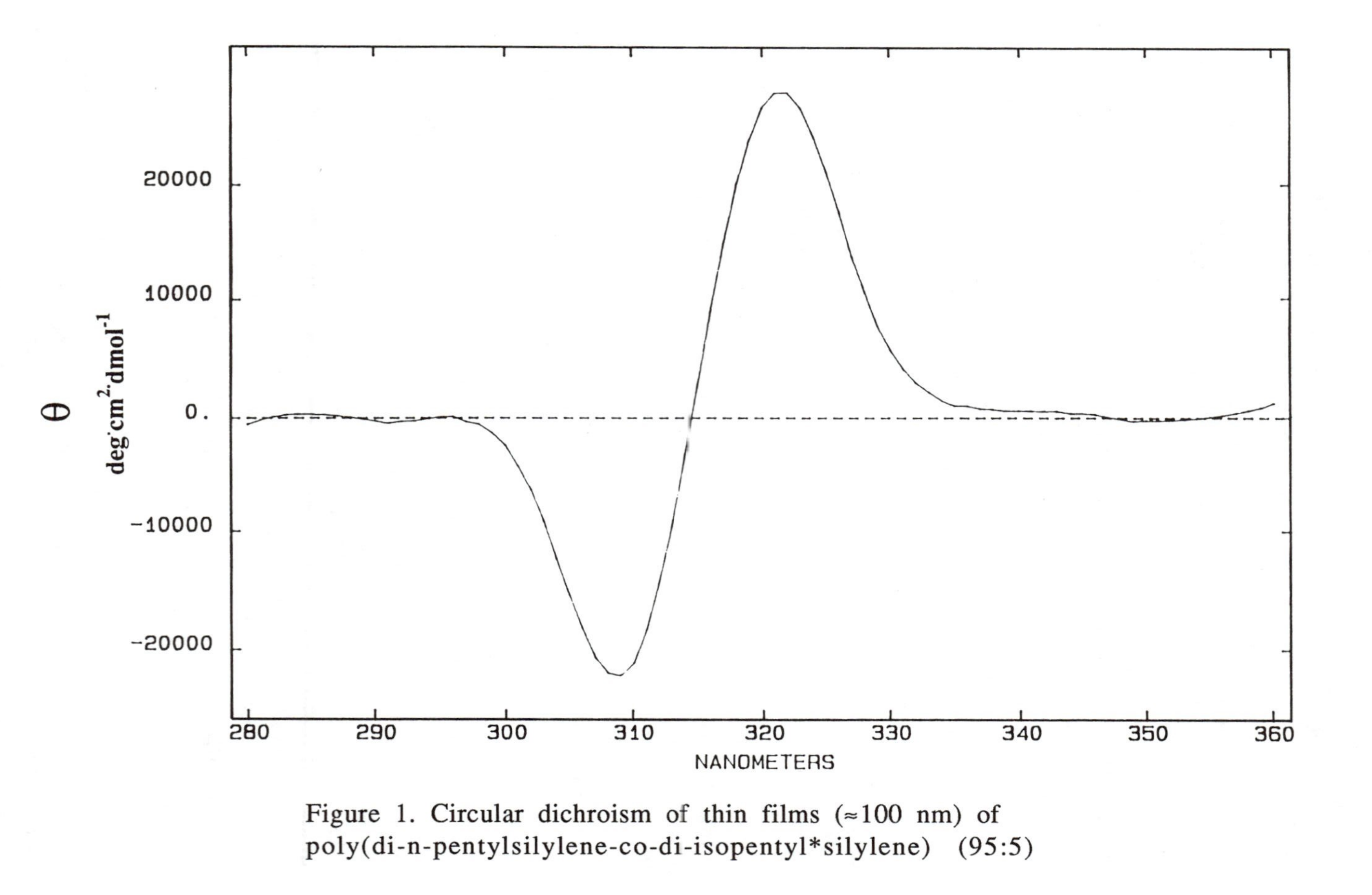

Figure 1. Circular dichroism of thin films (≈100 nm) of poly(di-n-pentylsilylene-co-di-isopentyl*silylene) (95:5)

pentylsilylene). These copolymers show relatively weak optical activities in solution and the optical rotation is approximately 50 times higher than predicted from the content of chiral substituents. The molar optical rotations increase with a decrease of temperature and become very large after addition of a non solvent (isopropanol). Polysilanes adopt a random coil structure in solution and there is only a small proportion of a helically oriented chromophores, especially at low proportion of chiral unit (every 10 or every 20 silicon atoms). However, at lower temperatures and at limited solubilities, a helical structure is formed which is accompanied by a dramatic increase in optical activity (over 100 times or 5,000 times based on optical activity of isolated chiral unit). This increase can be also partially ascribed to the intermolecular coupling of chromophores.

In agreement with this explanation, solid films of the copolymers with chiral groups exhibit strong optical activities and strong elipticities as determined by circular dichroism. (Figure 1).

Thus, as demonstrated in the above examples using sonochemical reductive coupling, ring-opening polymerization, and modifications, it is possible to prepare new interesting polysilanes with regular structures. Correct choice of initiators and catalysts is in these systems very important.

Acknowledgments.

This research has been sponsored by the Office of Naval Research, the Army Research Office and by the National Science Foundation. The author acknowledges support from Hoechst Celanese, Eastman Kodak, PPG Industries, and Xerox Corp. within the Presidential Young Investigator Award.

References

1. H. R. Allcock, *ACS Symp. Ser.*, 360, 251 (1988).
2. N. Matsumoto, K. Takeda, H. Teramae, M. Fujino, *Adv. Chem. Ser.*, 224, 515 (1990)
3. G. Raabe,J. Michl, *Chem. Rev.*, 85, 419 (1985)
4.D. Seyferth, J. M. Schwark, R. M. Stewart, *Organometallics*, 8, 1980 (1989)
5. M. Morton, A. Rembaum, E. E. Bostick, *J. Appl. Polym. Sci.* 8, 2707 (1964)

6. J. Chojnowski, M Mazurek, M. Scibiorek, L. Wilczek, *Makromol. Chem.*, 175, 3299 (1974)
7. J. Chojnowski, M. Scibiorek, P. Kowalski, *Makromol. Chem.*, 178, 1351 (1977)
8. Y. K. Godovsky, V. S. Papkov, *Adv. Polym. Sci.*, 88, 129 (1989)
9. G. Kogler, K. Loufakis, M. Moeller, *Polymer*, 31, 1538 (1990)
10. I. Yilgor, J. McGrath, *Adv. Pol. Sci.*, 86, 1 (1988)
11. D. P. Tate, *J. Polym. Sci. Symp.* 48, 33 (1974)
12. H. R. Penton, *ACS Symp. Ser.* 360, 277 (1988)
13. H. R. Allcock, *Chem. Eng. News* 63 (11), 22 (1985)
14. G. L. Hagnauer, N. S. Schneider, *J. Polym. Sci. A-2* 10, 699 (1972)
15. M. S. Sennett, G. L. Hagnauer, R. E. Singler, G. Davies *Macromolecules*, 19, 959 (1986)
16. R. H. Neilson, P. Wisian-Neilson, *Chem. Rev.* 88, 541 (1988)
17. J. S. Thayer, R. West, *Inorg. Chem.*, 3, 406 (1964)
18. E. P. Flindt, H. Rose, *Z. Anorg. Allg. Chem.*, 428, 204 (1977)
19. R. A. Montague, K. Matyjaszewski, *J. Am. Chem. Soc.*, 112, 6721 (1990)
20. R. H. Kratzer, K. L. Paciorek, *Inorg. Chem.*, 4, 1767 (1965)
21. R. West, *J. Organomet. Chem.*, 300, 327 (1986).
22. R. D. Miller and J. Michl, *Chem. Rev.*, 89, 1359 (1989).
23. S. Yajima, J. Hayashi, and M. Omori, *Chem. Lett.*, 1975, 931.
24. F. C. Schilling, F. A. Bovey, A. J. Lovinger, and J. M. Zeigler, *Adv. Chem. Ser.*, 224, 341 (1990).
25. R. D. Miller, *Adv. Chem. Ser.*, 224, 413 (1990).
26. R. West, L. D. David, P. I. Djurovich, K. S. V. Srinivasan, and H. Yu, *J. Am. Chem. Soc.*, 103, 7352 (1981).
27. M. A. Abkowitz, M. Stolka, R. J. Weagley, K. M. McGrane, and F. E. Knier, *Adv. Chem. Ser.*, 224, 467 (1990).
28. F. Kajzar, J. Messier, and C. Rosilio, *J. Appl. Phys.*, 60, 3040 (1986).
29. P. Trefonas III and R. West, *J. Polym Sci., Polym. Lett. Ed.*, 23, 469 (1985).
30. J. M. Zeigler, L. A. Harrah, A. W. Johnson, *ACS Polymer Preprints*, 28(1), 424 (1987).
31. J. F. Harrod, *ACS Symp. Ser.*, 360, 89 (1988).
32. H. G. Woo and T. D. Tilley, *J. Am. Chem. Soc.*, 111, 3757 (1989).
33. K. Sakamoto, K. Obata, H. Hirata, M. Nakajima, H. Sakurai, *J. Am. Chem. Soc.*, 111, 7641 (1989).
34. K. Matyjaszewski, Y. L. Chen, and H. K. Kim, *ACS Symp. Ser.*, 360, 78 (1988).
35. K. Matyjaszewski, J. Hrkach, H. K. Kim, and K. Ruehl, *Adv. Chem. Ser.*, 224, 285 (1990).

36. H. K. Kim and K. Matyjaszewski, *Polym. Bull.*, 22, 441 (1989).
37. H. K. Kim and K. Matyjaszewski, *J. Am. Chem. Soc.*, 110, 3321 (1988).
38. E. A. Ashby and J. Oswald, *J. Org. Chem.*, 53, 6068 (1988).
39. C. Chatgilialogu, K. U. Ingold, and J. C. Scaino, *J. Am. Chem. Soc.*, 105, 3292 (1982).
40. A. R. Wolff, I. Nozue, J. Maxka, and R. West, *J. Polym. Sci., Polym. Chem.*, 26, 701 (1988).
41. K. Matyjaszewski and Y. L. Chen, *J. Organomet. Chem.*, 340, 7 (1988).
42. Y. Gupta and K. Matyjaszewski, *ACS Polymer Preprints*, 31(1), 46 (1990).
43. F. Yenca, Y. L. Chen, and K. Matyjaszewski, *ACS Polymer Preprints*, 28 (2), 222 (1987).
44. H. Frey, M. Moeller, K. Matyjaszewski, *Macromolecules*, submitted.

RECEIVED November 15, 1991

Cationic Polymerization

ORDINARY CATIONIC POLYMERIZATION with proton and Lewis acid catalysts has been used for many years for olefins, vinyl ethers, epoxides, and other ring-opening monomers. Performance of these catalysts varies widely, depending on the catalyst's nature and with the monomer. In many cases, cationic systems are characterized by extensive chain transfer, rearrangements, and often the need to use very low temperatures to obtain high molecular weight. Some unique coordination cationic catalysts for vinyl ethers are described by Vandenberg in the Overview section. In the past 10 years, living cationic polymerization catalysts for olefins (such as isobutylene) and for vinyl ethers have been developed, and this field continues to expand rapidly. In this section, Kennedy and Price have used a combination of living cationic and living anionic polymerization to make and characterize a poly(methyl methacrylate)-*block*-isobutylene-*block*-poly(methyl methacrylate) ABA block copolymer. Allcock reviews his extensive mechanism studies on the thermal and catalyzed polymerization of cyclophosphazenes, concluding that a cationic process is preferred although "not fully established." Penczek et al. describe the mechanism aspects of the synthesis of poly(alkylene phosphates), analogs of important biopolymers, by the polyaddition of diepoxides to phosphorous acids.

Chapter 18

Mechanisms and Catalysis in Cyclophosphazene Polymerization

Harry R. Allcock

Department of Chemistry, The Pennsylvania State University, University Park, PA 16802

Ring-opening polymerization is one of the main access routes to heteroatom polymers such as polyphosphazenes or polysiloxanes. In the case of polyphosphazenes, a key reaction is the thermal ring-opening polymerization of $(NPCl_2)_3$ to poly(dichorophosphazene), $(NPCl_2)_n$, which is the starting point for a wide range of macromolecular substitution reactions that yield stable and useful organophosphazene high polymers. The mechanism of polymerization of $(NPCl_2)_3$ is discussed, together with an extension of the polymerization to organophosphazene trimers and to heterocyclic monomers that contain carbon or sulfur as well as phosphorus and nitrogen in the ring.

The broadest emphasis in polymer synthesis in the past 50 years has been placed on the polymerization of petrochemical monomers to organic polymers. However, the emerging field of inorganic-organic polymers is gaining increased attention because it provides an essential bridge between organic polymers and the broader areas of inorganic materials such as ceramics, metals, semiconductors, and superconductors.

Prominent among these newer polymer systems is the growing field of phosphazene macromolecules (*1*). These polymers have the general structure shown in **1**. They can be considered as being related to the older area of polysiloxanes (**2**) (2) and stucturally to polyethers such as polyaldehydes or polyketones (**3**).

Polyphosphazenes are unique macromolecules in a number of ways. First, the inorganic backbone provides an unusually high degree of molecular flexibility (comparable to polysiloxanes). Second, the presence of equi-elemental amounts of phosphorus and nitrogen in these polymers provides a built-in resistance to burning. Third, the backbone is transparent from the near infrared to 220 mu in the ultraviolet. Fourth, the methods of synthesis allow an almost unprecedented range of organic, organometallic, or inorganic side groups to be linked to the skeleton. Different side groups alter the solubility, solvent resistance, crystallinity, glass transition temperature, optical absorption, electrical conductivity, biomedical compatibility, refractive index, and non-linear-optical response. Thus, the opportunities for

0097–6156/92/0496–0236$06.00/0

property tuning and the development of structure-property correlations make this a powerful system for both fundamental macromolecular studies and technological development.

Overview of the Ring-Opening Polymerization / Macromolecular Substitution Route.

The most widely used synthesis route to these polymers, discovered and developed in our program (*1, 3-5*) involves a two-step sequence, as summarized in Scheme I. The first step is a thermal ring-opening polymerization of a cyclic inorganic monomer (**4**) (prepared on a manufacturing scale from phosphorus pentachloride and ammonium chloride), to an essentially linear high polymer (**5**) known as poly(dichlorophosphazene). This polymer, when dissolved in an organic solvent such as tetrahydrofuran, benzene, or toluene, serves as a reactive macromolecular intermediate. Nucleophilic reagents, such as alkoxides or amines, readily replace the chlorine atoms to generate a broad series of organophosphazene high polymers (**6, 7**), as shown in Scheme I. Polymers with two or more types of different side groups (**9, 10**) can be prepared by simultaneous or sequential cosubstitution reactions (Scheme I). Given the broad range of nucleophiles available, it will be clear that the number of accessible polymers is very large (*1, 3-31*). By late 1991 more than 300 different examples had been reported. More than 2000 papers and patents have been published on these reactions and their products, and roughly 200 new reports appear each year. Thus, it will be clear that the ring-opening polymerization step is a critical reaction in this field of polymers, and attempts to understand the polymerization mechanism are important for the development of the catalyzed processes needed for large scale reactions.

Approaches to Determining the Mechanism of $(NPCl_2)_3$ Polymerization

(a) **Electrical Conductivity of the Polymerization Mixture**. The ring-opening polymerization of **4** occurs when the pure molten trimer is heated to temperatures roughly 100°C above its 114°C melting point. Thus, temperatures of 210° or higher are needed in order to achieve useful rates of polymerization (70% conversion to polymer in 7-12 hours). At temperatures in the range of 275-300°C the rate of polymerization is too rapid to prevent the conversion of the polymer to a crosslinked derivative. The crosslinked form is not suitable for substitution reactions.

An early clue to the mechanism of polymerization was provided by measurements of the electrical conductivity of the molten trimer as the temperature was increased (*32*). At temperatures between the melting point and 200°C the conductivity is essentially zero. Between 203°C and 250°C the conductivity rises with temperature increases, and measurements of AC versus DC conductivity suggest an ionic mechanism for the conduction. The conductivity is ascribed to the ionization of chloride ions from phosphorus, as shown in Scheme II. The polarizability of the molten reaction mixture also rises in tandem with the conductivity, and this can be rationalized in terms of a conversion of cyclic molecules to linear species. These results formed the basis of the initial suggestion (*32*) that the ring-opening polymerization of **4** is a cationically initiated process, as shown in pathway (a) in Scheme II.

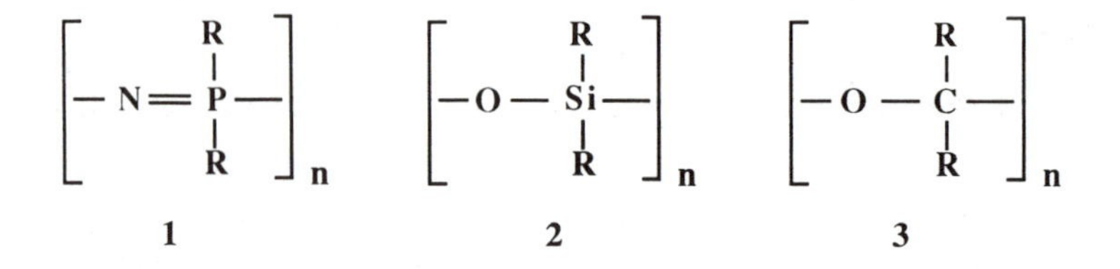
R
N
P
R
n
R
O
Si
R
n
R
O
C
R
n
1
2
3

Scheme I

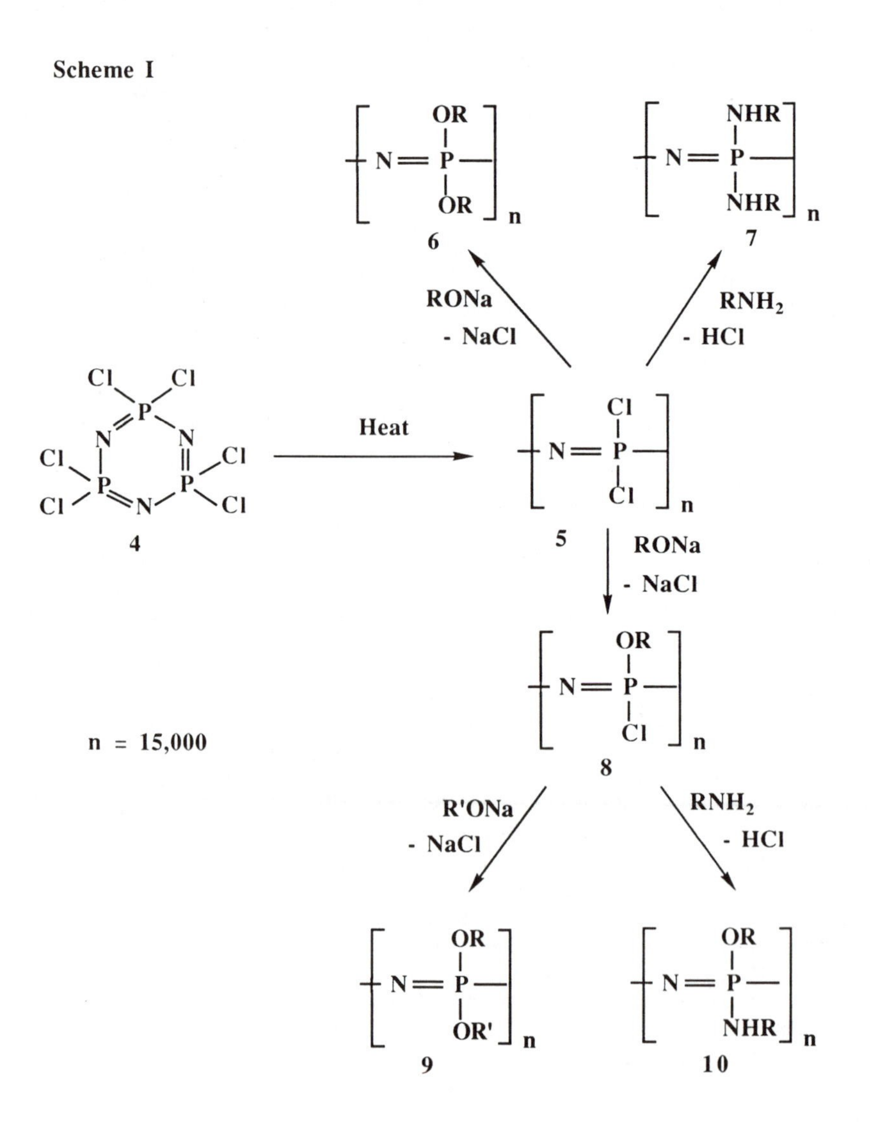
OR
N
P
OR
n
6
NHR
N
P
NHR
n
7
RONa
- NaCl
RNH2
- HCl
Cl
Cl
P
N
N
Cl
Cl
P
P
Cl
N
Cl
4
Heat
Cl
N
P
Cl
n
5
RONa
- NaCl
OR
N
P
Cl
n
8
n = 15,000
R'ONa
- NaCl
RNH2
- HCl
OR
N
P
OR'
n
9
OR
N
P
NHR
n
10

Scheme II

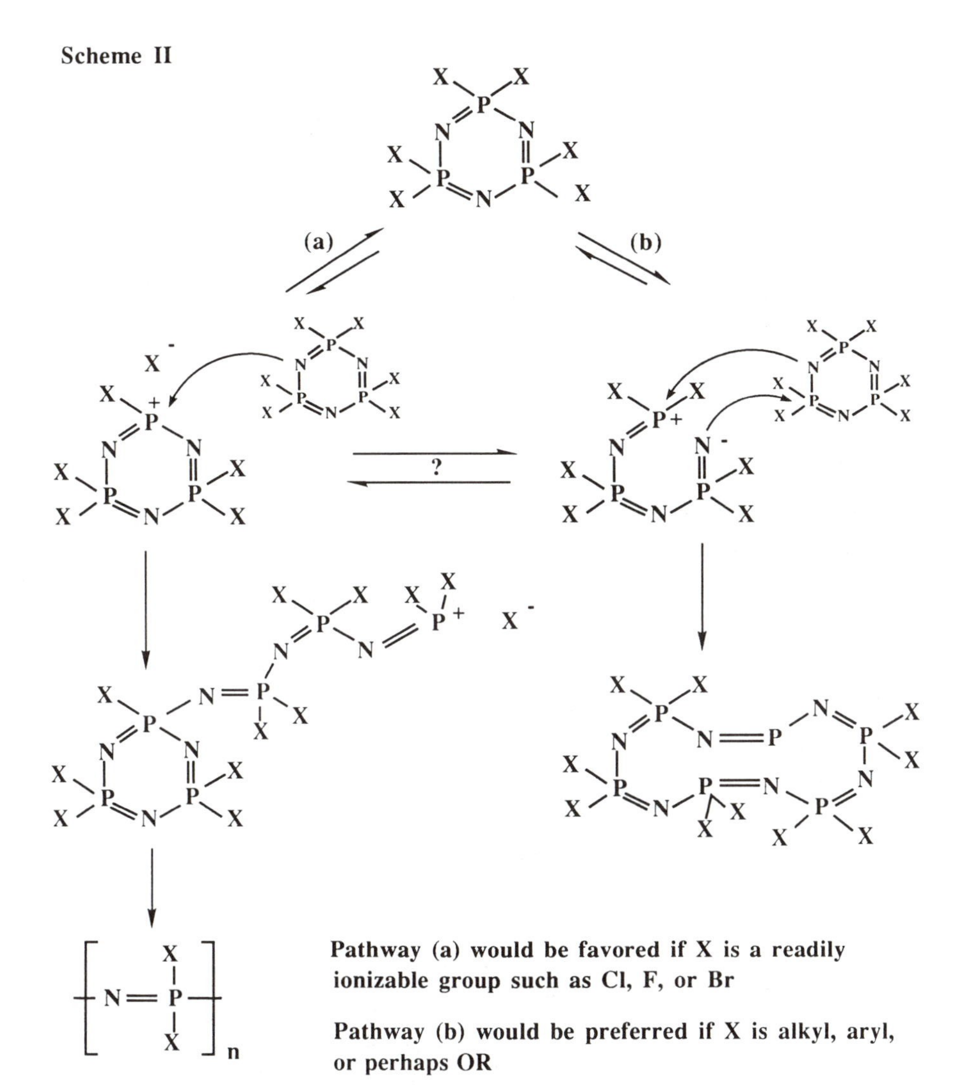
(a)
(b)
?
Pathway (a) would be favored if X is a readily ionizable group such as Cl, F, or Br
Pathway (b) would be preferred if X is alkyl, aryl, or perhaps OR

(b) Changes in the Halogen or Pseudohalogen Side Units. Cyclic phosphazenes are known that bear fluoro- (**11**), bromo- (**12**) , iodo-, isothiocyano- (**13**), and phosphazo (**14, 15**) side groups. With the exception of the iodo-derivatives, which decompose on heating, all of these different classes of cyclophosphazenes undergo ring-opening polymerization to the appropriate high polymer (*33-38*). However, the polymerization rates vary markedly with the type of side groups. A rough measure of polymerization reactivity can be obtained from the minimum temperature needed to generate polymer within a specific time (say 12 hours). These values are 300-350°C for $(NPF_2)_3$ (*34*), 220°C for $(NPBr_2)_3$ (39, 40), and 145°C for $[NP(NCS)_2]_3$ (*37, 41*), compared to about 240°C for $(NPCl_2)_3$.

It can be argued that the ease of separation of the side groups from phosphorus decreases in the order Br>Cl>F, and this provides support (albeit oblique) for the mechanism shown in Scheme II, pathway (a).

(c) Replacement of Chlorine or Fluorine Side Units by Organic Groups. The synthesis of organo-halogeno-phosphazene cyclic trimers, such as those shown in Chart 1, can be accomplished by a variety of organometallic halogen replacement reactions (*42*). It has been found that the replacement of one or two halogen atoms around the ring by organic units generally does not retard the polymerization process significantly (*43-54*). However, the higher the loading of organic side groups, the lower is the tendency for ring-opening polymerization. High loadings of organic groups favor ring-expansion reactions to cyclic tetramer, pentamer, hexamer, and so on, at the expense of polymerization. For example, compounds **16-19** undergo ring expansion when heated. Compound **20** neither polymerizes nor undergoes ring expansion. Specific details of this behavior are shown in Chart 1.

(d) Phosphazo Side Groups. A recent development that throws additional light on the mechanism is the polymerization of phosphazo-phosphazenes (*38*). Phosphazo-phosphazenes are cyclic trimers that bear one or two $-N{=}PCl_3$ side groups attached to a phophorus atom of the ring (**14, 15**). Both mono- and di-phosphazo-phosphazene trimers have been polymerized by heating (at 150°C for **15**). An important feature of these polymerizations is that the phosphazo units lower the temperature required for ring-opening polymerization compared to $(NPCl_2)_3$. Two possible explanations for this are, first, that the phosphazo units impose strain on the phosphazene ring which can be relieved by ring cleavage or, second, that the phosphazo units undergo P-Cl ionization more readily than can the PCl_2 units in $(NPCl_2)_3$. Thus, the possibility exists that the polymerization is controlled by the ionization-initiation process rather than by chain propagation.

(e) Polymerization of Halogen-Free Cyclic Phosphazenes. Although the presence of P-Cl bonds may be a prerequisite for the ring-opening-polymerization of most cyclic phosphazenes, two sets of circumstances exist where the requirement may be waived. First, it has been found that specific organophosphazenes such as **18** or **21**, although resistant to polymerization on their own, will nevertheless *co*polymerize with $(NPCl_2)_3$ (*43, 49, 53, 54*). Apparently the halogen-containing molecules provide cationic initiation sites by p-halogen ionization, and these initiate chain propagation of the organophosphazene species.

A second, and less expected, case is the recent report (*57*) that cyclic phosphazenes that bear both trifluoroethoxy and transannular ferrocenyl groups (**21**) polymerize not only in the presence of catalytic amounts of $(NPCl_2)_3$ but also in the

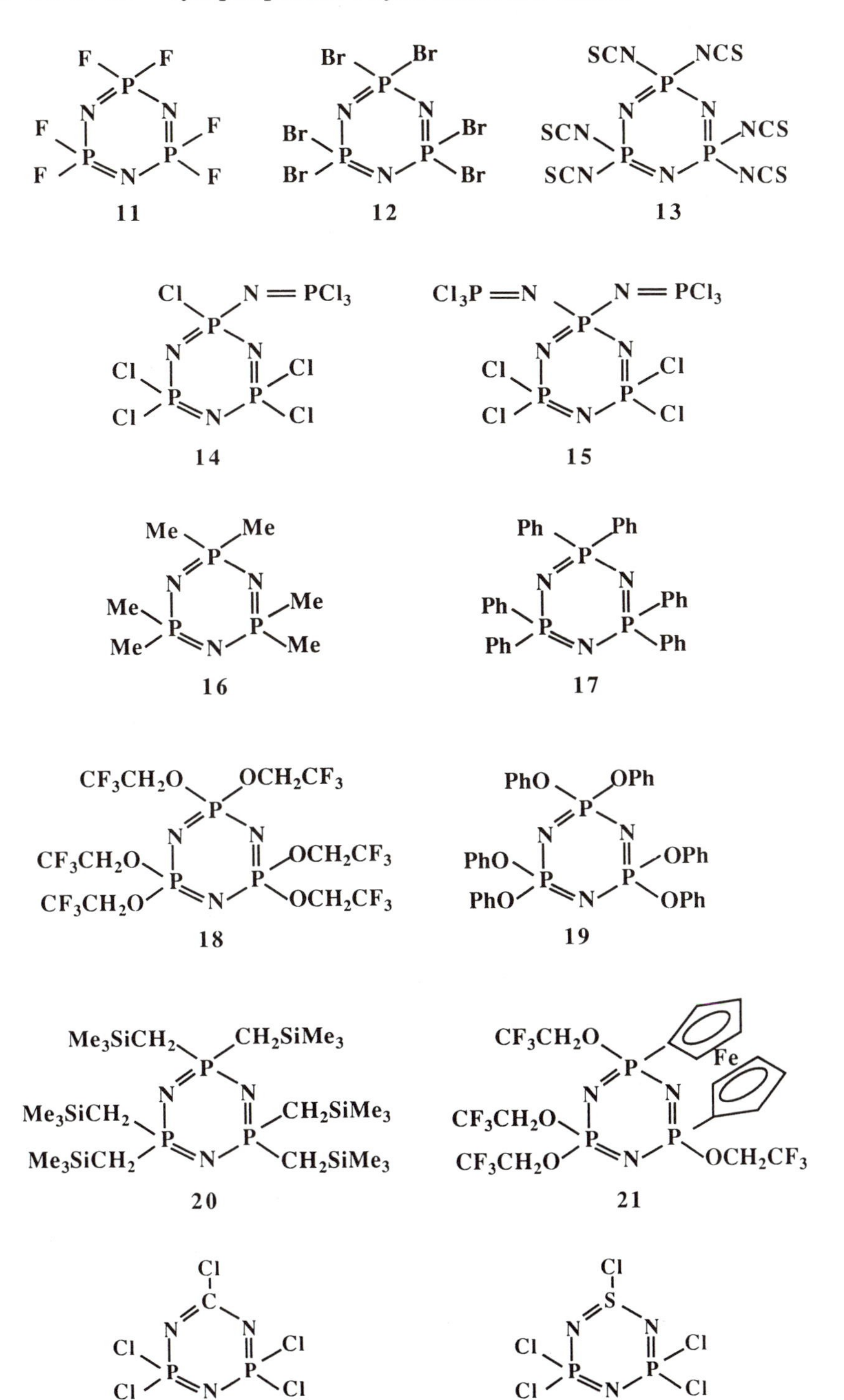

11
12
13
14
15
16
17
18
19
20
21
22
23

Chart 1

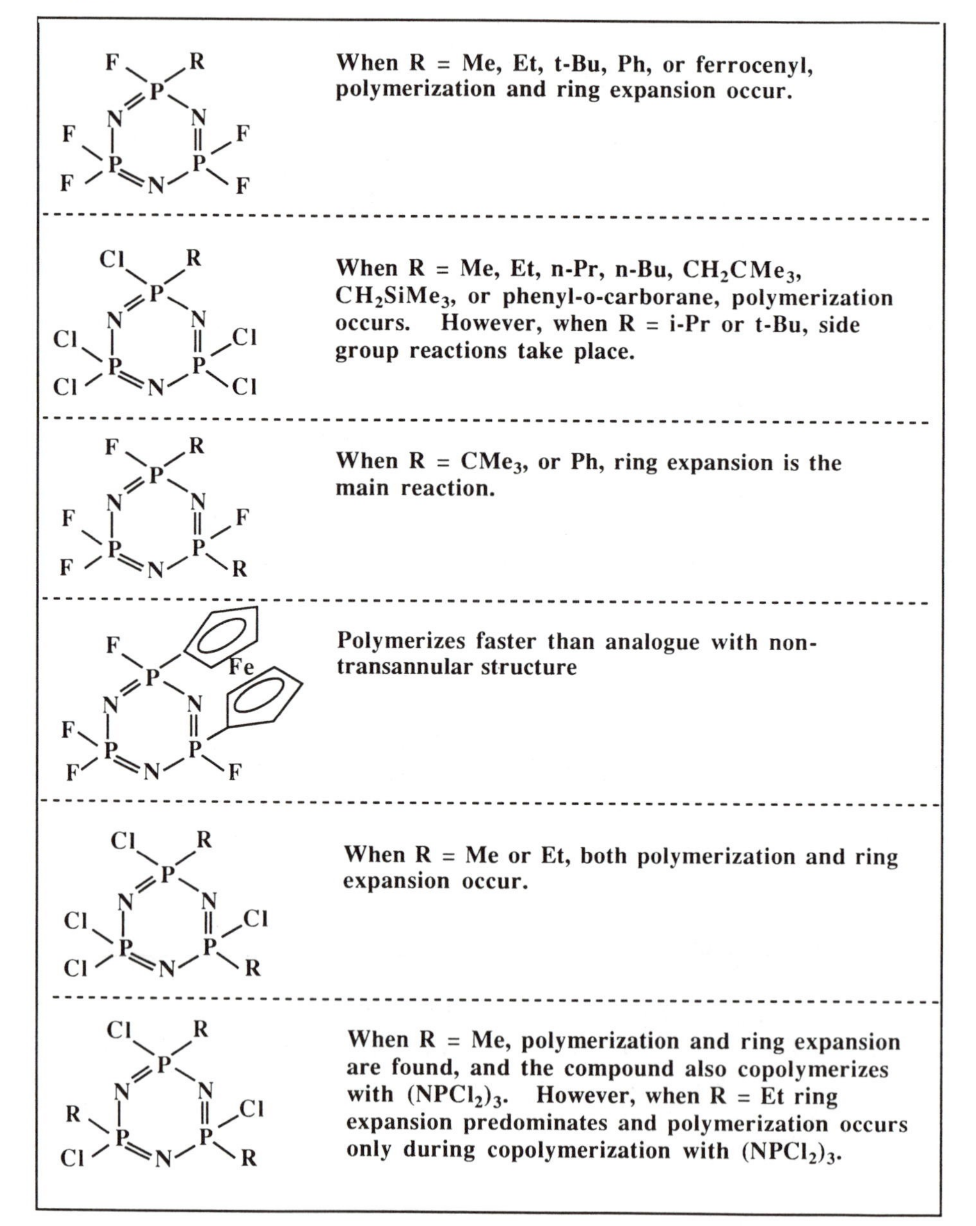
When R = Me, Et, t-Bu, Ph, or ferrocenyl, polymerization and ring expansion occur.
When R = Me, Et, n-Pr, n-Bu, CH_2CMe_3, CH_2SiMe_3, or phenyl-o-carborane, polymerization occurs. However, when R = i-Pr or t-Bu, side group reactions take place.
When R = CMe_3, or Ph, ring expansion is the main reaction.
Polymerizes faster than analogue with non-transannular structure
When R = Me or Et, both polymerization and ring expansion occur.
When R = Me, polymerization and ring expansion are found, and the compound also copolymerizes with $(NPCl_2)_3$. However, when R = Et ring expansion predominates and polymerization occurs only during copolymerization with $(NPCl_2)_3$.

absence of any P-halogen species. This behavior appears to be a consequence of phosphazene ring strain imposed by the transannular structure. Polymerization releases this strain.

(f) Polymerization of Related Inorganic Heterocyclic Molecules. Finally, we have recently reported the ring-opening polymerization of ring systems such as **22** and **23** (*58-60*). Here too, the ionization of chloride ion from phosphorus or sulfur may be a key component in the ring-opening-polymerization mechanism.

Possible Mechanisms

Two alternative mechanistic pathways are shown in Scheme II. In the first (Pathway a), the initiation step, as discussed above, is a thermally induced ionization of a halogen or pseudohalogen unit from phosphorus (*32, 61*). In the second (Pathway b), which is assumed to operate mainly when no P-halogen bonds are present (*57*), the initiation step is a heterolytic cleavage of a P-N skeletal bond to give a linear zwitterion. This then attacks another ring system to yield ring expansion products or linear intermediates. The possibility exists that medium or high molecular weight polymers could be generated by this process, and that some of these products may be giant macrocycles. It is not possible to say at this point if this mechanism proceeds via linear intermediates which subsequently cyclize, or if polymerization is really a ring-fusion process.

Possible Catalysts

If the foregoing arguments are correct, it should follow that reagents which assist the ionization of halogen or pseudohalogen anions from phosphorus should accelerate the initiation process. This provides a starting point for understanding the following results in which various reagents or experimental conditions have led to increases in the polymerization rates.

(a) Effect of Heat. Increases in temperature accelerate the ring-opening polymerization or ring-expansion reactions of every cyclophosphazene studied. This can be rationalized in terms of increases in the number of cationic centers that could initiate polymerization chains and, of course, to an increase in the rate of chain propagation. So far, no clear relationship has yet been established between the temperature of initiation and the degree of polymerization. Increased temperatures would also be expected to increase the equilibrium concentrations of zwitterions (Pathway b). However, temperature increases beyond a certain point may be detrimental to the polymerization process. For example, as mentioned earlier, the crosslinking of $(NPCl_2)_3$, $(NPF_2)_3$, and $[NP(NCS)_2]_3$ is also accelerated at elevated temperatures. The relative rates of polymerization versus crosslinking as the temperature is raised have not yet been established. Moreover, phosphazene ring-opening polymerizations, like most ring-opening polymerizations, are subject to ceiling temperature phenomena in which temperature increases favor an entropic-driven depolymerization to small molecule rings rather polymerization to high molecular weight chains.

(b) Influence of Free Radical Sources and High Energy Radiation. In general, no enhancement in the rate of polymerization has been reported for the melt polymerization following the addition of conventional free radical producing reagents. Moreover, the irradiation of molten $(NPCl_2)_3$ with X-rays or gamma-rays (*62, 63*) has not resulted in significant increases in the rates of polymerization. These results suggest that the melt polymerization mechanism does not follow a free radical pathway. However, it should be noted that crystalline $(NPCl_2)_3$ undergoes a low yield solid state polymerization reaction (*62, 63*), but this probably involves a different mechanism from the reaction in the molten state.

(c) Lewis Acids as Polymerization Accelerators. Lewis acids such as BCl_3 are used as accelerators for the molten state or solution state polymerization of $(NPCl_2)_3$ (*10, 61, 64, 65*). Presumably this is a consequence of the ability of BCl_3 to accept a Cl^- ion to form a BCl_4^- ion, thereby assisting the initiation process. In practice, the BCl_3 is used in the form of a complex with triphenyl phosphate $[Cl_3B{:}OP(OPh)_3]$ (*66*), which (as a solid) is easier to measure and manipulate than gaseous BCl_3. The average molecular weight of the polymer decreases with increases in catalyst concentration in the classical manner. However, the most useful feature of this initiator is that it allows a reduction in reaction temperature (to 210°C) (*10*), and favors polymerization over crosslinking. Thus, polymers prepared in the presence of this initiator are formed in higher yields with a reduced risk of crosslinking than in the pure melt polymerization process.

In theory other Lewis acids such as $AlCl_3$, PCl_3, or PCl_5 (PCl_6^- PCl_4^+) should also accelerate the ring-opening polymerization. So far the results are ambiguous. Addition of PCl_5 to molten $(NPCl_2)_3$ results in cleavage of the phosphazene rings to form linear end-capped, short chain species, but it does not appear to accelerate the polymerization (*67*). Aluminum chloride is reported to function in a similar way.

(d) $(NPCl_2)_3$ as a Polymerization Initiatior. If $(NPCl_2)_3$ polymerizes in the absence of an added initiator, will this molecule serve as an initiator for cyclic organo-phosphazenes that do not normally polymerize, and will it accelerate the polymerization of other phosphazenes? The answer to both questions is a qualified "yes". For example, $(NPClEt)_3$ or species **18** yield polymers when heated in the presence of $(NPCl_2)_3$, but they undergo ring expansion only in the absence of this reagent. Compound **21** polymerizes faster in the presence of $(NPCl_2)_3$ than in its absence.

(e) Water as an Initiator. Unless a careful control is exercised over the purity of $(NPCl_2)_3$ the rate of polymerization varies widely depending on the history of the sample. An investigation of this phenomenon led to the conclusion that traces of water in the system were at least partly responsible for the variations. Experiments in which measured amounts of water were added to the highly purified phosphazene trimer indicated that, in small amounts, water accelerates the polymerization (*67*). Water has also been reported to act as a "promoter" for the solution state polymerization of $(NPCl_2)_3$ when other protonic initiators are used as catalysts (*68*). However, at larger concentrations water decelerates the reaction and eventually functions as an inhibitor (*67*). An explanation for this behavior is that traces of water assist chloride ionization from phosphorus, while larger amounts bring about extensive conversion of P-Cl bonds to P-OH units which undergo intermolecular condensation to form crosslinked cyclomatrix side products, with concurrent liberation of HCl. Hydrogen chloride appears to inhibit the molten state ring-opening polymerization of $(NPCl_2)_3$ (*67*), although reports have appeared that HCl is a catalyst for the solution state polymerization (*68*).

(f) Other Potential Initiators. A number of other accelerators for this polymerization have been mentioned from time to time in the literature. These include sulfur, sulfamic acid, and sulfonic acids (*68*), alkylaluminum chlorides and hydrides (*69*), and a variety of metals. The mode of action of these additives is not known, although a cationic polymerization mechanism would appear to be consistent with the experimental observations.

(g) Effect of Solvents. Numerous attempts have been made to control the polymerization of $(NPCl_2)_3$ by carrying out the reaction in organic solvents. Linear aliphatic hydrocarbons and benzene are attacked by the cyclic trimer at elevated temperatures, and these side reactions disrupt the polymerization process, presumably by Friedel Crafts-type reactions. However, chlorinated aromatic solvents or carbon disulfide under pressure have been used successfully as polymerization solvents (*65, 68, 70*). Polymerizations of this type, especially those in chlorinated aromatic solvents catalyzed by BCl_3 or sulfamic acid (*66, 68*), provide control over the rate of polymerization and the polymer molecular weight, and they also lower the tendency for crosslinking in the final stages of the polymerization.

Conclusions

The mechanisms of cyclophosphazene ring-opening polymerization are by no means fully established, and other catalysts will undoubtedly be developed in the future. Moreover, the mechanisms of other reactions that lead to polyphosphazenes, such as the condensation-type polymerizations reported by Neilson and Wisian-Neilson (*71*) are also subject to catalysis (*72*) and this aspect will also be a fertile area for studies in the foreseeable future.

Literature Cited

1. Mark, J. E.; Allcock, H. R.; West, R. *Inorganic Polymers*; Prentice Hall, N.J., 1991, Chapter 3.
2. Mark, J. E.; Allcock, H. R.; West, R. *Inorganic Polymers*; Prentice Hall, N.J., 1991, Chapter 4.
3. Allcock, H. R.; Kugel, R. L. *J. Am. Chem. Soc.* **1965**, *87*, 4216.
4. Allcock, H. R.; Kugel, R. L.; Valan, K. J. *Inorg. Chem.* **1966**, *5*, 1709.
5. Allcock, H. R.; Kugel, R. L. *Inorg. Chem.* **1966**, *5*, 1716.
6. Allcock, H. R.; Mack, D. P. *J. Chem. Soc., Chem. Commun.* **1970**, *11*, 685.
7. Allcock, H. R.; Moore, G. Y. *Macromolecules* **1972**, *5*, 231.
8. Tate, D. P. *J. Polym. Sci., Polym. Symp.* **1974**, *48*, 33.
9. Singler, R. E.; Schneider, N. S.; Hagnauer, G. L. *Polym. Eng. Sci.* **1975**, *15*, 321.
10. Singler, R. E.; Sennett, M. S.; Willingham, R. A. In *Inorganic and Organometallic Polymers*, Zeldin, M.; Allcock, H. R.; Wynne, K. J., eds., 1988, ACS Symp. Ser. 360, 268.
11. Allcock, H. R.; Cook, W. J.; Mack D. P. *Inorg. Chem.* **1972**, *11*, 2584.
12. Allcock, H. R.; Fuller, T. J.; Mack, D. P.; Matsumura, K.; Smeltz, K. M. *Macromolecules* **1977**, *10*, 824.

13. Allcock, H. R.; Patterson, D. B.; Evans, T. L. *J. Am. Chem. Soc.* **1977**, *99*, 6095.
14. Allcock, H. R.; Wright, S. D.; Kosydar, K. M. *Macromolecules* **1978**, *11*, 357.
15. Allcock, H. R.; Patterson, D. B.; Evans, T. L. *Macromolecules* **1979**, *12*, 172.
16. Allcock, H. R.; Austin, P. E.; Neenan, T. X. *Macromolecules* **1982**, *15*, 689.
17. Allcock, H. R.; Scopelianos, A. G. *Macromolecules* **1983**, *16*, 715.
18. Blonsky, P. M.; Shriver, D. F.; Austin, P. E.; Allcock, H. R. *J. Am. Chem. Soc.* **1984**, *106*, 6854.
19. Allcock, H. R.; Austin, P. E.; Neenan, T. X.; Sisko, J. T.; Blonsky, P. M.; Shriver, D. F. *Macromolecules* **1986**, *19*, 1508.
20. Kim, C.; Allcock, H. R. *Macromolecules* **1987**, *20*, 1726.
21. Singler, R. E.; Willingham, R. A.; Lenz, R. W.; Furakawa, A.; Finkelmann, H. *Macromolecules* **1987**, *20*, 1727.
22. Allcock, H. R.; Connolly, M. S.; Sisko, J. T.; Al-Shali, S. *Macromolecules* **1988**, *21*, 323-334.
23. Allcock, H. R.; Brennan, D. J. *J. Organomet. Chem.*, **1988**, *341*, 231.
24. Allcock, H. R.; Kwon, S. *Macromolecules* **1988**, *21*, 1980.
25. Allcock, H. R.; Mang, M. N.; Dembek, A. A.; Wynne K. J. *Macromolecules* **1989**, *22*, 4179.
26. Allcock, H. R. In *Biodegradable Polymers as Drug Delivery Systems*; Langer, R.; Chasin, M., Eds., Marcel Dekker, N.Y., 1990, 163.
27. Dembek, A. A.; Allcock, H. R.; Kim, C.; Steier, W. H.; Devine, R. L. S.; Shi, Y. In *Materials for Nonlinear Optics*; Marder, S.; Sohn, J. E.; Stucky, G. S., eds., ACS Symp. Ser. 455; Washington, D.C., 1991, 258.
28. Allcock, H. R.; Chang, Ji Young, *Macromolecules* **1991**, *24*, 993.
29. Allcock, H. R.; Pucher, S. R. *Macromolecules* **1991**, *24*, 23.
30. Allcock, H. R.; Kim, C. *Macromolecules* **1991**, *24*, 2846.
31. Allcock, H. R.; Dembek, A. A.; Klingenberg, E. H. *Macromolecules*, **1991**, *24*, 5208.
32. Allcock, H. R.; Best, R. J. *Can. J. Chem.* **1964**, *42*, 447.
33. Seel, F.; Langer, J. *Z. Anorg. Allg. Chem.* **1958**, *295*, 316.
34 Allcock, H. R.; Patterson, D. B.; Evans, T. L. *Macromolecules* **1979**, *12* 172.
35. Allcock, H. R.; Connolly, M. S. *Macromolecules* **1985**, *18*, 1330.
36. Allcock, H. R.; Harris, P. J. *Inorg. Chem.* **1981**, *20*, 2844.
37. Allcock, H. R.; Rutt, J. S. *Macromolecules* **1991** *24*, 2852.
38. Allcock, H. R.; Ngo, D. C. *Macromolecules* (submitted).
39. Allcock, H. R. *Phosphorus-Nitrogen Compounds*, Academic Press, N. Y., 1972, p. 322.
40. Cordischi, D.; Site, A. D.; Mele, A. *J. Macromol. Chem.* **1966**, *1*, 219.
41. Otto, R. J. A.; Audrieth, L. F. *J. Am. Chem. Soc.* **1958**, *80*, 5894.
42. Allcock, H. R.; Desorcie, J. L.; Riding, G. H. *Polyhedron* **1987**, *6*, 119.
43. Allcock, H. R.; Moore, G. Y. *Macromolecules* **1975**, *8*, 377.
44. Ritchie, R. J.; Harris, P. J.; Allcock, H. R. *Macromolecules* **1979**, *12* 1014.
45. Allcock, H. R.; Ritchie, R. J.; Harris, P. J. *Macromolecules* **1980**, *13*, 1332.
46. Allcock, H. R.; Scopelianos, A. G.; O'Brien, J. P.; Bernheim, M. Y. *J. Am. Chem. Soc.* **1981**, *103*, 350.
47. Allcock, H. R.; Connolly, M. S. *Macromolecules* **1985**, *18*, 1330.
48. Allcock, H. R.; Brennan, D. J.; Graaskamp. J. M. *Macromolecules* **1988**, *21*, 1.
49. Allcock, H. R.; Riding, G. H.; Manners, I.; McDonnell, G. S.; Dodge, J. A.; Desorcie, J. L. *Polym. Prepr.* (ACS Div. Polym. Chem.) **1990**, *31*, 48.
50. Allcock, H. R.; McDonnell, G. S.; Desorcie, J. L. *Inorg. Chem.* **1990**, *29*, 3839.
51. Allcock, H. R.; Lavin, K. D.; Riding, G. H. *Macromolecules* **1985**, *18*, 1340.
52. Allcock, H. R.; Riding, G. H.; Lavin, K. D. *Macromolecules* **1987**, *20*, 6.

53. Manners, I.; Riding, G. H.; Dodge, J. A.; Allcock, H. R. *J. Am. Chem. Soc* **1989**, *111*, 3067.
54 Allcock, H. R.; Dodge, J. A.; Manners, I.; Riding G. H. *J. Am. Chem. Soc.* (in press).
55. Allcock, H. R.; Schmutz, J. L.; Kosydar, K. M. *Macromolecules* **1978**, *11*, 179.
56. Allcock, H. R.; McDonnell, G. S.; Desorcie, J. L. *Macromolecules* **1990**, *23*, 3873.
57. Manners, I.; Riding, G. H.; Dodge, J. A.; Allcock, H. R. *J. Am. Chem. Soc* **1989**, *111*, 3067.
58. Manners, I.; Renner, G.; Nuyken, O.; Allcock, H. R. *J. Am. Chem. Soc.* **1989**, *111*, 5478.
59. Allcock, H. R.; Coley, S. M.; Manners, I.; Renner, G.; Nuyken, O. *Macromolecules* **1991**, *24*, 2024.
60. Dodge, J. A.; Manners, I.; Allcock, H. R.; Renner, G.; Nuyken, O. *J. Am. Chem. Soc.* **1990**, *112*, 1268.
61. Lee, D. C.; Ford, J. R.; Fytas, G.; Chu, B.; Hagnauer, G. L. *Macromolecules* **1986**, *19*, 1586.
62. Caglioti, V.; Cordischi, D.; Mele, A. *Nature* (London) **1962**, *195*, 491.
63. Liu, H. Q.; Stannett, V. T. *Macromolecules* **1990**, *25*, 140.
64. Fieldhouse, J. W.; Graves, D. F. U.S. Patent, 4,226,840 (1980) (to Firestone Tire and Rubber Co.).
65. Sennett, M. S.; Hagnauer, G. L.; Singler, R. E.; Davies, G. *Macromolecules* **1986**, *19*, 959.
66. Levin, M. L.; Fieldhouse, J. W.; Allcock, H. R. *Acta Cryst.* **1982**, *B38*, 2284.
67. Allcock, H. R.; Gardner, J. E.; Smeltz, K. M. *Macromolecules* **1975**, *8*, 36.
68. Mujumdar, A. N.; Young, S. G.; Merker, R. L.; Magill, J. H. *Macromolecules* **1990**, *23*, 14.
69. Snyder, D. L.; Stayer, M. L.; Kang, J. W. U.S. Patent 4,123,503 (1978); German 606,802 (1975) (to Firestone Tire and Rubber Co.).
70. Scopelianos, A. G.; Allcock, H. R. *Macromolecules* **1987**, *20*, 432.
71. Neilson, R. H.; Wisian-Neilson, P. *Chem. Rev.* **1988**, *88*, 541.
72. Montague, R. A.; Matyjaszewski, K. *J. Am. Chem. Soc.* **1990**, *122*, 6721.

RECEIVED February 18, 1992

Chapter 19

Polyaddition of Epoxides and Diepoxides to the Acids of Phosphorus

Synthesis of Poly(alkylene phosphate)s

S. Penczek, P. Kubisa, P. Klosinski, T. Biela, and A. Nyk

Center of Molecular and Macromolecular Studies, Polish Academy of Sciences, 90–363 Lodz, Sienkiewicza 112, Poland

Addition of monoepoxides to the acids of phosphorus gives telechelic phosphorus containing polyethers. The relative rates of addition of ethylene oxide to the first acidic group in H_3PO_4, H_3PO_3 and to the only acidic group in $(C_2H_5O)_2P(O)OH$ are the same. The relative rates of addition to the second and third acidic groups in H_3PO_4 are higher than the rate of addition to the first one. The addition process is acid catalysed and either the protonated or the H-bonded monomer molecule actually reacts. Polyaddition of diepoxides to H_3PO_3 or H_3PO_4 leads to the linear and/or branched polymers. When gels are formed their hydrolysis gives water soluble poly(alkylene phosphate)s.

This paper gives a short account of the recent work from this laboratory directed towards synthesis of poly(alkylene phosphate)s by direct addition of diepoxides to the acids of phosphorus. The first high molecular weight polymers of this structure were prepared differently, namely by dealkylation of poly(alkylene phosphate)s, like poly(methyl ethylene phosphate). This particular parent polymer, prepared by catalytic ring-opening polymerization, first appeared in a patent authored by E.J. Vandenberg (*1*).

Poly(alkylene phosphate)s constitute an important class of novel polymers. Some of these polymers, having a sequence of atoms in the backbone similar to this occuring in biopolymers, like teichoic or nucleic acids, provide models for studying the behaviour of polyphosphate chains in solution, their interaction with other biopolymers (e.g. peptides) or the mechanism of transport of ions through membranes (teichoic acids are the components of the cell walls of certain bacteria). They can be considered also as potential biocompatible carriers of biologically active compounds. Synthetic methods for preparation of poly(alkylene phosphate)s can be divided into two groups:

- ring-opening polymerization of cyclic phosphates
- polycondensation of phosphoric acid with diols

Both approaches have been studied in this laboratory and it was shown

0097–6156/92/0496–0248$06.00/0

that under suitable conditions, high molecular weight polymers can be obtained by these methods. Both methods, however, have certain disadvantages. Preparation and purification of monomeric cyclic phosphates is difficult. These monomers polymerize only by ionic mechanism, thus free P-OH groups have to be blocked prior to polymerization and deblocked afterwards. This introduces an additional reaction step. Polycondensation method, on the other hand, requires more drastic reaction conditions and cannot be used if monomers contain fragile substituents. Thus, in order to develop a more convenient, simple method of polyphosphate synthesis we have looked into the possibility of forming the linear polyphosphate chain by polyaddition of acids of phosphorus to diepoxides. The simplified scheme of this process is shown below:

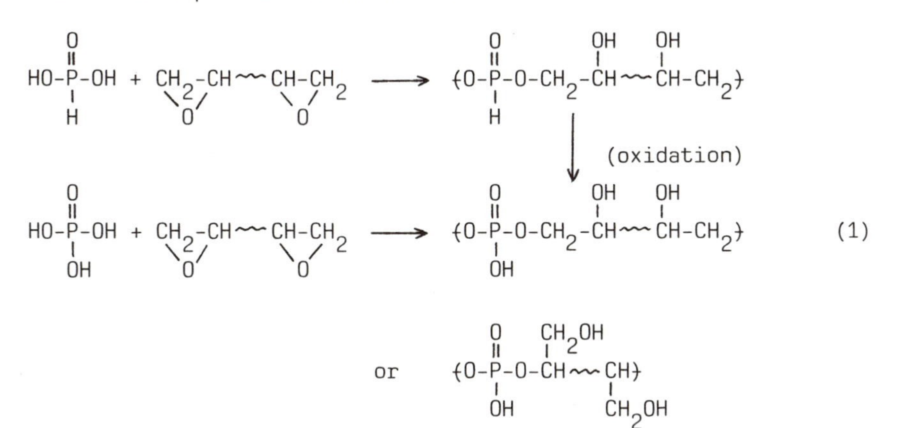

To estimate to what extent such a process can be controlled, two questions should be answered:

a. what is the relative reactivity of the first, second and third P-OH group in phosphoric acid i.e.: is it possible to convert two functions without affecting the third one.
b. what is the direction of the epoxide ring-opening in this reaction, i.e., is it possible to prepare products with predetermined structure.

In order to answer these questions we have studied the kinetics and the steric course of the addition of various oxiranes to different P-OH acids.

Kinetics of Addition of Oxiranes to the P-OH Acid

Ethylene Oxide

No kinetic and mechanistic data are available for the reactions of oxiranes with the acids of phosphorus. In order to compare the reactivities of subsequent P-OH groups in phosphorus containing acids we have studied addition of ethylene oxide (EO) to acids containing one, two, and three P-OH groups respectively (2).

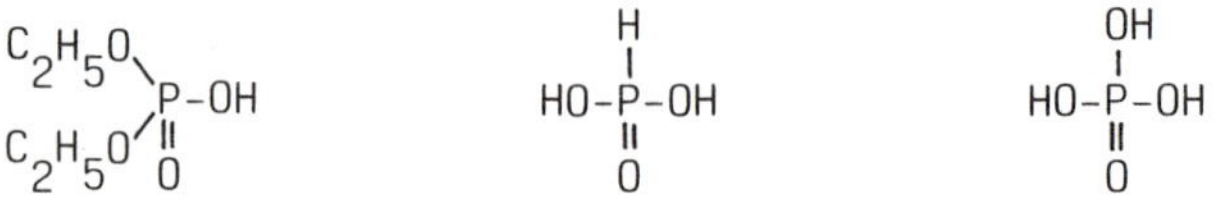

$(C_2H_5O)_2P(O)OH$	$H-P(O)(OH)_2$	$HO-P(O)(OH)_2$	(2)
diethyl phosphoric acid	phosphonic acid	phosphoric acid	
DEPA	PSA	PCA	

The kinetics of these reactions was followed by monitoring changes in EO concentration (^{1}H-NMR) and the concentration of P-OH groups (^{31}P-NMR). Analysis of the kinetics of the process led to the following conclusions (2):

- addition of oxiranes to the P-OH group is an acid catalysed reaction; in the absence of any "external" catalyst, this reaction is catalysed by the P-OH groups, playing thus a double role of reactant and catalyst
- P-OH groups catalyse also an addition of oxiranes to the C-OH groups of the product, thus, more than one oxirane unit may be introduced per one originally present P-OH group,
- in the absence of the "external" catalyst, reaction stops after the complete consumption of the P-OH groups, giving short chain oligomers; in the presence of the additional catalyst (e.g. $HBF_4 \cdot Et_2O$) reaction proceeds further and longer chain oligomers are formed,
- the catalysis involves both protonation of EO molecules (ionized form of the P-OH acid) and hydrogen bonding (unionized acid); the former reaction is much faster ($k = 2 \cdot 10\ mol^{-1} \cdot l \cdot s^{-1}$ in 1,4-dioxane at 25°), however, due to the low ionization degree of the P-OH acids under the conditions employed (ionization degree $\gamma \sim 4.8 \cdot 10^{-4}$ in ~1 $mol \cdot l^{-1}$ soln in 1,4-dioxane, 25°) reaction proceeds mainly (~ 85% of EO consumption) on much more abundant hydrogen bonded species.

The apparent second order rate constants of the addition reaction:

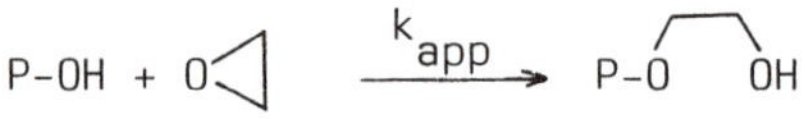

$$P{-}OH + O\!\triangleleft \xrightarrow{k_{app}} P{-}O\frown OH$$

are the following (Biela,T., Kubisa,P. *Makromol.Chem.*, in press).

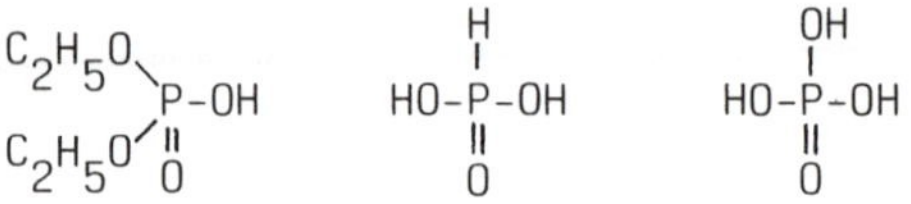

k (in $mol^{-1} \cdot l \cdot s^{-1}$)	$(C_2H_5O)_2P(O)OH$	$H-P(O)(OH)_2$	$HO-P(O)(OH)_2$
k_1	$2.6 \cdot 10^{-4}$	$2.7 \cdot 10^{-4}$	$2.4 \cdot 10^{-4}$
k_2	-	$5.5 \cdot 10^{-4}$	$4.6 \cdot 10^{-4}$
k_3	-	-	$9.7 \cdot 10^{-4}$

where k_1, k_2, and k_3 denote the apparent rate constants of the reaction of the first, second, and third P-OH group. These results show that:

- the reactivity of the different first P-OH groups is nearly the same, independent of the acid structure, i.e. k_1 values for all three acids studied are very close to each other
- for the di- and tribasic acids the reactivity increases with the degree of substitution i.e. for PCA $k_1 < k_2 < k_3$.

The similar reactivity of the first P-OH groups in all three acids studied is due to the similar acid strength. The known values of the dissociation constants in aqueous solution do not differ considerably (*3*). The increase of the reactivity with increasing the degree of substitution is probably due to the formation of the internal hydrogen bond:

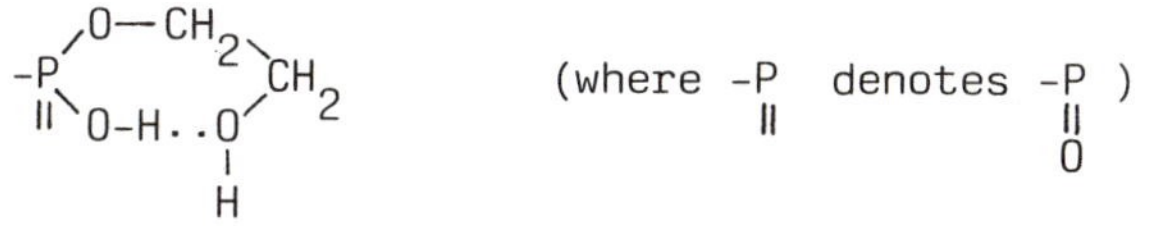

```
  ,O—CH2
-P       CH2        (where -P  denotes -P )        (3)
 ‖`O-H..O'                 ‖           ‖
        |                              O
        H
```

loosening the P-OH bond and increasing thus the nucleophilic character of the oxygen atom in the P-OH group.

In the course of reaction, some P-OH groups are converted into the P-O⌒OH units, whereas the unreacted P-OH groups are still present in the system. This results in further reaction: the oxirane molecule, activated (through protonation or hydrogen bonding to the P-OH groups), reacts with the terminal hydroxyl group (C-OH), extending the oligoether chain.

$$-P-O\frown OH + H-\overset{+}{O}\triangleleft \longrightarrow -P-O\frown O\frown OH + "H^+"$$

This reaction resembles closely the propagation step in the Activated Monomer (AM) polymerization of oxiranes, a process studied extensively in this laboratory (*4,5*). The kinetics of oxirane -POH addition is, however, much more complex than the simple kinetics of the AM polymerization, due to the possibility of formation of the intramolecular hydrogen bonds.

Thus, it was observed for the PCA-EO system that the rate constant of addition of the activated EO molecule to the P-O⌒OH group is close to the rate of addition to the P-OH group ($k = 6.5 \cdot 10^{-4}$ $mol^{-1} \cdot l \cdot s^{-1}$) whereas the apparent rate constant of addition to the P-O⌒O⌒OH unit is much lower, leading to the negligibly low concentrations of oligomers, containing more than one EO unit.

This can be attributed to the enhanced reactivity (nucleophilicity) of the C-OH group, involved in formation of the intramolecular hydrogen bond with the phosphoryl oxygen. The probability of cyclization, when one EO unit is attached (7 membered ring), appears to be higher than for two attached units (10 membered ring).

```
  ,O··H—O                    ,O···H-O-CH2
—P       |          vs     —P            CH2       (4)
 |`O-CH2-CH2                |`O-CH2-CH2-O'
```

The knowledge of the apparent rate constants in the series of consecutive-parallel reactions allows one to predict the composition of the reaction mixture at any stage of reaction, as illustrated by the scheme below:

```
     O                 O                       O                          O
     ||       EO*      ||            EO*       ||             EO*         ||
HO-P-OH  ------>  HO-P-EO-OH  ------>  HO-P-EO-OH  ------>  HO-EO-P-EO-OH
     |                 |                       |                          |
     OH                OH                      EO-OH                      EO-OH

(0.32,0.31)       (0.18,0.16)           (0.08,0.057)           (0.02,0.027)

                       |  EO*                  |  EO*                     |  EO*
                       v                       v                          v
                       O                       O                          O
                       ||          EO*         ||            EO*          ||
                  HO-P[EO]2OH  ------>  HO-P[EO]2OH  ------>  HO-EO-P[EO]2OH
                       |                       |                          |
                       OH                      EO-OH                      EO-OH

                  (0.13,0.146)          (0.09,0.105)           (0.06,0.041)

                                               |  EO*                     |  EO*
                                               v                          v
                                               O                          O
                                               ||            EO*          ||
                                        HO-P[EO]2OH  ------>  HO-EO-P[EO]2OH
                                               |                          |
                                               [EO]2OH                    [EO]2OH

                                        (0.04,0.048)           (0.05,0.068)

                                                                          |  EO*
                                                                          v
                                                                          O
                                                                          ||
                                                               HO[EO]2P[EO]2OH
                                                                          |
                                                                          [EO]2OH

                                                               (0.03,0.034)
```

In this scheme EO* denotes activated EO molecule and the numbers in parenthesis give the measured (from ^{31}P-NMR) and calculated (on the basis of the derived set of the rate constants) molar fractions of individual oligomers at the certain stage of reaction.

The results of the kinetic analysis indicate, that, due to the increasing reactivity of the subsequent P-OH group in reaction of PCA with oxiranes, it is not possible at any stage of reaction to achieve esterification of two P-OH groups without involving the third one. The final product, at complete conversion of the P-OH groups, is the mixture of the four triesters shown in the last vertical row of the scheme.

This result shows also, that the C-OH groups, formed in the reaction, may also react with oxiranes in the presence of P-OH, the rate of this reaction may, however, strongly depend on the structure of related species (e.g. whether C-OH groups may participate in formation of the hydrogen bonds with the phosphoryl oxygen).

Steric course of addition of the substituted oxiranes to the P-OH groups in phosphonic and phosphoric acid.

Unsymetrically substituted oxiranes, reacting with the P-OH groups, can give two products, due to two possible directions of the ring-opening:

$$\text{P-OH} + \text{O}\lhd \longrightarrow \text{P-O}\diagup\!\!\overline{\;\;\;}\!\!\diagdown\text{OH}\ (\alpha)\ \text{or}\ \text{P-O}\diagup\!\!\overline{\;\;\;}\!\!\diagdown\text{OH}\ (\beta) \quad (5)$$

Due to the high sensitivity of the chemical shift of the phosphorus nucleus in ^{31}P-NMR to the chemical environment, these isomeric products can be observed separately. Assignments of the pertinent signals in the ^{31}P-NMR were based on the comparison with the spectra of the suitable models (when available) and the multiplicity of the phosphorus signal arising from the P-H coupling.
Using this approach, the steric course of addition of various substituted oxiranes to DEPA, PSA, and PCA was studied (Biela,T., Kubisa, P., Penczek,S., unpublished data). In some systems, in order to detect the expected influence of the reaction conditions on the direction of the ring-opening, the products obtained in different solvents and different temperatures, were analysed. The results are summarized in Table I.

Table I. Proportions of products of α-ring-opening (attack on the less substituted carbon atom of the oxiranes ring) in the reaction of various oxiranes with phosphorous acid (PSA) at <50% conversion of P-OH groups

Solvent	t°C	Fraction of α-ring-opening for				
		CH_3-oxirane	C_2H_5-oxirane	CH_2Cl-oxirane	C_6H_5-oxirane	$(CH_3)_2$-oxirane
1,4-dioxane	25	0.50	0.41	0.66	0.63	0.89
$(C_2H_5)_2O$	25	0.50	0.28	0.62	0.66	0.89
$CH_3O(CH_2CH_2O)_3CH_3$	25	0.50	0.38	0.66	0.60	0.89
$CH_3O(CH_2CH_2O)_3CH_3$	-18	0.49	0.40	0.60	0.61	0.88

Analysis of the products at different conversions of the P-OH groups indicates, that in both PSA and PCA the fraction of the α-ring-opening does not practically change with the degree of conversion, i.e. is the same for the reaction of the first P-OH groups in PCA as for the reaction of the third group in the diester. This indicates that the steric crowding around phosphorus is not important enough to affect the direction of the ring-opening in the subsequent additions. Results of Table I indicate, that the direction of the ring-opening is not affected by reaction conditions. Fractions of α- and β-ring--openings in the studied reactions are close to these, observed for acid catalysed solvolysis of the corresponding oxiranes. Thus, for the monosubstituted oxiranes, reaction is not regiospecific, although depending on the electronic and steric effect of substituents, slight preferences for α- or β-ring-opening are observed. Only for the disubstituted oxirane: isobutylene oxide the α-ring-opening is clearly preferred, apparently due to the prevailing steric effect and only in this case reasonably high selectivity can be obtained.

Addition of the Acids of Phosphorus to Diepoxides

The results described above for phosphorus acids addition to mono-epoxides allow one to better understand the conditions needed to obtain high polymers by polyaddition of these acids to the diepoxides. First, the phosphonic acid (the tautomeric form of the phosphorous acid) was applied in order to obtain linear polymers (*6*) (Equation (6)).

$$HO{-}P(=O)(H){-}OH + \underset{\backslash O /}{CH_2{-}CH}{-}R{-}\underset{\backslash O /}{CH{-}CH_2} \longrightarrow {-}\!\!(O{-}P(=O)(H){-}O{-}CH_2{-}CH(OH){-}R{-}CH(OH){-}CH_2)\!\!{-} \qquad (6)$$

Diglycidyl ethers of di- and triethylene glycols (-R- = = $-CH_2O(CH_2CH_2O)_2CH_2-$, and $-CH_2O(CH_2CH_2O)_3CH_2-$ respectively) were used as the diepoxy compounds. The polyadditions were performed at room temperature in 1,4-dioxane solvent. However, the reaction mixture after the complete conversion of the diepoxide contained unreacted acid and its monoesters. This resulted from the reaction of the C-OH groups formed with the epoxy groups. After blocking in statu nascendi of the C-OH groups only polymeric diesters were detected (^{31}P-NMR). Acetic anhydride and 3,4-dihydro-2H-pyran (DHP) were the most efficient. Acetic anhydride was used in the stoichiometric amount, whereas 2-3 fold excess of the 3,4-dihydro-2H-pyran was applied.

$$HO{-}P(=O)(H){-}OH + \underset{\backslash O /}{CH_2{-}CH}{-}R{-}\underset{\backslash O /}{CH{-}CH_2} + (CH_3CO)_2O \longrightarrow {-}\!\!(O{-}P(=O)(H){-}O{-}CH_2{-}CH(OCOCH_3){-}R{-}CH(OCOCH_3){-}CH_2)\!\!{-} \qquad (7)$$

$$HO{-}P(=O)(H){-}OH + \underset{\backslash O /}{CH_2{-}CH}{-}R{-}\underset{\backslash O /}{CH{-}CH_2} + \text{DHP} \longrightarrow {-}\!\!(O{-}P(=O)(H){-}O{-}CH_2{-}CH(O\text{-THP}){-}R{-}CH(O\text{-THP}){-}CH_2)\!\!{-} \qquad (8)$$

Oxidation of thus obtained poly(alkylene phosphonate)s with N_2O_4 and deblocking of the OH groups (by simple aging in H_2O) gave the required poly(alkylene phosphate)s (*6*):

$$\text{-(O-}\underset{\text{H}}{\overset{\text{O}}{\text{P}}}\text{-O-CH}_2\text{-}\underset{}{\overset{\text{OCOCH}_3}{\text{CH}}}\text{-R-}\underset{\text{OCOCH}_3}{\text{CH}}\text{-CH}_2\text{-)} \xrightarrow[\text{2.deblocking}]{\text{1. } N_2O_4} \text{-(O-}\underset{\text{OH}}{\overset{\text{O}}{\text{P}}}\text{-O-CH}_2\text{-}\overset{\text{OH}}{\text{CH}}\text{-R-}\overset{\text{OH}}{\text{CH}}\text{-CH}_2\text{-)} \quad (9)$$

Similarly to the addition of the monoepoxides, both ring-openings (α and β) occurred with probabilities close to 1:1.
Application of the blocking reagents increased the proportion of the cyclization processes, namely formation of cyclic end-groups, like 2-hydro-2-oxo-1,3,2-dioxaphospholanes:

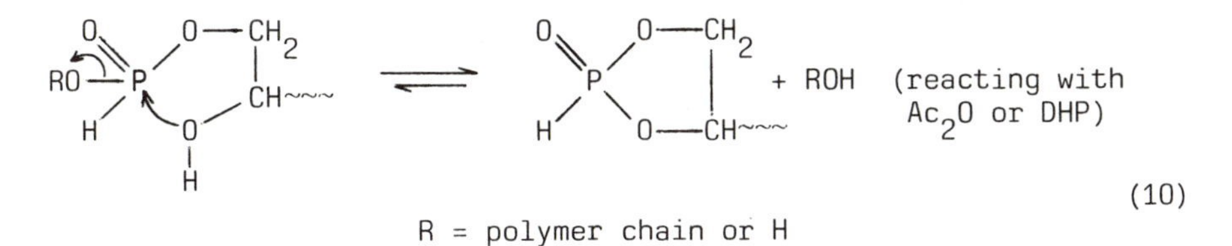

R = polymer chain or H

$\overline{M}n$ of these polymers, measured by vpo, reached $3 \cdot 10^4$. Since the found proportion of the cyclic end-groups (assuming two end-groups per macromolecule) would only give $\overline{M}n \sim 1.5 \cdot 10^3$, then the most probable structure of the obtained products is the branched chain with at least 1 branch for every 5 linear units (*6*).
Branching has been further suppressed by using disilyl methyl esters of the phosphoric acid instead of the acid itself (Nyk,A., Klosinski, P., Penczek,S., unpublished data):

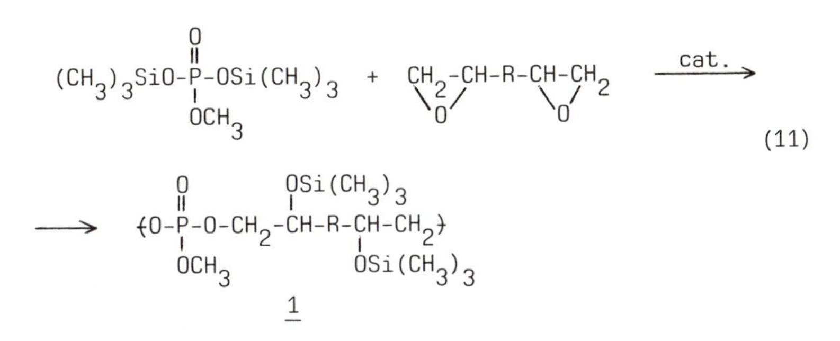

This polyaddition was performed at 80°, without solvent. $SnCl_2$, methylimidazole or 4-dimethylaminopyridine were used as catalysts. $\overline{M}n$ of the obtained polymers ranged from 3 to $9 \cdot 10^3$. The structures of 1 (where R = $CH_2O(CH_2CH_2O)_2CH_2$ or a simple chemical bond) were studied by ^1H, $^{31}P\{^1H\}$, $^{13}C\{^1H\}$ and $^{29}Si\{^1H\}$NMR (Nyk,A., Klosinski, P., Penczek,S., unpublished data). Also in these polymers cyclic end-groups were found by ^{31}P NMR. Their formation is visualized below, in Equation (12). The side product, hexamethyldisiloxane, was detected in concentration corresponding to the concentration of the end-groups formed.

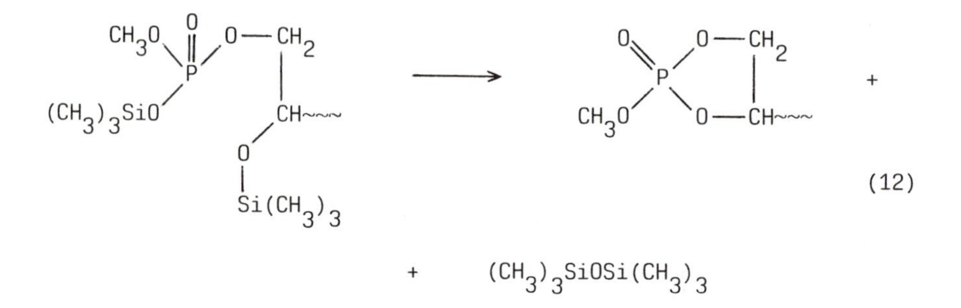

The regioselectivity of polyaddition (α or β ring-opening) is highly dependent on the structure of the catalyst used.
The best results (80% of the secondary blocked hydroxyls formed) were obtained with 4-dimethylaminopyridine.
Further transformation of polyphosphates 1 by desilylation with water and followed by dealkylation of the CH_3O groups with trimethylamine yields polyphosphates with unsubstituted -OH groups - close analogs of teichoic acid:

$$\left(O-\underset{OCH_3}{\overset{O}{\overset{\|}{P}}}-O-CH_2\overset{OSi(CH_3)_3}{CH}-R-\underset{OSi(CH_3)_3}{CH}-CH_2\right) \xrightarrow[2.\,(CH_3)_3N]{1.\,H_2O} \left(O-\underset{O^-\,\overset{+}{N}(CH_3)_4}{\overset{O}{\overset{\|}{P}}}-O-CH_2-\overset{OH}{CH}-R-\overset{OH}{CH}-CH_2\right) \tag{13}$$

1

Direct polyaddition of phosphoric acid to diepoxy compounds (e.g. in 1,4-dioxane solvent) provided products insoluble in the reaction medium (*7*). When placed in water, these gels are first becoming highly swollen and then slowly dissolve, by breaking the triester knots and providing poly(alkylene phosphate)s (polydiesters), relatively resistant to further hydrolysis. It is known (*8*,*9*), that the ratio of the rate constants of hydrolysis of triesters to diesters may be as high as 10^2 at pH~2, i.e. at pH of the 5-10% water suspension/solution of these polymers.
The original degree of swelling approaches 10^3%. The polymeric materials, obtained after hydrolysis, were characterized by NMR, titration, and molecular weight measurements. $\overline{M}n$ (measured by vpo) of soluble, highly branched polymers reached several thousands (*7*).

Literature Cited

1. Vandenberg,E.J. US Patent, **1970**, 3,520,489.
2. Biela,T., Kubisa,P. *Makromol.Chem.*, **1991**,*192*,473.
3. Kirby,A.J., Warren,S.G. "*The Organic Chemistry of Phosphorus*" in "*Reaction Mechanism in Organic Chemistry* " Monograph 5, Eaborn,C., Chapman,N.B. Eds., Elsevier, Amsterdam, London, New York **1967**, p.24.
4. Penczek,S., Kubisa,P., Matyjaszewski,K., Szymanski,R. *Pure and Appl.Chem.* **1984**,140.
5. Biedron,T., Bednarek,M., Kubisa,P., Penczek,S. *Makromol.Chem., Macromol.Symp.* **1990**,*32*,155.
6. Klosinski,P., Penczek,S. *Makromol.Chem., Rapid Commun.* **1988**,*9*,159.
7. Nyk,A., Klosinski,P., Penczek,S. *Makromol.Chem.* **1991**,*192*,833.
8. Bunton,C.A., Mhala,M.M., Oldhani,K.G., Vernon,C.A. *J.Chem.Soc.* **1960**,2670.
9. Barnard,P.W.C., Bunton,C.A., Llewellyn,D.R., Vernon,C.A., Welch, V.A. *J.Chem.Soc.* **1961**,2670.

RECEIVED November 15, 1991

Chapter 20

Poly(methyl methacrylate)-*block*-polyisobutylene-*block*-poly(methyl methacrylate) Thermoplastic Elastomers

Synthesis, Characterization, and Some Mechanical Properties

Joseph P. Kennedy and Jack L. Price[1]

Department and Institute of Polymer Science, The University of Akron, Akron, OH 44325–3909

Poly(methyl methacrylate-*b*-isobutylene-*b*-methyl methacrylate) (PMMA-PIB-PMMA) triblock copolymers have been synthesized by a combination of living cationic and living anionic polymerizations. The total synthesis involved five steps: 1.) The living cationic polymerization of isobutylene (IB) leading to well defined, narrow molecular weight distribution, α,ω-dichloropolyisobutylenes (^{t}Cl-PIB-^{t}Cl). 2.) The alkylation of toluene with ^{t}Cl-PIB-^{t}Cl using $AlCl_3$ at -78°C which yields α,ω-p-ditolylpolyisobutylene (CH_3-Ø-PIB-Ø-CH_3). 3.) The quantitative lithiation of the latter to α,ω-dibenzyllithiumpolyisobutylene ($LiCH_2$-Ø-PIB-Ø-CH_2Li) by using a complex of *s*-butyllithium (*s*-BuLi) and N,N,N',N'-tetramethylethyldiamine (TMEDA) (TMEDA/*s*-BuLi = 2) in *n*-hexane at 0°C for 48 hrs. 4.) The reaction of $LiCH_2$-Ø-PIB-Ø-CH_2Li with 1,1-diphenylethylene. 5.) The living anionic polymerization of methyl methacrylate (MMA) with the latter intermediate. An inventory of various PMMA-PIB-PMMA have been prepared, characterized, and some of their mechanical properties studied. The molecular weight distribution of the triblocks was nearly identical to the PIB center block which indicates good blocking efficiency. DSC and DMTA showed two transitions (-61 and 105°C for PIB and PMMA, respectively) which indicates a two phase microarchitecture. Stress-strain measurements were performed on clear cast sheets and the data analyzed in terms of compositional information of these new thermoplastic elastomers.

[1]Current address: 3M Austin Center, Austin, TX 78769

0097–6156/92/0496–0258$06.00/0

The discovery of the living cationic polymerization of vinyl ethers (*1,2*) and the synthesis of block copolymers using these techniques (*3*) was rapidly followed by the discovery of living cationic polymerization of vinyl hydrocarbon monomers (*4,5*) which paved the way for the development of poly(styryl-*b*-isobutylene-*b*-styryl) thermoplastic elastomers (*6*) by sequential monomer addition much in the same manner that the discovery of living anionic polymerization by Szwarc (*7*) led to the development of thermoplastic elastomers (*8,9*). Living cationic polymerization of IB produces telechelic PIB's with narrow molecular weight distribution ($\overline{M_w} / \overline{M_n}$ < 1.1) in molecular weights ranging from oligomers to $\overline{M_n}$ ~ 125,000 (*4,10*). Chainend functionalization studies of ^{t}Cl-PIB-^{t}Cl have led to a variety of new materials including allyl-, epoxy-, and hydroxyl-terminated PIB's (*11,12*), α,ω-diaryl-PIB's (*13*), and lithium (*14*) and potassium (*15*) chainend-functionalized PIB's.

The development of living anionic polymerization of MMA to narrow molecular weight polymer has been more arduous than that of the polymerization of hydrocarbon monomers. Several undesirable reactions compete with propagation above -65°C and the difficulty of purifying the monomer has been the bane of controlled MMA polymerizations. The development of highly controlled MMA polymerizations was reviewed by Van Beylen *et al.* (*16*). Alkyllithium initiators, in THF, attack the olefin and carbonyl groups of the monomer. Treatment of alkyllithium initiators with 1,1-diphenylethylene was found to prevent the addition to the carbonyl due to steric hindrance and directed the addition of the alkyllithium exclusively to the olefin (*17*). Polymerization of MMA in the range -78 to -65°C in THF yields well-defined narrow molecular weight distribution polymer ($\overline{M_w} / \overline{M_n}$ ~ 1.16) (*18*).

Further improvement in the polymerization of MMA could be achieved by the scrupulous purification of the monomer by using an alkylaluminum (*19*). Polymerization of MMA purified by Et_3Al gave a narrow molecular weight distribution ($\overline{M_w} / \overline{M_n}$ ~ 1.05) PMMA when initiated with 1,1-diphenylhexyllithium in THF at -78°C.

Chainend transformation has been used to synthesize block copolymers by changing the nature of the propagating species (*20*). Chainend transformations must be quantitative or reduced blocking efficiency will result. In the case of the synthesis of triblock, reduced blocking efficiency would result in homopolymer and/or diblock with an attendant reduction in mechanical properties. A notable example of chainend transformation of living anionic to living cationic polymerization yielded poly(styrene-*b*-tetramethylene oxide); the synthesis was performed by terminating the living anionic polymerization of styrene with an excess of bromine or xylene dibromide to yield the bromide or benzylbromide chainend, respectively, which upon treatment with AgPF6 in the presence of THF initiated the living cationic polymerization of THF to yield the target block copolymer (*21,22,23*).

The use of chelating agents [e.g., N,N,N',N'-tetramethylethylenediamine (TMEDA)] has greatly enhanced the yield of lithiation of hydrocarbons. Langer (*24*) reported that benzene did not react with *n*-butyllithium (*n*-BuLi) up to 100°C; however, complete lithiation occured in one hour with the *n*-BuLi-TMEDA complex in benzene at 50°C. The *n*-BuLi-TMEDA complex was found to metalate toluene quantitatively within minutes at 25°C to yield benzyllithium; however, *n*-BuLi in THF or diethylether produced negligible conversions. The lithiation of toluene to benzyllithium has been studied and a variety of conditions have been found for quantitative reaction (*25*). For example, the lithiation of toluene occurs with a 15.2% solution of *n*-BuLi in *n*-hexane and a 10% excess of toluene with a 1/1 ratio of *n*-BuLi-TMEDA at 66°C for 36 min. Lithiations of various compounds has been reviewed by Mallan and Rebb (*26*).

The strategy for the synthesis of PMMA-PIB-PMMA is presented in Scheme 1. The initial step involves IB polymerization by a difunctional initiator to produce ^{t}Cl-PIB-^{t}Cl (**1**) (*27*). The sterically hindered bifunctional initiator (**I**) was chosen to ensure quantitative terminal functionality of the PIB. Termination results in the *t*-chlorine chainend which quantitatively para-alkylates toluene in the presence of $AlCl_3$ to yield CH_3-Ø-PIB-Ø-CH_3 (**2**) (*13*). Treatment of CH_3-Ø-PIB-Ø-CH_3 with *s*-BuLi-TMEDA yields $LiCH_2$-Ø-PIB-Ø-CH_2Li (**3**). Addition of 1,1-diphenylethylene converts the chainend to the sterically hindered gem diphenyl lithium derivative (**4**). The synthesis is completed by the addition of MMA, followed by termination with methanol to yield the target triblock copolymer (**5**).

Experimental

Materials. *n*-Hexane (Fisher) was prepared by refluxing for 48 hours with conc. H_2SO_4 (10% v/v) followed by washing with distilled water until neutral. Preliminary drying occurred by storing over $CaCl_2$ and final drying by refluxing overnight with CaH_2 followed by fractional distillation under nitrogen. THF (Fisher) was distilled from CaH_2 onto sodium-benzophenone and freshly distilled as required. Toluene (Fisher) was fractionally distilled over CaH_2. TMEDA (Aldrich) was stirred with CaH_2 overnight and vacuum distilled. The molarity of *n*-, *t*- and *s*-BuLi (Aldrich) was determined by using the double titration method of Gilman (*28*) and then used without further treatment. Degassed methanol (Fisher) was prepared by distilling over CaH_2 under N_2. p-*t*-Butyltoluene (TBT) (Aldrich) was refluxed over CaH_2 and fractionally distilled. Trimethylchlorosilane (TMCS) (Aldrich) was distilled over CaH_2 prior to use. 1,1-Diphenylethylene was synthesized

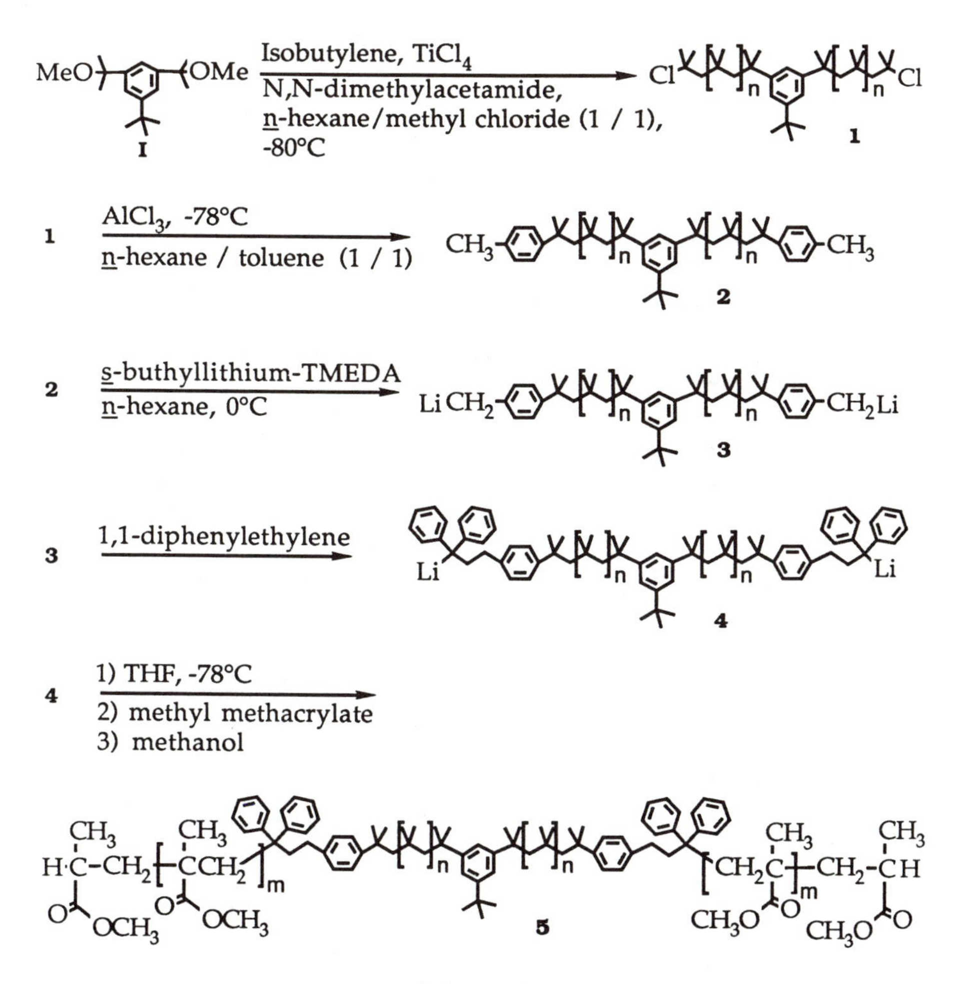

MeO
OMe
I
Isobutylene, TiCl4
N,N-dimethylacetamide,
n-hexane/methyl chloride (1 / 1),
-80°C
Cl
n
n
Cl
1
1
AlCl3, -78°C
n-hexane / toluene (1 / 1)
CH3
CH3
2
2
s-buthyllithium-TMEDA
n-hexane, 0°C
Li CH2
CH2Li
3
3
1,1-diphenylethylene
Li
Li
4
4
1) THF, -78°C
2) methyl methacrylate
3) methanol
CH3
CH3
H·C-CH2
CCH2
m
O
OCH3
OCH3
5
CH2C
CH2-CH
CH3O
O

Scheme 1.

by reacting benzophenone (Aldrich) with methylmagnesium iodide in ether followed by dehydration of the resultant alcohol by refluxing in hexane with P_2O_5. 1,1-Diphenylethylene was distilled over CaH_2 with vacuum, treated with sufficient *n*-BuLi to produce the characteristic red color, and again distilled. MMA (Aldrich) was distilled over CaH_2, stored at -20°C in a flask closed with a rubber serum cap, and aliquots were removed by a syringe. An aliquot was syringed into a vacuum distillation apparatus and the monomer was rigorously degassed by several freezing-thawing cycles. Triethylaluminum (Aldrich) in *n*-hexane was slowly added under a blanket of dry nitrogen until a bright yellow color developed. Distillation of the monomer completed the purification process (*19*).

Instrumentation. Gas chromatography was performed by using a Perkin-Elmer 8410 gas chromatograph with a flame ionization detector equipped with a 10 ft. stainless steel column, 1/8 in. dia. packed with 1% dimethylsiloxane (DuPont SE 30) on Chromabsorb. The injector and oven temperature was 250°C. and the column was at 200°C. The flow rate of the carrier gas (helium) was 25 mL/min.

GPC analysis was performed with a Waters Model 6000A Pump and a series of five μ-Styragel columns (10^5, 10^4, 10^3, 500 and 100 Å). Elutions were conducted at 30°C with THF as the solvent with a flow rate of 1 cc/min. The molecular weights were determined by the retention volume and based on polyisobutylene standards produced in these labs.

^{1}H-NMR spectra were taken with a Varian 200 MHz Spectrometer using $CDCl_3$ as the solvent with tetramethyl silane (TMS) as internal standard.

Tensile measurements were made on an Instron Mechanical Tester equiped with an extensometer guage. Microdumbells were cut from solution cast films and extended at 50 cm/min. at ambient temperature.

Differential scanning calorimetry (DSC) was performed using a DuPont 1090 Thermal Analyzer. The samples were scanned from -100°C to 150°C at a rate of 20°C/min.

Dynamic mechanical thermal analysis (DMTA) was performed on a Polymer Laboratories 9122 instrument in a single cantalever configuration at 4°C/min.

Procedures. The polymerization of IB to telechelic ^{t}Cl-PIB-^{t}Cl has been described (*10,27*).

Model lithiations and anionic polymerizations were performed in Teflon-valve sealed reactors (*29*) under dry nitrogen. The reactors and syringes were dried in an oven overnight at 120°C and assembled while still hot. Teflon-coated magnetic stir bars were used during the lithiation; however, vigorous manual shaking/swirling was necessary during the MMA polymerization. A flow of dry nitrogen was maintained when the Teflon stopper was removed from the reactor.

Model lithiation studies were performed using TBT as the model for the CH_3-Ø-PIB-Ø-CH_3 chainends. A typical model experiment was performed by syringing 10 mL dry *n*-hexane, 0.1 mL TBT (5.80×10^{-2}M), and 0.18 mL TMEDA (1.39×10^{-1}M) into a series of 50 mL reactors and cooling to 0°C. After equilibration *s*-BuLi (6.96×10^{-2}M) was syringed into the reactors. At periodic intervals 0.20 mL TMCS was added and allowed to react until the color disappeared and a bright white precipitate formed (at least 1/2 h.). The solution was poured into a separatory funnel and washed three times with water. The *n*-hexane fraction was separated and immediately analyzed by GC. The extent of the reaction was followed by observing the disappearance of the TBT peak. Figure 1. shows a series of GC traces of charges quenched with TMCS at various times. The first peak at lowest elution time, ~2.7 min. corresponds to *n*-hexane. TBT appears at ~4.8 min. and the silylated TBT (TBT-Si) at ~6.6 min. A higher boiling product appears at ~8.1 min. which is probably a disilylated TBT that arises by lithiation of the TBT-Si by excess *s*-BuLi or lithiatied TBT followed by silylation during quenching. Figure 1. shows that the concentration of TBT decreases as the products increase with reaction time. The extent of reaction was calculated by comparing the area of the TBT peak with the total area of the peaks.

The CH_3-Ø-PIB-Ø-CH_3 was synthesized by charging a reactor equipped with a mechanical stirrer and N_2 inlet with 100 g of ^tCl-PIB-^tCl, 100 mL *n*-hexane and 100 mL toluene. In the case of high molecular weight ($\overline{M_n}$ > 5,000) ^tCl-PIB-^tCl, such as 50,000, sufficient solvent was added, to reduce the viscosity of the charges to aid stirring. The reactor was placed in a Dry Ice-isopropanol bath, thermoequilibrated, $AlCl_3$ (1.2 M) was added, and the charge was stirred for 5 h. The reaction was quenched by the slow addition of cold MeOH.

The CH_3-Ø-PIB-Ø-CH_3 was separated from aluminum-containing residues and toluene by precipitation into acetone and reprecipitation from *n*-hexane into acetone. The polymer was redissolved in *n*-hexane and filtered. The *n*-hexane was removed by a rotovap and the flask with the CH_3-Ø-PIB-Ø-CH_3 was placed in a vacuum oven at ~50°C for 7 days to remove traces of acetone. Dry *n*-hexane was added to make the 15% v/v stock solution.

The synthesis of a PMMA-PIB-PMMA triblock, for example $\overline{M_n}$ = 10,000-53,200-10,000, was initiated by syringing 33 mL (1.88×10^{-4} moles of chainends) of the 15% CH_3-Ø-PIB-Ø-CH_3 solution into a reactor and cooled to 0°C. The polymer solution was then purified by adding 0.2 mL *s*-BuLi and stirring for 1h. One drop of 1,1-diphenylethylene was added and allowed to react for 1h; if the solution remained colorless, the procedure was repeated. If a red or yellow color persisted, TMEDA, 0.14

mL (7.52×10^{-4} moles), was syringed in, followed by *s*-BuLi, 0.69 mL (3.76×10^{-4} moles), and the charge was stirred at 0°C for 48 h.

After lithiation of CH_3-Ø-PIB-Ø-CH_3, 1,1-diphenylethylene, 0.13 mL (7.54×10^{-4} moles), was added and the charge stirred for at least 1 h. Dry THF, 150 mL, was syringed into the reactor and the temperature was lowered to -78°C. The Teflon valve of the reactor was replaced with a rubber serum cap. Then purified MMA (4.0 mL) was taken up into a syringe followed by an equal volume of THF, the syringe was gently rocked to mix the system, and the solution was slowly dripped into the vigorously agitated reactor. The charge was vigorously shaken for at least 0.5 h after addition of the MMA solution. The reaction was terminated by the addition of degassed methanol.

The triblock was precipitated by slowly pouring the charge into water or methanol, and the crumb was dried and weighed to determine overall yield. The triblock was extracted for approximately 20 h. with acetone (to remove unblocked PMMA) and, subsequently, with *n*-pentane (to determine the amount of unblocked PIB). The triblock and the extract were dried and weighed.

The composition of the triblocks was determined by ^{1}H-NMR spectroscopy by comparing the -OCH_3 resonance of PMMA (δ = 3.6) with the -CH_2- resonance of the PIB (δ = 1.4) (see Figure 2.). The PIB centerblock was the internal standard since its molecular weight was accurately known from GPC. The stereoregularity of the extracted PMMA was determined by ^{1}H-NMR spectroscopy (*30,31*). The PMMA was found to be ~75% syndiotactic, which is similar to values obtained by earlier investigators (32).

The triblocks were cast from 20% solutions of either THF or 1/1 (v/v) THF/toluene onto taut cellophane membranes using a glass cylinder ~4 in. diameter as the form. The glass cylinder was tightly covered with aluminum foil and the samples were dried for several days. The solid films were dried to constant weight at room temperature in vacuo. Clear smooth rubber discs were produced by this technique. The test pieces slowly became yellow at room temperature or more quickly if heated, possibly due to oxidation or to traces of TMEDA in the product.

Results and Discussion

Preliminary Experiments. Model studies to determine the conditions necessary for the quantitative lithiation of CH_3-Ø-PIB-Ø-CH_3 were carried out using TBT. Initially, following the literature (25), lithiations were performed at 60°C. However, we found that under these conditions quantitative conversions did not occur. By increasing the concentration of *s*-BuLi-TMEDA the conversions increased only to ~80% (see Figure 3.). Reducing the temperature to 0°C, however, improved the results dramatically . Figure 4. shows that lithiation of TBT with *s*-

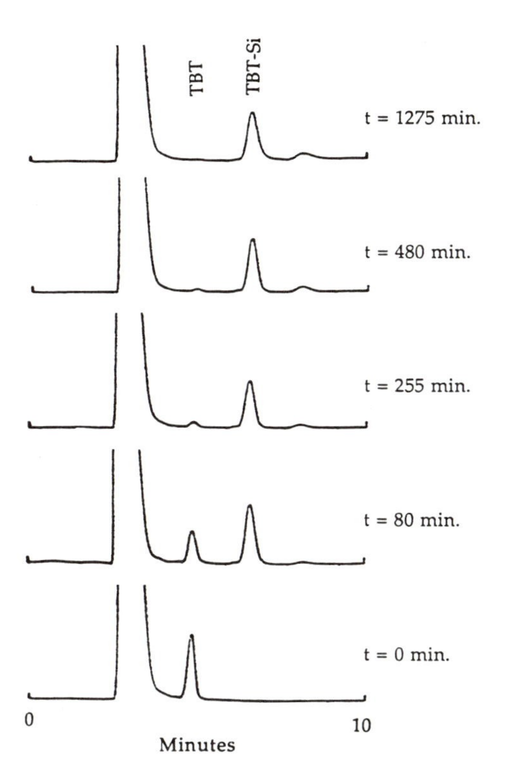

Figure 1. GC traces of the TBT and *s*-BuLi-TMEDA reaction; quenching times on the right.

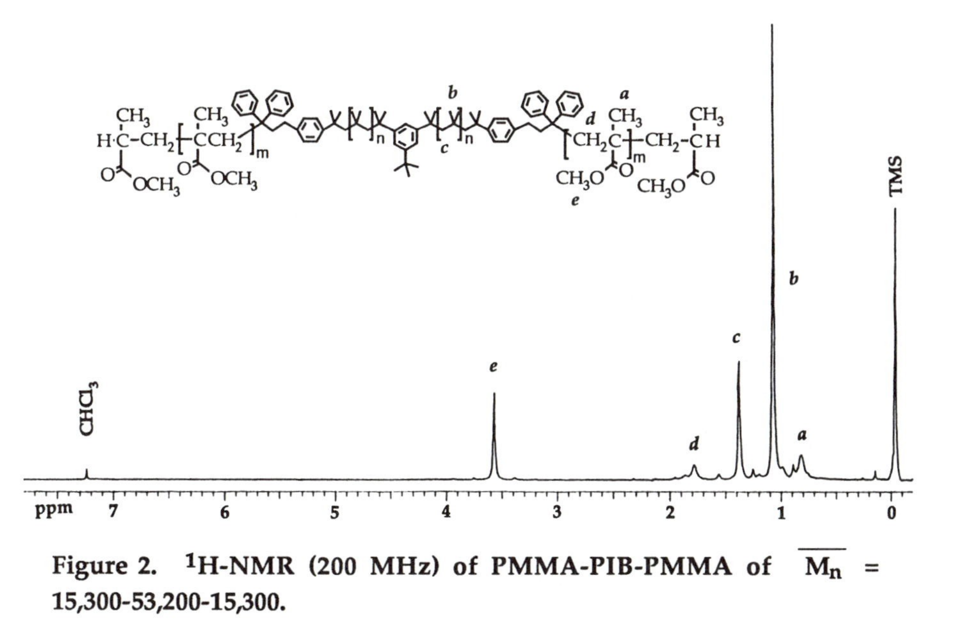

Figure 2. ^{1}H-NMR (200 MHz) of PMMA-PIB-PMMA of $\overline{M_n}$ = 15,300-53,200-15,300.

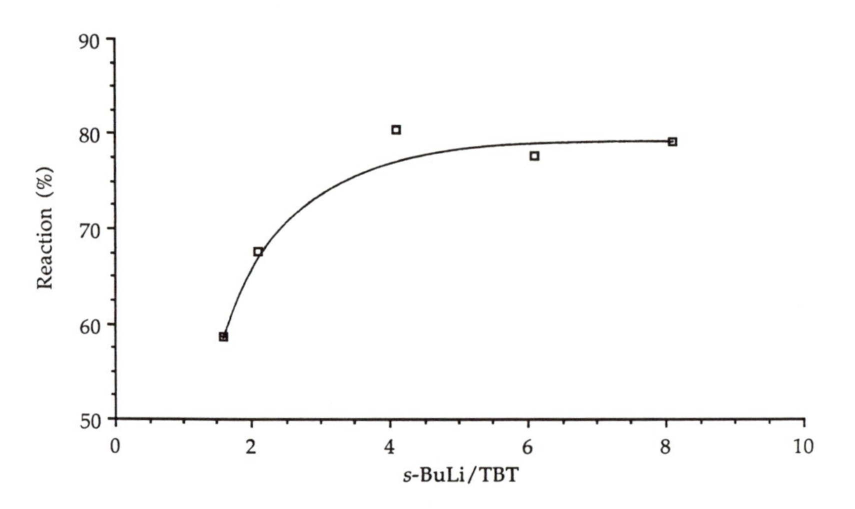

Figure 3. The effect of increasing *s*-BuLi concentrations on the metalation of TBT in *n*-hexane. [TBT] = 5.8 x 10^{-2} M, *s*-BuLi/TMEDA = 1, 60°C, t = 1380 min.

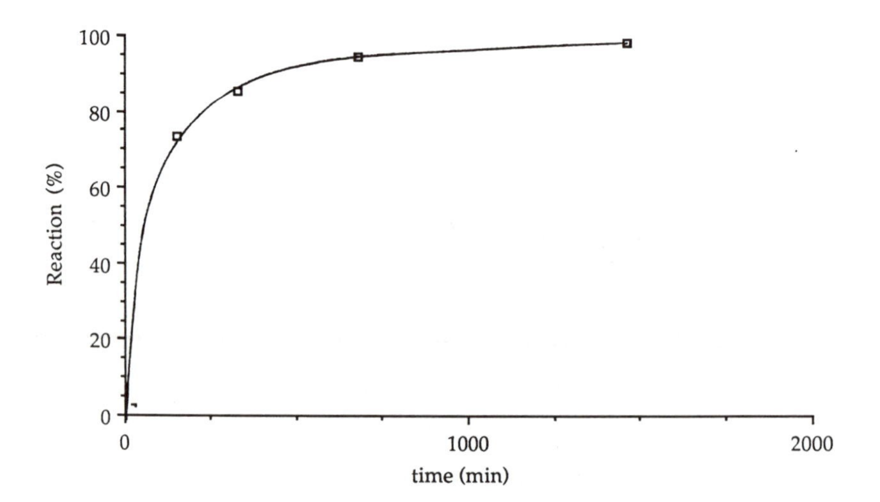

Figure 4. The litiation of TBT by *s*-BuLi in *n*-hexane. [TBT] = 5.8 x 10^{-2}M, [*s*-BuLi] = [TMEDA] = 6.69 x10^{-2}M, 0°C.

BuLi-TMEDA goes to completion at 0°C. Lower conversions at 60°C are probably due to lithiation of the TMEDA and the Teflon stir bar. At low concentrations these undesirable reactions occur at rates comparable with the lithiation of CH_3-Ø-PIB-Ø-CH_3; however, at high concentrations of CH_3-Ø-PIB-Ø-CH_3 the rates of these undesirable reactions are negligible. It appears that at 0°C the lithiation of TMEDA or Teflon is greatly reduced or even eliminated. While a darkening of the Teflon stir bar was still evident, the same results were obtained by the use of glass-coated stir bars or by not using stir bars. In any event, the glass-covered stir bars were too fragile to withstand the vigorous stirring needed for the polymerization of MMA.

Figure 5. shows the extent of lithiation of TBT with time using complexes of *n*-, *t*- and *s*-BuLi with TMEDA at 0°C. It is evident that *n*- and *t*-BuLi -TMEDA complexes do not lithiate TBT quantitatively even after long reaction times, while lithiation with *s*-BuLi-TMEDA rapidly yields complete reactions. While *s*-BuLi is known to be more reactive than *n* - or *t*-BuLi they were explored because they might produce fewer side reactions. It was found, however, that *n* and *t*-BuLi reacted too slow to be practical for this application.

The effect of TMEDA/*s*-BuLi ratio on the rate of reaction is shown in Figure 6. Lithiations were started by the addition of *s*-BuLi to TBT in hexane and the charges were quenched with TMCS after 1h. The rate of lithiation is seen to increase up to TMEDA/*s*-BuLi ≈ 2 after which it remains constant. While the TMEDA is known to chelate a lithium cation an additional TMEDA molecule may aid in the solvation of the cation resulting in an increased rate; however, additional TMEDA molecules may be prohibited from interacting due to steric factors.

Figure 7. shows the rate of lithiation at two TBT concentrations representing chainend concentrations equivalent to $\overline{M_n}$ = 5000 and $\overline{M_n}$ = 50,000. In line with this information, the lithiation of low molecular weight (<5,000) PIB's was carried out for 24 h., while the higher molecular weight products (>5,000) were reacted for 48 h. Figure 7. further illustrates that the concentration of the products remain constant even after long reaction times.

Blocking Experiments. Table I. shows the various triblocks synthesized. The shortest triblocks (the first two entries in Table I.) are viscous liquids whereas the rest are solids. The low molecular weight triblocks were synthesized to facilitate the chainend reactions and analysis by ^{1}H-NMR spectroscopy. The extent of reaction was determined by observing the disappearance of the resonance of the terminal CH_3-group (δ = 2.3 ppm) of the CH_3-Ø-PIB-Ø-CH_3. Under the conditions developed for the lithiation of TBT (see section on Preliminary Experiments), this resonance has completely disappeared in the low molecular weight triblocks.

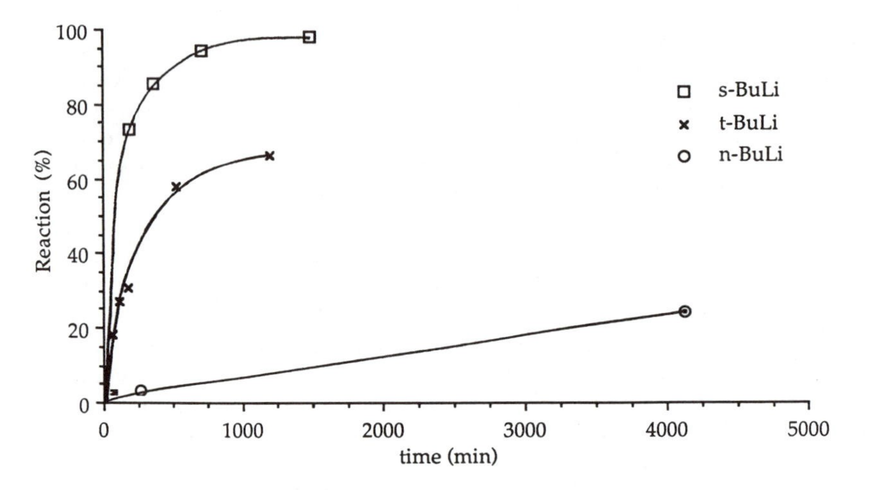

Figure 5. Lithiation of TBT with *n*-, *s*- and *t*-BuLi-TMEDA. [TBT] = 5.8 x10^{-2} M, [TMEDA] = 6.96 x 10^{-2} M, [BuLi] = 6.96 x 10^{-2} M, 0°C.

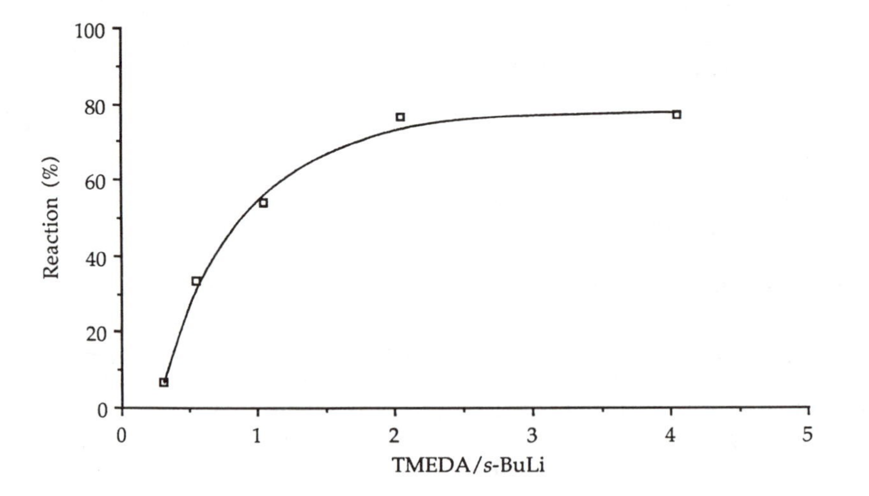

Figure 6. Lithiation of TBT in 1 h at various ratios of TMEDA/*s*-BuLi. [TBT] = 5.8 x 10^{-2}M, 0°C.

The molecular weights of the triblocks were determined by ^{1}H-NMR using PIB as the internal standard since its molecular weight is well known by GPC. The composition values were calculated from the amount of PMMA found in the triblock by ^{1}H-NMR. The GPC values are shown for comparison with the ^{1}H-NMR values. The difference between the GPC and NMR molecular weight measurements are small at low molecular weights and increase with total molecular weight and PMMA content. The triblock, $\overline{M_n}$ = 1,130-4,020-1,130, has a 6.8% difference between the GPC and NMR values and contains 38% PMMA while the triblock, $\overline{M_n}$ = 15,300-53,200-15,300, has a 16.5% difference between the GPC and NMR values and contains 36% PMMA. The difference of the GPC values from the ^{1}H-NMR values are expected due to the difference in hydrodynamic volume of the triblock compared with the PIB standards used for calibration.

Table I. PMMA-PIB-PMMA triblock copolymers

Center Block		Triblock			
$\overline{M_n}$	$\overline{M_w}/\overline{M_n}$	$\overline{M_n}$ (GPC)	$\overline{M_n}$ (NMR)	PMMA (%)	Composition
4020	1.14	4530	4460	12	250-4020-250
4020	1.14	4700	4680	17	350-4020-350
4020	1.14	6050	6490	38	1130-4020-1130
12300	1.23	15300	16500	25	2100-12300-2100
41500	1.23	46300	46800	12	2530-41500-2530
53200	1.17	59800	62100	14	4500-53200-4500
53200	1.17	65100	72600	27	9700-53200-9700
53200	1.17	69900	83700	36	15300-53200-15300

The high molecular weight triblocks were extracted with *n*-pentane (good solvent for PIB, nonsolvent for PMMA), and the composition of the extracts was analyzed by ^{1}H-NMR spectroscopy. Table II. shows the results. The extracts contained small amounts of PMMA indicating that the PIB segments pulled PMMA blocks into solution, i.e., even the pentane-soluble fractions contained block copolymer. The amount of the extract and the percent PMMA decrease with increasing triblock molecular weight because the molecular weights of the endblock increase and thus their solubility decreases. The extraction of the triblocks with low molecular weight PMMA blocks could not be carried out because the triblocks passed through the thimble and formed an emulsion in the solvent pot.

Table II. Results of extraction of high molecular weight triblocks

n-Pentane Extract		Triblock
Extracted (%wt)	PMMA (%)	
13.79	6.90	4500-53200-4500
8.25	0.87	9700-53200-9700
7.80	0.73	15300-53200-15300

The polymerization of the PMMA was living, and, as shown by the data in Table III., the molecular weights of the triblocks were within 10% of the expected values.

Table III. Molecular weight control in the triblock synthesis ($\overline{M_n}$ x 10^{-3})

Expected			Actual		
Triblock	$\overline{M_n}$	PMMA (%)	Triblock	$\overline{M_n}$ (NMR)	PMMA (%)
5.0-53.2-5.0	63.2	15.8	4.5-53.2-4.5	62.1	14.3
10.0-53.2-10.0	73.2	27.3	9.7-53.2-9.7	72.6	26.7
15.0-53.2-15.0	83.2	36.1	15.3-53.2-15.3	83.7	36.4

Figure 8. shows the results of a representative GPC experiment. As shown by the RI traces, the molecular weight of the triblock has considerably shifted to the left after blocking MMA from the center block. Further, the shape of the trace of the triblock is quite similar to that of the initial (center) block indicating that diblock or homopolymer contamination of the former is negligble.

Glass transition temperatures were determined by DSC. According to the data shown in Figure 9., the Tg of PIB and PMMA appear at -61.5°C and 105°C, respectively. The two distinct Tg's confirm the two phase nature of the triblocks. DMTA (Figure 10.) shows the two Tg's in the tan δ plot and also shows the high modulus of the triblock even in the melt, possibly indicating considerable phase separation even in the melt. A similar behavior has been observed for poly(styrene-*b*-butadiene-*b*-styrene) where the melt viscosity of the triblock was greater than either of the homopolymers or of the random copolymer (*33,34*). The two phase structure still exists in the melt and a large amount of energy is required to move the molten polystyrene blocks through the molten polybutadiene matrix.

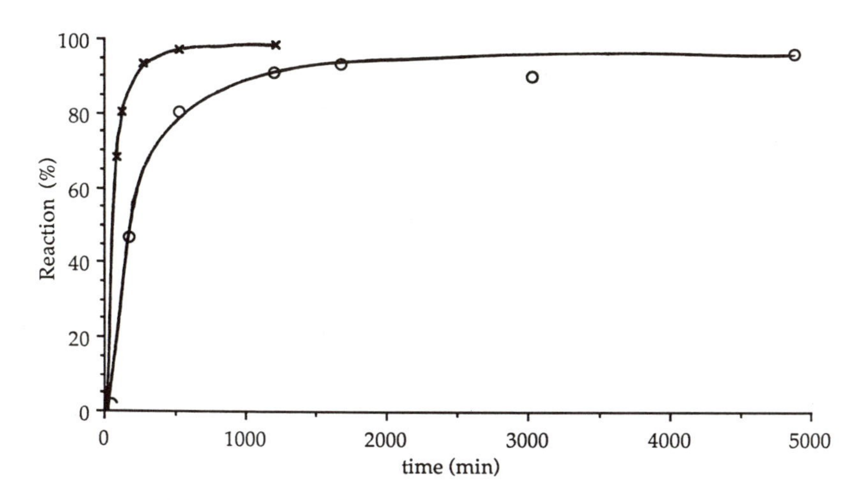

Figure 7. Lithiation of TBT at two concentrations, equivalent to high and low molecular weight PIB. TMEDA/*s*-BuLi = 2.0, 0°C; x, [TBT] = 5.80×10^{-2}M, [*s*-BuLi] = 6.96×10^{-2}M; o, [TBT] = 4.35×10^{-3}M, [*s*-BuLi] = 5.63×10^{-3}M.

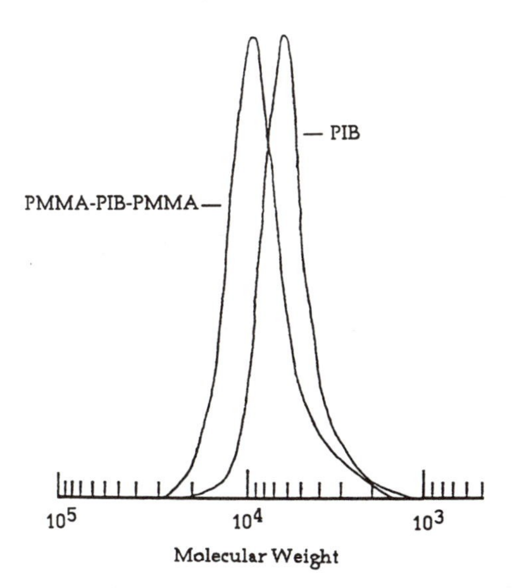

Figure 8. GPC traces of PIB and PMMA-PIB-PMMA, $\overline{M_n}$ = 15,300-53,200-15,300.

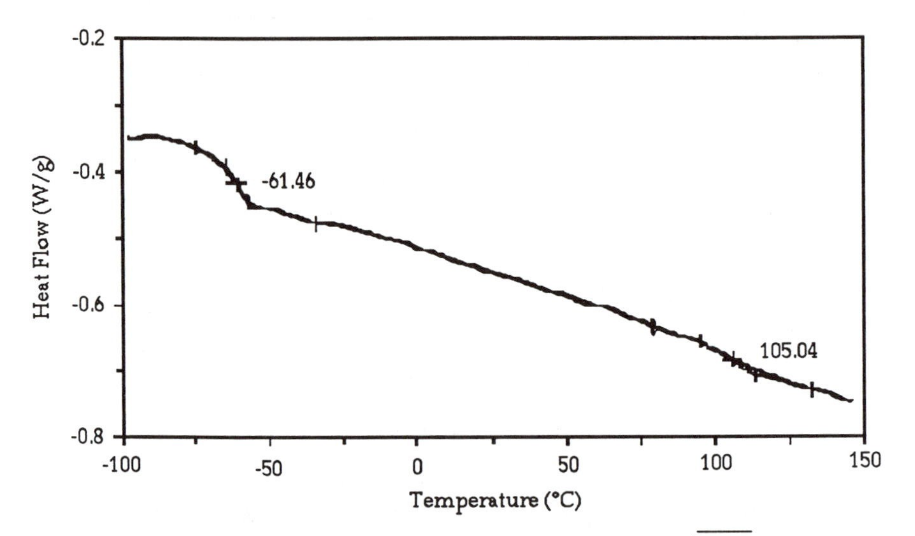

Figure 9. DSC trace of extracted PMMA-PIB-PMMA ($\overline{M_n}$ = 15,300-53,200-15,300).

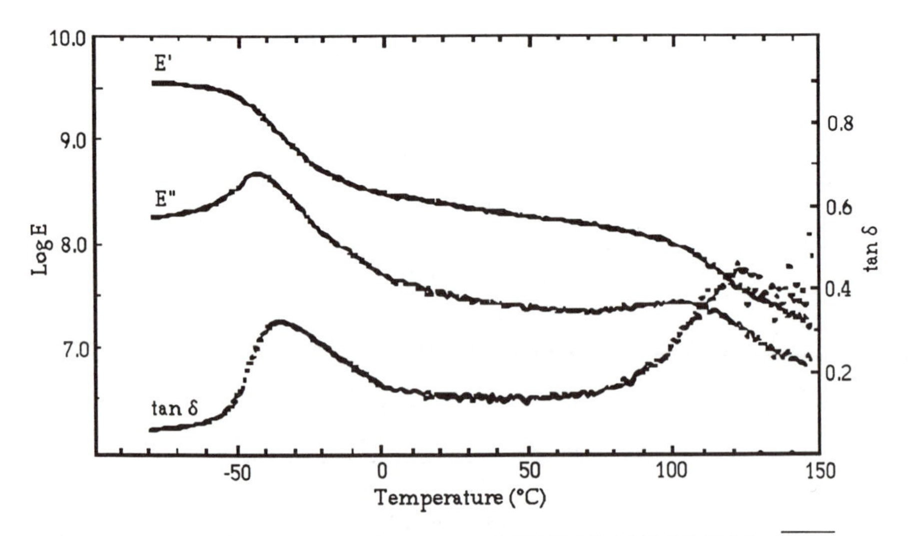

Figure 10. DMTA trace of extracted PMMA-PIB-PMMA ($\overline{M_n}$ = 15,300-53,200-15,300).

Figure 11. shows the stress-strain traces for triblocks cast from THF. Most materials exhibit a high initial yield followed by a rubbery extension. The 15,300-53,200-15,300 triblock shows a distinct yield point followed by a short extension implying a continuous PMMA phase. The tensile properties of the triblocks are low until the PMMA block reaches $\overline{M_n}$ = 9700. Morton (*35*) found that with poly(styrene-*b*-isoprene-*b*-styrene) triblocks a signifigant reduction in tensile strength occurred if the $\overline{M_n}$ of the polystyrene block was below 10,000 and that with polystyrene below $\overline{M_n}$ = 6,000 phase separation did not even occur. Above the critical molecular weight ($\overline{M_n}$ ~ 10,000) however, the tensile strength was independent of the amount of the polystyrene molecular weight.

The hysteresis of PMMA-PIB-PMMA has been investigated. Figure 12. shows a representative set of observations with the 9,700-53,200-9,700 triblock. Evidently the initial extention to ~300% disturbs some continuous PMMA domains and leads in the second extention to a more discrete domain structure with relatively little hysteresis. The phase separation of triblock copolymers has been described for other systems (*36*) as spheres, cylinders or lamellae depending on the volume ratio of the two phases and the casting solvent. The THF casting solvent in this case apparently is a good solvent for both PMMA and PIB presumably resulting in the PMMA fraction producing a cylindrical architecture.

Figures 13. and 14., respectively show and contrast stress-strain traces of the triblocks 15,300-53,200-15,300 and 9,700-53,200-9,700 cast from THF and a 1/1 (v/v) mixture of THF/toluene. The triblock 15,300-53,200-15,300 cast from THF/toluene exhibits a lower yield point followed by an extension where the test piece necks. After necking is completed a rubbery extension occurs and the specimen breaks at ~14 MPa and 700% elongation. The triblock 9,700-53,200-9,700 cast from THF/toluene behaves as a lightly crosslinked rubber with a low stress extension to ~200% elongation followed by an extension with increasing stress as the chains reach the extension limit. The ultimate properites of this material are very similar to those of the 15,300-53,200-15,300 triblock. The triblocks exhibit a large increase in stress and strain by casting them from a poor solvent for the PMMA which causes the glassy phase to form more discrete domains. Evidently the THF evaporates from the casting before the toluene so that in effect toluene becomes the casting medium. However, toluene is a much poorer solvent for PMMA than for PIB, which forces the PMMA blocks to form discrete domains dispersed in the PIB matrix.

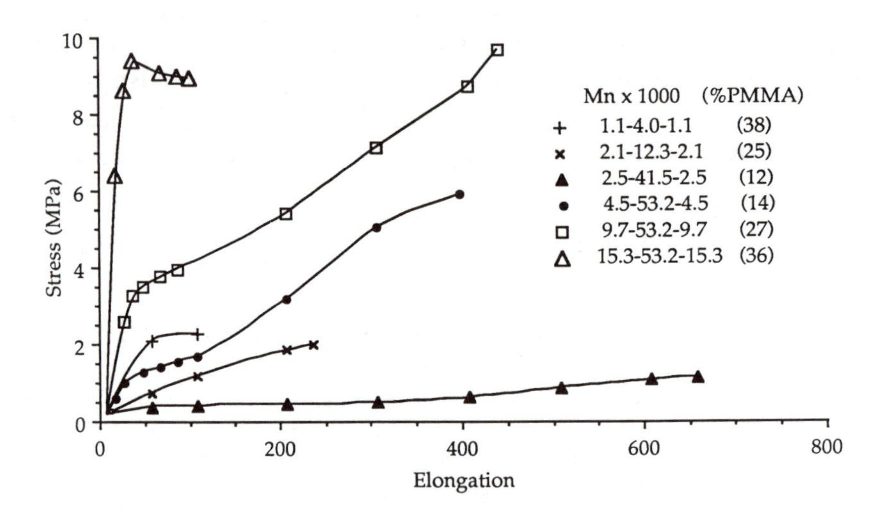

Figure 11. Stress-strain plots for triblocks cast from THF.

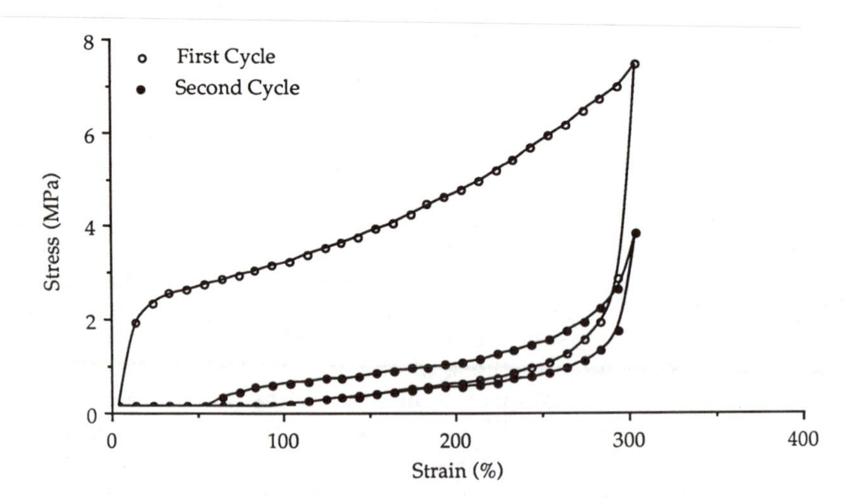

Figure 12. Hysteresis of PMMA-PIB-PMMA, $\overline{M_n}$ = 9,700-53,200-9,700, over two cycles.

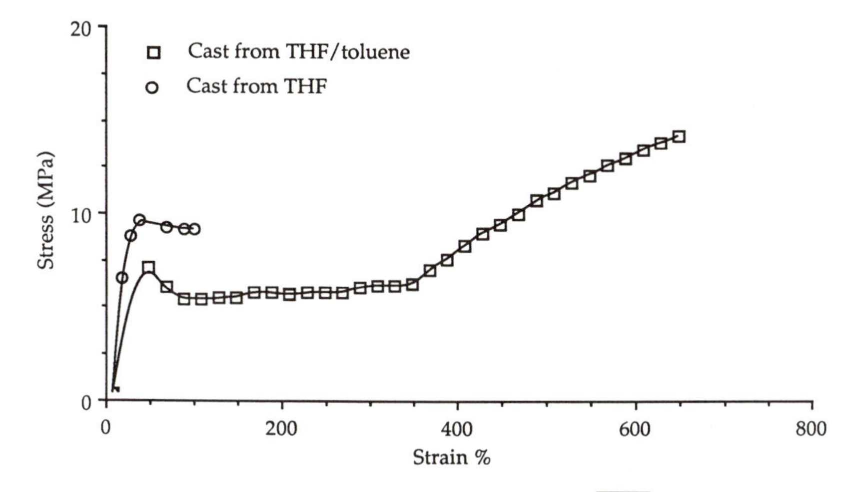

Figure 13. Comparison of PMMA-PIB-PMMA, $\overline{M_n}$ = 15,300-53,200-15,300, cast from two solvent systems.

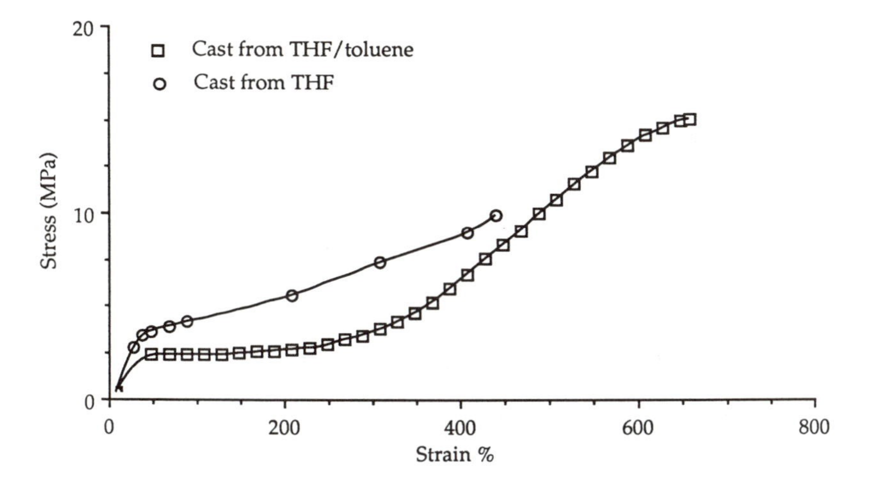

Figure 14. Comparison of PMMA-PIB-PMMA, $\overline{M_n}$ = 9,700-53,200-9,700, cast from two solvent systems.

Conclusions

The synthesis of the triblock copolymer PMMA-PIB-PMMA was performed by combining living cationic and living anionic polymerization with a chainend transformation of telechelic PIB. Conditions were developed for the efficient chainend transformation of ^{t}Cl-PIB-^{t}Cl, produced by living cationic polymerization, to $LiCH_2$-Ø-PIB-Ø-CH_2Li permitting the subsequent living anionic polymerization of MMA. The higher molecular weight triblocks cast from THF/toluene exhibit good mechanical properties comparable with poly(styrene-*b*-isobutylene-*b*-styrene) (*6*).

The PMMA-PIB-PMMA combines in one molecule a nonpolar hydrocarbon polymer and a polar PMMA moiety. The all hydrocarbon backbone of this molecule should provide excellent thermal and hydrolytic stability. Diblock polymers of this composition may be excellent compatabilizers, surfactants or adhesives due to their polar/nonpolar nature.

In addition to the synthesis of triblocks described in this paper, the $LiCH_2$-Ø-PIB-Ø-CH_2Li intermediate may be useful for the preparation of other block copolymers that can utilize the excellent property package of PIB, such as outstanding chemical and weather resistance, low gas permeability, high tack and damping.

While we have obtained good blocking efficiency and molecular weight control, the syringe technique employed is probably at the limit of its capability. Higher molecular weight center blocks would probably require high vacuum techniques (*37*).

Literature Cited

1. Miyamoto, M.; Sawamoto, M.; Higashimura, T. *Macromol.* **1984,** *17,* 265.
2. Miyamoto, M.; Sawamoto, M.; Higashimura, T. *Macromol.* **1984,** *17,* 2228.
3. Miyamoto, M.; Sawamoto, M.; Higashimura, T. *Macromol.* **1985,** *18,* 123.
4. Faust, R.; Kennedy, J. P. *J. Polym. Sci., Part A, Polym. Chem.* **1990,** *25,* 1847-1869.
5. Kennedy, J. P.; Faust, R. U.S. 4,910,321.
6. Kennedy, J. P. *Makromol. Chem., Makromol. Symp.* **1990,** *32,* 119-129.
7. Szwarc, M.; Levy, M.; Milkovich, R. *J. Am. Chem. Soc.* **1956,** *78,* 2656.
8. Legge, N. R. *Rubber Chem. Technol.* **1987,** *60,* G83.
9. Fetters, L. J. *J. Polymer Sci.:Part C.* **1969,** *26,* 1.
10. Kaszas, G.; Puskas, J. E.; Kennedy, J. P. *Polym. Bull.* **1987,** *18,* 123-130.
11. Ivan, B.; Kennedy, J. P. *J. Polym. Sci., Part A., Polym. Chem.* **1990,** *28,* 89-104.
12. Ivan, B.; Kennedy, J. P.; Chang, V. S. C. *J. Polym. Sci., Polym. Chem. Ed.* **1980,** *18,* 3177-3191.

13. Kennedy, J. P.; Hiza, M. *J. Polym. Sci., Polym. Chem. Ed.* **1983**, *21*, 3573-3590.
14. Kennedy, J. P.; Peng, K.; Wilczek, L. *Polym. Bull.* **1988**, *19*, 441-448.
15. Nemes, S.; Peng, K. L.; Wilczek, L.; Kennedy, J. P. *Polym. Bull.* **1990**, *24*, 187-194.
16. Van Beylen, M.; Bywater, S.; Smets, G.; Swarc, M.; Worsfold, D. *Adv. Polym. Sci.* **1988**, *86*, 87.
17. Wiles, D. M.; Bywater, S. *Trans. Fara. Soc.* **1965**, *61*, 150.
18. Anderson, B. C.; Andrews, G. D.; Arthur Jr., P.; Jacobson, H. W.; Melby, L. R.; Playtis, A. J.; Sharkey, W. H. *Macromol.* **1981**, *14*, 1601.
19. Allen, R. D.; Long, T. E.; McGrath, J. E. *Polym. Bull.* **1986**, *15(2)*, 127.
20. Rempp, P.; Franta, E.; Herz, J. E. *Adv. Polym. Sci.* **1988**, *86*, 145-173.
21. Burgess, F. J.; Cunliffe, A. V.; MacCallum, J. R.; Richards, D. H. *Polymer*, 1977, *18*, 719.
22. Burgess, F. J.; Cunliffe, A. V.; MacCallum, J. R.; Richards, D. H. *Polymer*, **1977**, *18*, 726.
23. Burgess, F. J.; Cunliffe, A. V.; Dawkins, J. V.; Richards, D. H. *Polymer*, **1977**, *18*, 733.
24. Langer Jr., A. W. *New York Acad. of Sci. Trans.* **1965**, *27*, 741.
25. Smith, W. N. In *Polyamine-Chelated Alkali Metal Compounds*; Langer, A. W., Ed. Adv. Chem. Ser. 130; American Chemical Society: Washington D.C., 1974.
26. Mallan, J. M.; Bebb, R. L. *Chem. Rev.*, **1969**, *69*, 693.
27. Wang, B.; Mishra, M. K.; Kennedy, J. P. *Polym. Bull.* **1987**, *17*, 205.
28. Gilman, H.; Haubein, A. H. *J. Am. Chem. Soc.* **1944**, *66*, 1515.
29. Shriver, D. F. *The Manipulation of Air-Sensitive Compounds*; McGraw-Hill Book Company: New York, N.Y., 1969.
30. Bovey, F.A.; Tiers, G.V.D. *J. Polym. Sci.* **1960**, *44*, 173.
31. Tessier, M.; Marechal, E. *Eur. Polym. J.* **1986**, *22(11)*, 889.
32. Hatada, K.; Kitayama, T.; Ute, K. *Prog. Polym. Sci.* **1988**, *13*, 189.
33. Holden, G.; Bishop, E. T.; Legge, N. R. *J. Polym. Sci.* **1969**, *C26*, 37.
34. Futamura, S.; Meinecke, E. A. *Polym. Eng. Sci.* **1977**, *17(8)*, 563.
35. Morton, M. *Rubber Chem. Technol.* **1983**, *56(5)*, 1096.
36. Molau, G. E. In *Block Copolymers*. S. L. Aggarwal, Ed. Plenum Press: New York, N.Y., 1970.
37. Morton, M.; Fetters, L. J. *Rubber Chem. Technol.* **1975**, *48(3)*, 359.

RECEIVED February 10, 1992

Indexes

Author Index

Allcock, Harry R., 236
Barazzoni, L., 56
Bell, Andrew, 121
Biela, T., 248
Booth, Brian L., 78
Cheng, H. N., 157
Chiellini, E., 56
Chien, James C. W., 27
Condé, Ph., 149
D'Antone, S., 56
Denger, Ch., 88
Fink, G., 88
Galambos, Adam, 104
Heinrichs, A., 88
Hocks, L., 149
Inoue, Shohei, 194
Jejelowo, Moses O., 78
Kaminsky, W., 63
Kashiwa, N., 72
Kennedy, Joseph P., 258
Kioka, M., 72
Klosinski, P., 248
Kubisa, P., 248
Lau, Suk-fai, 170
Le Borgne, A., 205
Marchessault, R. H., 182
Masi, F., 56
Matyjaszewski, Krzysztof, 215
Menconi, F., 56
Mizuno, A., 72
Möhring, V., 88
Möller-Lindenhof, N., 63
Niedoba, S., 63
Nyk, A., 248
Parris, J. M., 182
Penczek, S., 248
Price, Jack L., 258
Rabe, O., 63
Solaro, R., 56
Spassky, N., 205
Tait, Peter J. T., 78
Teyssié, Ph., 149
Tirrell, David A., 136
Tsutsui, T., 72
Vandenberg, Edwin J., 2,
Vincens, V., 205
Warin, R., 149
Wicks, Douglas A., 136
Wolkowicz, Michael, 104
Zeigler, Robert, 104

Affiliation Index

Arizona State University, 2
Carnegie Mellon University, 215
Enichem S.p.A., 56
Hercules Advanced Materials and Systems Company, 121,157,170
Himont Research and Development Center, 104
Max–Planck Institut für Kohlenforschung, 88
McGill University, 182
Mitsui Petrochemical Industries, 72
The Pennsylvania State University, 236
Polish Academy of Sciences, 248
Université Pierre et Marie Curie, 205
University of Akron, 258
University of Hamburg, 63
University of Liége, 149
University of Manchester Institute of Science and Technology, 78
University of Massachusetts, 27,136
University of Pisa, 56
University of Tokyo, 194

Subject Index

A

Acetone, 4
Alkylaluminoxanes
 applications, 11–12
 development, 10,11*f*
 polymerization mechanism studies, 13–14
Alkylcyclopentadienylzirconium chloride–methylaluminoxane catalyst system, effect of alkyl group on rate of ethylene polymerization, 78–86
Aluminoxane
 cocatalysts with soluble zirconium complexes, use as catalysts, 78
 effect on polymerization activity, 64
Aluminum complexes as catalysts for oxirane oligomerization, 205–214
 experimental procedure, 206
 methoxymethyloxirane oligomerization using (tetraphenylporphinato)-aluminum chloride, 207,208*f*,209*t*,210*f*
 oxirane oligomerization using aluminum complexes of Schiff's bases, 209,211*t*,212*f*,213
 Schiff's bases, 209,211–214
 stereoelective oligomerization of methyloxirane, 213*t*,214
 structure, 206
Aluminum porphyrins
 block copolymer synthesis, 198*t*
 effect of stereochemistry of ring opening of epoxide, 200,202,203*t*
 end-reactive polymer synthesis, 198,199*t*
 reactions, 195–197
Amine–epoxy resin, calorimetric study of cure behavior at ambient conditions, 170–181
Anionic catalyst, role in determination of structure of final product, 193

B

Bifunctional oligoethers, 205
Bis(cyclopentadienyl)titanium(IV), use in ethylene polymerization, 63
Block copolymers, use of metalloporphyrin catalysts in polymerization, 198*t*
1-Butene polymerization with ethylene-bis(1-indenyl)zirconium dichloride and methylaluminoxane catalyst system, 72–76

C

Calorimetric study of cure behavior of amine–epoxy resin at ambient conditions, 170–181
Catalysis
 coordination copolymerization of tetrahydrofuran and oxepane with oxetanes and epoxides, 17–18
 cyclophosphazene polymerization, 236–247
 early polymerization catalyst studies, 3–5
 epoxide polymerization, 10–14
 hydroxy polyether synthesis, 14–17
 Ziegler catalyst polymerization, 5–10
Catalysts
 CH-type, *See* CH-type catalysts
 CW-type, *See* CW-type catalysts
 for structural control of inorganic polymers, 215–233
 metalloporphyrin, for control of polymerization, 194–203
 methylaluminoxane–substituted $ZrCl_2$ systems, 72–76,78–87
 nickel(0)–phosphorane, in α-olefin polymerization, 88–103
 optically active, for isotactic olefin polymerization, 63–71

Catalysts—*Continued*
soluble zirconium complexes with aluminoxane cocatalysts, 78
substituted $ZrCl_2$–methylaluminoxane systems, 72–76,78–87
tungsten phenoxide, for dicyclopentadiene polymerization, 121–133
Ziegler–Natta, based on titanium halides and organoaluminum sulfates, 56–62
Ziegler–Natta, $MgCl_2$-supported, 27–54
Catalyzed polymerization of phosphoranimines, 220–222
Cationic polymerization, development of catalysts, 235
Cellulose synthetic analog, X-ray fiber diffraction study of poly[3,3-bis(hydroxymethyl)oxetane], 182–191
Chain-end transformation, use in synthesis of block copolymers, 259–260
Characterization, poly(methyl methacrylate)-*block*-polyisobutylene-*block*-poly(methyl methacrylate) thermoplastic elastomers, 258–275
Chelating agents, effect of lithiation of hydrocarbons, 260
Chlorinated polyethers, kinetics of nucleophilic displacement reactions, 136–147
CH-type catalysts
active site determination, 36,37–38
active site structure, 50,52–53
activity, 28,35,36*t*
chain propagation, 39,40*t*
chain transfer, 40*t*
comparison to CW-type catalysts, 49*t*,50,51*t*
deactivation, 41,42–43*f*
effect of ester structures, 35
effect of Lewis base at active site, 41,44–48
mechanism for formation, 32
oxidation state and EPR of Ti, 48*t*,49
CH-type catalysts—*Continued*
powder X-ray diffraction pattern, 30–34
preparation, 28–38
stereospecificity, 28,35,36*t*
^{13}C-NMR analysis of polyether elastomers, 157–169
computerized analytical approaches, 166*t*,167,168*t*
contribution of catalytic sites to overall polymer yield, 168*t*
epichlorohydrin–allyl glycidyl ether copolymer, 162,163*f*,164
ethylene oxide–epichlorohydrin–allyl glycidyl ether terpolymer, 164,165*f*
ethylene oxide–epichlorohydrin copolymer, 159,160*f*,161
evidence for multisite nature of catalyst system, 166*t*,167
experimental procedure, 158
poly(allyl glycidyl ether), 161–162,163*f*
polyepichlorohydrin, 158–159,160*f*
Computerized analytical approaches, ^{13}C-NMR analysis of polyether elastomers, 166*t*,167,168*t*
Coordination polymerization, oxirane, soluble polynuclear μ-oxometal alkoxide aggregates, 149–156
Copolysilanes with chiral substituents, synthesis, 229,230*f*,231
Cumene hydroperoxide, 3–4
Cure behavior of amine–epoxy resin, calorimetric study, 170–181
advancement related to apparent activation energy, 175,177*f*
advancement related to time and temperature, 173,175
experimental procedure, 171–172
glass transition temperature vs. conversion, 177,179*f*
heat generation vs. time during curing, 172,174*f*
rate equation, 172–173
reaction time vs. reciprocal temperature plot, 175,176*f*

Cure behavior of amine–epoxy resin, calorimetric study—*Continued*
reduced reaction rate plot, 173,174*f*
resin cure behavior, 172–173,174*f*,175
time–temperature–conversion diagram, construction and application, 175–181
CW-type catalysts
active site determination, 36,37–38
active site structures, 50,52–53
activity, 36*t*
chain propagation, 39,40*t*
chain transfer, 40*t*
comparison to CH-type catalysts, 49*t*,50,51*t*
deactivation, 41,42–43*f*
ester structures, 35
Lewis base at active site, 41,44–48
oxidation state and EPR of titanium, 48*t*,49
preparation, 28
stereospecificity, 36*t*
Cyclopentadienyl ring substituents, effect on rate of ethylene polymerization, 78–87
butyl substituent, 79,81*f*
concentration vs. duration of polymerization, 82,83*f*,85*t*,86
electronic and steric effects, 82,85–86
experimental procedures, 86
methyl substituent, 79,80*f*
propagation rate coefficient vs. duration of polymerization, 82,84*f*,85*t*,86
propyl substituent, 79,80*f*
schematic representation, 79*t*,81*f*
Cyclophosphazene polymerization, 236–247
approaches for mechanism determination, 237–243
catalysts, 243–245
changes in halogen or pseudohalogen side units, 240–241
effect of heat, 243
effect of $(NPCl_2)_3$ as polymerization initiator, 244
Cyclophosphazene polymerization—*Continued*
effect of solvents, 245
electrical conductivity, 237,239
free radical sources related to high-energy radiation, 243–244
Lewis acids as polymerization accelerators, 244
mechanisms, 243
phosphazo side groups, 240
replacement of chlorine or fluorine side units by organic groups, 240–242
synthesis via ring-opening polymerization–macromolecular substitution, 237–238
water as initiator, 244–245

D

Dialkylzinc–water catalyst system, description, 194
Dicumyl peroxide, synthesis, 4
Dicyclopentadiene polymerization using well-characterized tungsten phenoxide complexes, 121–133
applications, 122
calculation procedure, 125
cyclic voltammetric procedure, 124
electrochemical characterization of procatalysts, 126–127,128*f*
experimental procedure, 123–124
polymerization activity of procatalyst, 127,129–130,131*f*
polymerization procedure, 124
reaction conditions related to properties, 130*t*
schematic representation, 122–125
synthetic routes to procatalysts, 124–127
Diepoxides, polyaddition to acids of phosphorus, 254–256
Disproportionated rosin soap, reaction times for emulsion polymerization, 3

E

End-reactive polymers, use of metalloporphyrin catalysts, 198,199*t*
Epichlorohydrin
elastomers, 12–13
polymerization catalysts, 10,11*f*,12
Epichlorohydrin–allyl glycidyl ether copolymer, ^{13}C-NMR analysis, 162,163*f*,164
Epoxide(s)
coordination copolymerization of tetrahydrofuran and oxepane, 17–18
development of catalysts for ring-opening coordination polymerization, 135
polyaddition to acids of phosphorus, 248–257
polymerization, 10–14
stereochemistry of ring opening, 200,202,203*t*
Epoxide–episulfide copolymerization, control by light with zinc porphyrin, 199–200,201*f*
Ester structure, effect on $MgCl_2$-supported Ziegler–Natta catalysts, 35
Ethylene oxide, addition to P–OH acid, 247–254
Ethylene oxide–epichlorohydrin–allyl glycidyl ether terpolymer, ^{13}C-NMR analysis, 164,165*f*
Ethylene oxide–epichlorohydrin copolymer, ^{13}C-NMR analysis, 159,160*f*,161
Ethylene polymerization
effect of cyclopentadienyl ring substituents on rate, 78–86
effect of hydrogen, 5,6*t*
Ziegler–Natta catalysts based on titanium halide and organoaluminum salts, 57–60*t*
Ethylenebis(1-indenyl)zirconium dichloride, effect of hydrogen addition on 1-butene polymerization, 72–76
Ethylenebis(4,5,6,7-tetrahydroindenyl)-zirconium dichloride, use in isotactic olefin polymerization, 64–70

F

Functional groups, polymer-bound, structural effects and reactivity, 136–148

H

Halogenated polyethers, factors affecting reaction rate, 136–137
Halogen-free cyclic phosphazenes, polymerization, 240,243
Heterogeneous catalysts, improvements for production of isotactic polypropylene, 25
High-molecular-weight polymers from monosubstituted epoxides, 12*t*
Highly stereoregular syndiotactic polypropylene from homogeneous catalysts, 104–120
crystal forms, 116–117,118*f*,119
estimation of thermodynamic parameters, 113,116
evidence for figure-8 crystal form, 116–117,118*f*,119
evidence for planar all-trans crystal form, 116–117,118*f*,119
experimental materials, 105–106
molecular characterization, 106–107
optical micrographs, 113,114*f*
SAXS results, 110,111*f*
scanning electron micrographs, 113,115*f*
solid-state ^{13}C-NMR results, 110,112*f*,113,114*f*
solid-state structure, 107
thermal behavior, 105,108*f*,*t*,109*f*,110
WAXS results, 110,111*f*
Homogeneous catalysts
developments, 25–26

Homogeneous catalysts—*Continued*
to produce highly stereoregular syndiotactic polypropylene, 104–120
Homogeneous systems, reasons for study, 63
Hydrogen, effect on ethylene polymerization using Ziegler–Natta catalyst, 5,6*t*
Hydrogen addition, effect on 1-butene polymerization with ethylene-bis(1-indenyl)zirconium dichloride–methylaluminoxane catalyst system, 72–76
catalytic activity, 73,74*t*
chain-transfer reaction with H, 76
chain-transfer reaction without H, 74
^{13}C-NMR spectra, 73–75
DSC procedure, 73
intrinsic viscosity, 73–74
melting temperature, 76
polymerization procedure, 73
Hydroxy polyethers, 14–17
Hydroxyoxetane polymers
commercialization, 17
poly[3,3-bis(hydroxymethyl)-oxetane], 16
poly[3-ethyl-3-(hydroxymethyl)]oxetane, 17
poly(3-hydroxyoxetane), 14–16
poly[3-methyl-3-(hydroxymethyl)]oxetane, 16–17

I

Immortal polymerization, 197
Initiators for structural control of inorganic polymers, 215–233
polyphosphazene synthesis, 218–233
polysilane synthesis, 223–231
polysiloxane synthesis, 218
Inorganic heterocyclic phosphazenes, polymerization, 243
Inorganic polymers, catalysts and initiators for structural control, 215–231
Insertion polymerization, 9–10
Isotactic olefin polymerization with optically active catalysts, 63–70
alkyl side chain length related to monomer, 66,67*t*,68
catalyst, 64
density, 66,67*t*
glass transition point, 66,67*t*
isotacticity, 66,67*t*
mean molecular mass, 66,67*t*
melting point, 66,67*t*
molecular rotation vs. oligomer, 69*t*,70
monomer related to reaction rate, 64,65*t*
product molecular weight, 70*f*
properties vs. polyolefin, 66,69*t*
specific optical rotation of oligomer vs. wavelength, 68,69*t*,70
temperature related to properties, 64,66
Isotactic polymer chain, stereochemistry, 64,65*f*
Isotactic polypropylene
commercial development, 8–9
discovery, 5–6
high-yield synthesis using $TiCl_3 \cdot nAlCl_3$, 6,7–8*f*
improvements in technology for commercial success, 104–105
production using heterogeneous catalysts, 25
worldwide consumption, 6
Isotactic propylene polymerization with ethylenebis(1-indenyl)zirconium dichloride and methylaluminoxane, effect of hydrogen addition, 72–76

K

Kinetics, nucleophilic displacement reactions on chlorinated polyethers, 136–148

L

Lewis base, effect at active site of
$MgCl_2$-supported
Ziegler–Natta catalysts
electron paramagnetic resonance, 48
site stereoselection, 44,46*t*,47
stereoelective polymerization of racemic olefins, 44,45*t*
Linear high-density polyethylene, 4–5
2,ω-linkage, α-olefins, 88–103
Living cationic polymerization, 259

M

Magnesium chloride supported Ziegler–Natta catalysts, 27–55
advantages, 27–28
comparison between CH- and CW-types, 49*t*,50,51*t*
ester structures, 35
kinetics, 36–43
Lewis base at active sites, 41,44–48
oxidation state and EPR of titanium, 48*t*,49
preparation, 28–38
structures of active sites, 50,52–53
Mechanical properties, poly(methyl methacrylate)-*block*-polyisobutylene-*block*-poly(methyl methacrylate) thermoplastic elastomers, 258–275
Mechanisms, cyclophosphazene polymerization, 236–247
Metallocenes, effect on polymerization activity, 64
Metalloporphyrin catalysts for control of polymerization, 194–204
applicability to ring-opening and addition polymerizations, 195–197
block copolymer synthesis, 198*t*
effect on stereochemistry of ring opening of epoxide, 200,202,203*t*
end-reactive polymer synthesis, 198,199*t*
Metalloporphyrin catalysts for control of polymerization—*Continued*
epoxide–episulfide copolymerization, 199–200,201*f*
experimental procedure, 202–204
Metathesis systems, 26
Methoxymethyloxirane oligomerization using (tetraphenylporphinato)aluminum chloride
^{13}C-NMR spectra, 205,208*f*,209,210*f*
experimental procedure, 207
FABMS, 207,208*f*
molecular weights of different fractions, 207,209*t*
nature of terminal hydroxyl group, 207
structure and functionality, 207
structure of dimer and trimer, 209,210*f*
Methyl methacrylate, development of living anionic polymerization, 259
Methylaluminoxane–substituted zirconium chloride catalyst systems
cyclopentadienyl ring substituent related to rate of ethylene polymerization, 78–86
hydrogen addition related to 1-butene polymerization, 72–76
Methyloxirane, stereoelective oligomerization using aluminum complexes of Schiff's bases, 213*t*,214
4-Methyl-1-pentene, polymerization using Ziegler–Natta catalysts based on titanium halide and organoaluminum salts, 58,59–60*t*,61
Migratory nickel(0)–phosphorane catalyst, α-olefin polymerization by 2,ω-linkage, 88–103
Morphology, highly stereoregular syndiotactic polypropylene produced by homogeneous catalysts, 104–120

N

Nickel(0)–phosphorane catalyst, α-olefin polymerization by 2,ω-linkage, 88–103

Nucleophilic displacement reactions on chlorinated polyethers, 136–148
benzoate ion related to second-order kinetic plots, 141,142*f*,143
benzoate ion related to spin–lattice reaction times, 147*t*
calculated rate constant ratios, 146*t*
conversion vs. time for various temperatures, 140,142*f*
determination of kinetics of substitution by benzoate ion, 140,141*t*
deviation from second-order kinetics, 143
effect of secondary chloride, 143
effect of solvent, 143
experimental materials, 138
fit of second-order kinetics for high and low conversion data, 144,145*f*
goodness of fit of rate constants, 144,145*f*
initiator preparation procedure, 138
kinetic method, 140
measurement procedure, 139–140
preparation procedure for tetra-butylammonium benzoate, 138–139
preparation procedure for various polyoxiranes, 138
second-order kinetic plots of low conversion data, 141,142*f*
sensitivity of rate constant ratios, 144,146*t*

O

α-Olefin(s), interest in polymerization, 56
Olefin polymerization, isotactic, *See* Isotactic olefin polymerization with optically active catalysts
α-Olefin polymerization by 2,ω-linkage using nickel(0)–phosphorane catalysts, 88–103
activation energy vs. C number, 99,100*f*
activation parameters, 97*t*
amount of HD found vs. amount of catalyst used, 93,95*f*

α-Olefin polymerization by 2,ω-linkage using nickel(0)–phosphorane catalysts—*Continued*
degree of polymerization vs. C number, 93,95*f*
Fischer projection of possible configuration of two methyl groups for general case, 100,101*f*
Fourier transform spectra of 1-hexene polymerization, 93,96*f*,97
kinetic model, 97,98*f*,99
2,ω-linkage formation, 90,91*f*
migration mechanism, 90,92*f*,93
migration of nickel catalysts via ß-elimination–addition, 93,94*f*
molecular weight vs. pressure inside reaction vessel, 93,96*f*
rate constant vs. pressure, 97,98*f*
structures resulting from alternative combination of R and S monomers, 100,102*f*,103
α-Olefin polymerization using Ziegler–Natta catalysts based on titanium halide and organoaluminum salts, 56–62
effect of temperature, 60
ethylene polymerization procedure, 57,58*t*
experimental conditions, 57,60
4-methyl-1-pentene polymerization procedure, 58,59*t*
number of methyl groups branching, 60*t*
polymeric product characterization procedure, 57
propylene polymerization procedure, 58*t*
Oligomerization of oxiranes with Al complexes as initiators, 205–214
Optically active catalysts, isotactic olefin polymerization, 63–70
Organoaluminum sulfates, polymerization of C_2–C_6 α-olefins, 56–61
Organometallic polymers, lack of definition, 216
Oxepane, coordination copolymerization with tetrahydrofuran using oxetanes and epoxides, 17–18

Oxetanes, coordination copolymerization of tetrahydrofuran and oxepane, 17–18
Oxirane(s)
addition to P–OH acid, 248–254
oligomerization using aluminum complexes of Schiff's bases, 205–214
polymerization by polynuclear μ-oxometal alkoxides aggregates, 149–155
μ-Oxometal alkoxide aggregates as catalysts for oxirane polymerization, 149–156

P

Phenol, synthesis, 4
Phosphazene macromolecules, *See* Polyphosphazenes
Phosphinites, formation of polyphosphazenes, 222–223
Phosphonites, formation of polyphosphazenes, 222–223
Phosphorane–migratory nickel(0) catalyst, α-olefin polymerization by 2,ω-linkage, 88–103
Phosphoranimines, catalyzed polymerization, 220–222
Polar monomers, 9–10
Polyaddition of epoxides and diepoxides to acids of phosphorus, 248–257
Poly(alkylene phosphate)s, synthetic methods, 248–249
Poly(alkylene phosphate) synthesis using polyaddition of epoxides and diepoxides, 248–257
addition of diepoxides to acids, 254–256
kinetics of ethylene oxide addition, 250–253
reaction, 249
steric course of addition of substituted oxiranes to P–OH groups in acids, 253*t*,254
structures of acids, 249–250
Poly(allyl glycidyl ether), ^{13}C-NMR analysis, 161–162,163*f*
Poly[3,3-bis(hydroxymethyl)oxetane]
commercialization, 17
properties, 182–183
synthesis, 16
X-ray fiber diffraction study, 184–190
Poly(dimethylsiloxane), synthesis, 218
Polyepichlorohydrin
^{13}C-NMR analysis, 158–159,160*f*
microstructure studies, 157–158
Polyethers
chlorinated, kinetics of nucleophilic displacement reactions, 136–148
elastomers, ^{13}C-NMR analysis, 157–168
Poly[3-ethyl-3-(hydroxymethyl)-oxetane], 17
Polygermanes, inorganic polymer, 216
Polyglycidol, synthesis, 14
Poly(3-hydroxyoxetane), 14–17
Polymer(s) with inorganic backbones, controlling structure, 215–233
Polymer-bound functional groups, structural effects and reactivity, 136–148
Polymerization
1-butene, 72–76
C_2–C_6 α-olefins in presence of modified Ziegler–Natta catalysts based on titanium halides and organoaluminum sulfates, 56–62
catalyzed, phosphoranimines, 220–222
cationic, development of catalysts, 235
control using metalloporphyrin catalysts, 194–204
cyclophosphazene, 236–247
dicyclopentadiene, 121–133
epoxides, 10–14
ethylene, 63,78–86
inorganic heterocyclic phosphazenes, 243
insertion, 9–10
isotactic olefins, 63–70
living cationic, 259
α-olefins by 2,ω-linkage using nickel(0)–phosphorane catalysts, 88–103

Polymerization—*Continued*
α-olefins using Ziegler–Natta catalysts based on titanium halide and organoaluminum salts, 56–62
oxiranes by polynuclear μ-oxometal alkoxides aggregates, 149–155
poly(alkylene phosphate) synthesis using polyaddition of epoxides and diepoxides, 248–257
Poly[3-methyl-3-(hydroxymethyl)oxetane], 16–17
Poly(methyl methacrylate)-*block*-polyisobutylene-*block*-poly(methyl methacrylate) thermoplastic elastomers
n-BuLi, *t*-BuLi, and *s*-BuLi related to lithiation, 267,268*f*
s-BuLi concentration related to lithiation, 264,266*f*
p-*tert*-butyltoluene concentration related to lithiation rate, 267,271*f*
evidence for phase separation, 270,272*f*
experimental procedure, 262–264,265*f*
extraction of high-molecular-weight triblock copolymers, 269,270*t*
GC traces, 263,265*f*
glass transition temperatures, 270,272*f*
GPC traces, 270,271*f*
^{1}H-NMR spectra, 264,265*f*
hysteresis, 273,274*f*
instrumentation, 262
molecular-weight control in triblock copolymer synthesis, 270*t*
solvent related to stress–strain, 273,275*f*
stress–strain plots, 273,274*f*
synthetic scheme, 260–261
temperature related to lithiation, 264,266*f*,267
tetramethylethyldiamine:*s*-BuLi ratio related to reaction rate, 267,268*f*
triblock copolymers synthesized, 267,269*t*
Polynuclear μ-oxometal alkoxide aggregates as catalysts for oxirane polymerization, 149–156
^{13}C-NMR chemical shifts for complex aggregate, 151*t*,152
Polynuclear μ-oxometal alkoxide aggregates as catalysts for oxirane polymerization—*Continued*
experimental procedure, 150–151
fate of monomer, 153,154*t*
mechanism, 154–155
schematic representation of main features of complex aggregate, 152,153*f*
structure of complex aggregate, 151*t*,152,153*f*
Poly-α-olefins
^{13}C-NMR spectra, 90,91*f*
properties, 50,51*t*
Polyoxetanes, general composition, 182,183*f*
Poly(phenylmethylsilylene), modification, 228–229
Polyphosphazenes
properties, 216
structure, 236,238
synthesis via direct reaction of silyl azides with phosphonites and phosphinites, 222–223
synthesis via high-temperature condensation of *N*-silyl-phosphoranimines, 219–220
synthesis via ring-opening polymerization of hexachloro-cyclotriphosphazene, 218–219
Polypropylene
isotactic, *See* Isotactic polypropylene
syndiotactic, produced by homogeneous catalysts, 104–120
Polysilanes
commercial importance, 223
copolysilanes with chiral substituents, 229,230*f*,231
description, 223
modification of poly(phenylmethyl-silylene), 228–229
properties, 223
synthesis via reductive coupling disubstituted dichlorosilane with alkali metals, 223–227

Polysilanes—*Continued*
synthesis via ring-opening polymerization, 227–228
Polysiloxanes, synthesis, 218
Polysulfur, inorganic polymer, 216
Porphinatoaluminum derivative initiators, applications, 205–206
Prepreg lot, effect of resin on tack and flow properties, 170–171
Procatalyst, definition, 122
Propylene
polymerization using Ziegler catalyst, 5–6
polymerization using Ziegler–Natta catalysts based on titanium halide and organoaluminum salts, 58–60*t*
Propylene oxide copolymers, 13

R

Reactivity, polymer-bound functional groups, 136–148
Redox free radical polymerization, mechanism studies, 3–4
Ring opening of epoxide, stereochemistry, 200,202,203*t*
Ring-opening polymerization
coordination, epoxides, 135
metathesis, cyclic olefins, 121
synthesis of polysilanes, 227–228

S

Silyl azides, formation of polyphosphazenes, 222–223
Sodium stearate, emulsion polymerization with potassium persulfate, 3
Soluble polynuclear μ-oxometal alkoxide aggregates, coordination polymerization, 149–156
Soluble zirconium complexes with aluminoxane cocatalysts, use as catalysts, 78
Sonochemical reductive coupling process, synthesis of polysilanes, 224–227
Stereochemistry, isotactic polymer chain, 64,65*f*
Stereoregular syndiotactic polypropylene produced by homogeneous catalysts, 104–120
Stereoselectivity ratio, definition, 214
Structural effects, polymer-bound functional groups, 136–148
Structure
highly stereoregular syndiotactic polypropylene produced by homogeneous catalysts, 104–120
polymers with inorganic backbones, 236–247
Syndiotactic polypropylene produced by homogeneous catalysts, 104–120
Synthesis
poly(alkylene phosphate)s, 248–257
poly(methyl methacrylate)-*block*-polyisobutylene-*block*-poly(methyl methacrylate) thermoplastic elastomers, 258–275
Synthetic analog of cellulose, X-ray fiber diffraction study of poly[3,3-bis(hydroxymethyl)oxetane], 182–191

T

Tetrahydrofuran, coordination copolymerization with oxepane using oxetane and epoxides, 17–18
(Tetraphenylporphinato)aluminum chloride, use in methoxymethyloxirane oligomerization, 207,208,209*t*,210*f*
Thermoplastic elastomers, poly(methyl methacrylate)-*block*-polyisobutylene-*block*-poly(methyl methacrylate), 258–275
∂-$TiCl_3 \cdot 0.33AlCl_3$, advantages for use in polypropylene synthesis, 27–28
$TiCl_3 \cdot nAlCl_3$, use in high-yield synthesis of isotactic polypropylene, 6,7–8*f*
Time–temperature–superposition technique, description, 171

Titanium, EPR spectroscopy, 48–49
Titanium halides, polymerization of C_2–C_6 α-olefins, 56–61
Titanium trichloride system, 63
Transition metal catalyzed coordination polymerization of olefins, 25–26
Trialkylaluminum–H_2O catalysts
applications, 11–12
polymerization mechanism studies, 13–14
Trialkylaluminum–H_2O–chelate catalysts
applications, 11–12
polymerization mechanism studies, 13–14
Tungsten phenoxide complexes, use in dicyclopentadiene polymerization, 121–131

V

Vandenberg catalysts
description, 194
use in polyether elastomer synthesis, 157
Vinyl ether coordination catalysts, mechanism of polymerization, 9,10*f*

X

X-ray fiber diffraction study of poly[3,3-bis(hydroxymethyl)oxetane], 182–191
backbone conformation, 186
experimental materials, 183
X-ray fiber diffraction study of poly[3,3-bis(hydroxymethyl)oxetane]—*Continued*
hydroxymethyl conformation, 186,187*t*,188*f*
initial refinement, 187,189,190*f*
molecular conformations, 184,186*f*,*t*
structural determination, 187
unit cell determination, 184
X-ray fiber diffraction pattern, 183–185

Z

Ziegler catalyst polymerizations
commercial development, 8–9
ethylene and propylene, 5–6
$TiCl_3 \cdot nAlCl_3$ as catalyst, 6,7–8*f*
vinyl ethers as catalysts, 9–10
Ziegler–Natta catalysts
for α-olefin polymerization, based on titanium halide and organoaluminum salts, 56–62
$MgCl_2$-supported, 27–55
Zinc porphyrins
control of copolymerizability by light for epoxide–episulfide copolymerization, 199–200,201*f*
effect of stereochemistry of ring opening of epoxide, 200,202,203*t*
Zirconium chloride, substituted–methylaluminoxane catalyst system
cyclopentadienyl ring substituent related to rate of ethylene polymerization, 78–86
hydrogen addition related to 1-butene polymerization, 72–76

Production: Margaret J. Brown
Indexing: Deborah H. Steiner
Acquisition: Anne Wilson and A Maureen Rouhi
Cover design: Amy Hayes

Printed and bound by Maple Press, York, PA